# FROM NERVE TO MIND

# FROM NERVE TO MIND

A Volume prepared in Honor
of Professor Wade Marshall

*Edited by*

## ROBERT G. GRENELL

*Section of Neurobiology, Psychiatric Institute,
University of Maryland Hospital, Baltimore,
Maryland 21201, U.S.A.*

GORDON AND BREACH SCIENCE PUBLISHERS
New York          London          Paris

These papers (except the last) were originally published in *Intern. J. Neuroscience.* **1**, Nos. 5 and 6; **2**, Nos. 1–5; **3**, No. 4.

# Contents

# INTRODUCTION

Professor Wade Hampton Marshall

The papers in this volume were prepared as a tribute to Wade Hampton Marshall on the occasion of his retirement on July 10, 1970, as Chief, Laboratory of Neurophysiology, National Institute of Mental Health. The papers have been submitted by scientists who have collaborated with Dr. Marshall or who have been members of his Laboratory.

Wade Marshall was born December 17, 1907, in Pittsburg, Pennsylvania. After graduating from Beloit College in 1930, he prepared himself for a career in physiology at the University of Chicago. Under the sponsorship of Ralph Gerard, he was awarded the degree of Master of Arts in 1931 and the degree of Doctor of Philosophy in 1934. His doctoral thesis was entitled 'Production of reflex

pressor and depressor response by selective electrical stimulation of the vagus.' While a graduate student he developed ingenious methods for recording from brain nuclei, and the findings by him and his colleagues resulted in the publication of a unique brain and electroencephalographic atlas.

After receiving his degree Dr. Marshall served for two years as an Instructor in Physiology at George Washington University Medical School and then moved to the Johns Hopkins Medical School where he worked until 1943. His research at Hopkins (with Samuel A. Talbot, Clinton N. Woolsey, Philip B. Bard and others) was concerned primarily with mapping electrically the cortical representation of the somatic and visual systems. The publications on this work have become standard references. His interest in how the brain processes visual information led him in the late 30's to describe a mechanism for binocular interaction which became noted for its heuristic value.

During World War II he applied his technical ingenuity in instrumentation to the development of weapons at the Applied Physics Laboratory at the John Hopkins, with publications on subjects as remote from neurophysiology as 'Preliminary Tests on the 1.6 Mach Number Supersonic Jet.'

Dr. Marshall became a member of the Intramural Research Program of the National Institute of Mental Health in 1947. In 1954 he was appointed Chief of the Laboratory of Neurophysiology. His interest in cortical function led him to investigate the so-called suppressor action of various cortical areas, and he showed that many of the reported findings were attributable to the phenomenon of spreading depression originally described by Leaó. In an extensive critical review of this subject in 1959 (*Physiological Reviews*) he concluded that spreading depression represents a pathological process that may result from a variety of injuries to the brain that in each case lead to the alteration of the composition of fluid in the spaces around the nerve cells. Specifically, he postulated that the generalized depression of neural function was the result of injury to the blood-brain barrier.

Since then, he and his collaborators have examined the mechanisms by which the blood-brain barrier participates in regulating the chemical environment of the extracellular spaces surrounding neurons. Attention has focused particularly on the role of the potassium ion, the $CO_2$ content of the blood, and the relation of blood flow in altering the electrical potential of the brain.

Much of the success of these experiments resulted from the development of instrumentation by Dr. Marshall and his staff. Dr. Marshall was personally responsible for constructing a pore electrode for recording direct current potentials from a cortical region of less than 1 mm. He and his staff devised special pH electrodes for determining changes in acidity of the cerebral cortex. The Bak amplifier, which was developed under his aegis, became standard equipment for intracellular recording in many laboratories throughout the world.

Perhaps one of Wade Marshall's greatest contributions to neurophysiology has been the training of young scientists. In his 23 years at the National Institute of Mental Health more than 60 workers have received postdoctoral training in a stimulating and constructively critical atmosphere. Many of them doubtless regard Dr. Marshall's support and soft-spoken encouragement as pivotal factors in their pursuit of a career in neurophysiology.

From its beginning his Laboratory has been one in which a wide variety of approaches to the study of the nervous system has been encouraged. The resulting range of interests is reflected in the articles in this volume dedicated to him.

PAUL D. MacLEAN, M.D.
*Chief, Laboratory of Brain*
  *Evolution and Behavior*
*National Institute of Mental*
  *Health*
*Bethesda, Maryland 20014*

DAVID O. CARPENTER M.D.
*Laboratory of Neurophysiology*
*National Institute of Mental*
  *Health*
*Bethesda, Maryland 20014*

# THE RELATION OF AXON SHEATH THICKNESS TO FIBER SIZE IN THE CENTRAL NERVOUS SYSTEM OF VERTEBRATES†

GEORGE H. BISHOP, MARGARET H. CLARE, and WILLIAM M. LANDAU

*Department of Neurology*
*Washington University School of Medicine, St. Louis, Missouri* 63110

In a series of land vertebrates sheath thicknesses of myelinated nerve fibers in a variety of central tracts have been measured, and plotted against axon diameters. Preparations include monkey, cat, rat, mouse, mole, hedgehog, bullfrog, and green frog. Tracts examined include dorsal column cuneate and gracilis bundles, pyramidal tract, optic nerve and tract, trigeminal and auditory nerves, and for comparison, the saphenous nerve. Maximal sizes of axons (inside sheath diameter) in different preparations vary widely, from 12 $\mu$ or more to 2 $\mu$, and most tracts have axon diameter ranges down to 0.3 $\mu$. In the cat auditory nerve the axon size range is restricted (80% between 2 and 3 $\mu$). All fibers in the mammalian central tracts are myelinated, with a wide scatter of values of the ratio of sheath thickness to axon diameter, as measured in a single cross section. The ratio varies with fiber diameter, becoming larger with decreasing diameter. Average ratios attain the value of 1/5 in different preparations at different axon diameters, generally from 3 $\mu$ or less in central tracts (in some preparations ratios remain just below 1/5 in the smallest fibers present) as compared to 5 $\mu$ in the cat saphenous nerve. In certain tracts (e.g., dorsal columns of spinal cord), the ratio is approximately 1/5 over the whole size range of the fibers present. In the frog optic nerve the myelinated fibers tend to have a uniform sheath thickness throughout the diameter range.

## INTRODUCTION

The study here reported comprises a survey of the ratio of myelin sheath thickness to nerve fiber diameter, in a variety of central tracts and peripheral nerves. A series of vertebrates, including the frog and a wide spectrum of mammals, were studied, with special interest in the characteristics of central tracts as compared to those of peripheral nerves.

In spite of the technical problems occasioned by the irregular contours of fibers and sheaths as seen by electron microscopy, it seems certain that sheath thickness is one of the important biophysical parameters for central axons.

In their early study of osmicated cross sections of myelinated peripheral nerves from a number of vertebrates, Donaldson and Hoke (1905) concluded that the areas of axon (axis cylinder) and of myelin sheath are equal in general. Later studies have

†This work was supported in part by Contract Nonr-816 (16) with the Office of Naval Research, by a grant from the Supreme Council Thirty-Third Degree Scottish Rite, Northern Jurisdiction, U.S.A., through the National Association for Mental Health, and by National Institutes of Health Program Project Grant NB-04513.

shown that their conclusion is not universally valid. Schmitt and Bear (1937) measured inside and outside sheath diameter, and termed the ratio the *g* factor. That factor varied directly with fiber size. The ratio of sheath thickness to axon diameter varies inversely with fiber size. Axon and sheath areas are equal when the sheath thickness is approximately one-fifth of axon diameter.

Earlier work had dealt with inner and outer sheath diameters of peripheral nerve fibers, from which sheath thicknesses can be figured, but have not usually been emphasized. The variation of configuration of a peripheral fiber along its length has been studied by Hursh (1939) and by Sunderland and Roche (1958), who provide an extensive review. The latter authors also found that for fibers of a given diameter, sheath thicknesses may vary by as much as $\pm 25\%$ of the mean values.

In spite of the uncertainty involved in the estimation of fiber diameter from measurements of a single cross section, the classic Gasser and Erlanger studies (1927) indicate a distinct correlation of peripheral fiber size with electrical threshold, conduction rate, and related physiological parameters. Their conclusions have become axioms of neurophysiological thought. Myelin sheath thick-

1

c

ness has interested many investigators because of the inferred effect of sheath insulation and capacitance on nerve impulse conduction.

Rushton (1951), in a theoretical review, dealt with the physics of conduction in myelinated fibers of varying size, and the relationship of velocity of conduction to axon diameter, sheath thickness, and internode length. He considered the relationship of myelinated and unmyelinated fiber conduction velocities in both vertebrate and invertebrate peripheral axons, and concluded that for diameters less than 1 micron, there is a biological advantage for the axon to abandon its myelin sheath. Indeed, this may be the case in mammalian peripheral nerve. However, his generalization fails in the mammalian central nervous system paths where fibers are myelinated down to diameters of a third of a micron (Bishop and Smith, 1964). Furthermore electron microscopic studies of central fibers using a variety of fixation techniques, have generally failed to show the uniform round configurations as observed by optical microscopy of peripheral nerves fixed in 1 % osmic acid. Nevertheless, the general relationship of fiber size to physiological parameters does hold for central tracts (Bishop, et al., 1953; Landau, et al., 1968).

Rushton's (1951) theoretical rationalization of the structural and related electrical properties of the axon connotes confidence in Gasser's aphorism that 'the nerve is a gentleman'. One may, in fact, infer that a given fiber, however it may vary in diameter, in internodal length, and perhaps in sheath thickness and nodal resistance, can be rationalized as one with a statistical average of a group of axons of similar measurements, so that all would have closely similar physiological values. These latter properties are traditionally measurable on whole nerves or on multifibered tracts. One or more critical fibers can be identified as having a given set of values of threshold, conduction rate, etc., but it is usually not feasible to obtain the actual structural dimensions throughout a significant distance of conduction.

Several species of available vertebrates of different phylogenetic development and size were studied including the squirrel monkey (Saimiri sciurea), cat, mouse, rat, European hedgehog (Erinaceus europaeus), mole (Scalopus aquaticus, subspecies Mississippiensis), bullfrog (Rana catesbiana), and green frog (Rana clamitans).

METHODS

Tissues have been fixed by immersion and by perfusion in attempts to obtain cross sections of nerve fibers that were as nearly round as possible. Many fixatives were employed: Immersion in 1 and 2 % osmic acid in water and in Tyrode solution, in 3 to 10 % buffered formalin for one to several days, in glutaraldehyde, and in glutaraldehyde and formalin, all followed by osmication; perfusion with glutaraldehyde and formalin, and with formalin in a variety of buffers. Nerves were treated with osmic acid for at least two hours, up to 24 hours. Tissues were run through graded alcohols, toluene and Epon (two successive plastic solutions containing catalyst), and hardened at 60°C for at least 12 hours. Sections were cut at 0.1 $\mu$ or less, and usually restained with lead acetate.

By light microscopy we confirmed that Donaldson and Hoke's (1905) technique of simply immersing peripheral nerve tissues in 1 % osmic acid provides remarkably uniform round fiber contours. However, examination of the same tissue by electron microscopy showed such marked sheath splitting and fragmentation that valid measurements could not be made. Immersion of C.N.S. blocks in 1 % osmic acid was similarly unsatisfactory.

With all these variations no better results were obtained in electron microscope pictures than by immersion in 10 % buffered formol followed by Dalton's fluid containing 1 or 2 % osmic acid. Sample areas for study were selected at random from cross sections. The chief criteria of a satisfactory preparation were the clean fixation of myelin and the absence of obvious shrinkage of axoplasm from the myelin sheath. As to the fixation techniques in general, we have arrived at the conclusion that either there are yet no satisfactory fixatives for myelinated nerve fibers as studied by electron microscopy, or nerve fibers in vivo must vary widely from a uniform tubular model.

Electron micrographs at 5,000 to 30,000 diameters show irregularities in shape not conspicuous in light micrographs at 1,000 diameters. Measurements of sheath thickness were generally made on fibers of nearly round cross section. Only one-third of the fibers of a given area may fulfill that criterion. In general, the shapes of the larger axons deviate more from round cross sections than do the shapes of smaller ones. In many preparations none of the large fibers were even approximately circular in outline, and equivalent circles had to be estimated as to fiber size. Measurements on enlarged photographs were made to the nearest 0.5 mm for fiber diameters, and estimated to the nearest 0.2 mm for sheath thickness. Inside sheath diameter was considered to be equivalent to axis cylinder (axon)

diameter. Considering these and other complications, there is a variable error of measurement,
but a relatively small one as compared to scatter
of sheath thicknesses per fiber diameter. Thus we
have plotted the measurements of individual fibers
instead of giving the averages of all those of a
given diameter.

Ovoid cross sections are obviously cut at an
angle to the vertical. In other clusters fibers are
packed so closely together that their shapes are
generally distorted as if by overall pressure. Still
others are crenelated and where a sharp curvature
exists the myelin sheath is frayed as if a shrinkage
of the whole during dehydration caused local
buckling in a sheath stiffened in fixation before or
more rigidly than the enclosed protoplasm. In a
large fiber the neuroplasm may be retracted from
the sheath at one point and the sheath itself frayed
at another with a clear space beneath it. On the
other hand in areas where the large fibers are not
densely packed their contours approximate closer
to circles. When judged to be sufficiently round to
make a reasonable estimate of an equivalent circle,
the sheaths of such fibers have been measured at
their thinnest regions.

## RESULTS

Prior studies involving measurements of inside and
outside sheath diameters have been carried out
chiefly on peripheral nerves where the smallest
myelinated fibers are 2 $\mu$ in diameter. The sheath
of a 4 $\mu$ myelinated fiber is at the limit of resolution
of the light microscope, and central tracts contain
myelinated fibers as small as 0.3 $\mu$ outside diameter.
Measurements upon electron micrographs of thin
osmicated sections from a variety of mammalian
nerves and tracts indicate that the sheaths have a
wide scatter of values for a given axon diameter;
and that the average sheath thickness is usually
one-fifth or less this axon diameter. Fibers of
reasonably circular contour were selected for
measurement, and only a sufficient number were
plotted to indicate the general trend for each
preparation. That is, minimal and maximal values
for axon diameters and for sheath thickness are
indicated, although the plots in Figures 1–8 do not
necessarily correlate to the relative numbers of
fibers of all diameters in the nerve or tract. In the
mammalian central tracts examined, all the fibers
were myelinated and fiber diameters of 0.3 $\mu$ or
less were present in many, with a maximum of
numbers at 1 $\mu$. However, exceptions to this dis-

tribution will be noted below. The maximum fiber
sizes in the tracts examined varied from 2 to 10
or 12 $\mu$, and in a given preparation the number of
fibers usually diminished rapidly as their diameters
increased. The few fibers of large size together with
the number which were distorted in outline com-
plicated the determination of average sheath
thickness to axon ratios.

In the figures presenting sheath thickness vs.
inside sheath diameters, we have plotted the sheath
measurements at 5 times the scale of axon diameters.
This relieves crowding of the plotted points. Taking
a 5 $\mu$ axon as an example, a 1 $\mu$ sheath value is at
the same ordinate distance above 0 as the axon
diameter is along the abscissa. A 45° diagonal
line through zero then is the locus of all sheath
values which are one-fifth of their respective
axon diameters (Figures 1 and 2). This serves to

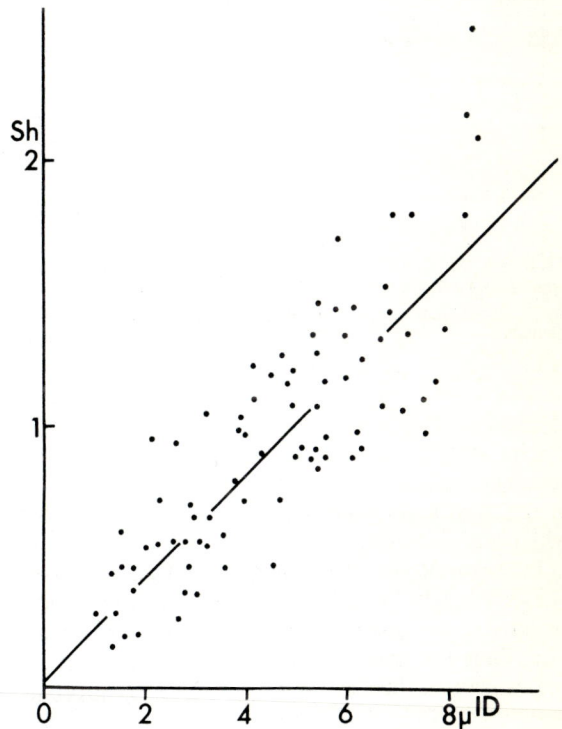

FIGURE 1 Cat cervical dorsal column, gracilis area.
Perfused with glutaraldehyde with cacodylate buffer, followed
by 2% osmic; electron microscope × 5500. Plot of sheath
(ordinate) vs. axon diameter (abscissa) shows no significant
deviation from scatter about the diagonal line of 1/5 ratio.
Only selected round fibers were counted, and relatively
few very small fibers were present.

emphasize the scatter of sheath thickness values.
A dash line that locates the approximate average

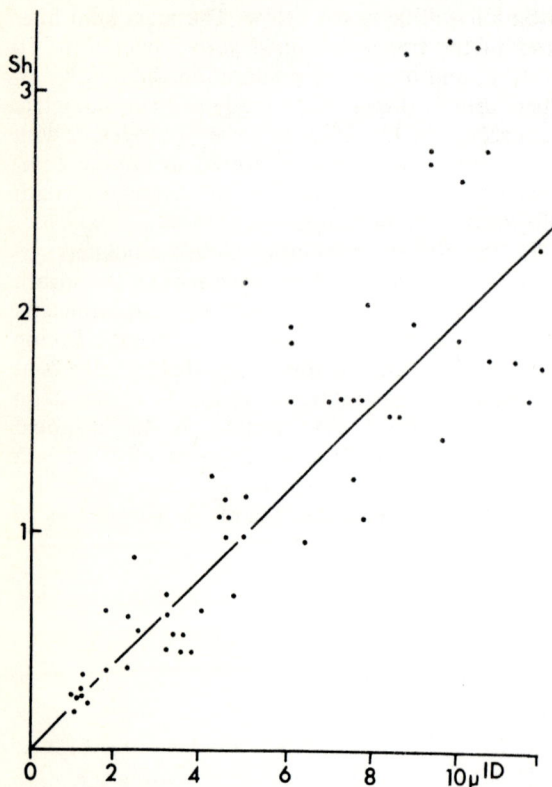

FIGURE 2 Same preparation as Figure 1, but cuneate area. Larger fibers than in gracilis, but same constant ratio. Some small fibers appear to enter the area from the region of the comma tract. Distribution denotes constant *g* factor.

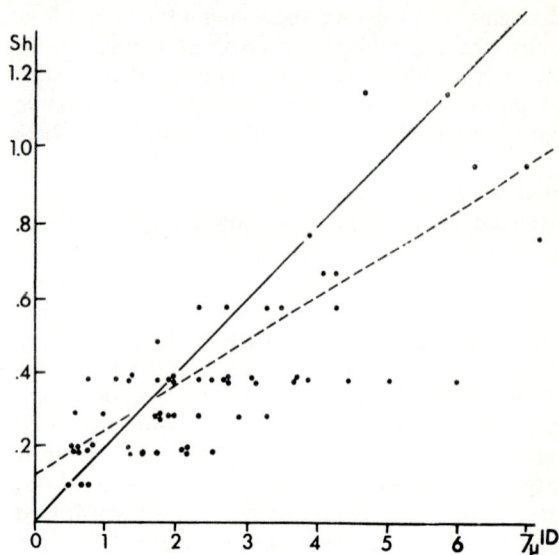

FIGURE 3 Cat optic tract. 10% formol followed by Dalton osmic. E.M. × 5200. Dash line, approximate average of sheath thickness scatter. Slant of this line to standard 1/5 ratio indicates relative increase in ratio in smaller fiber range. Thinner sheaths than in similar sections of optic nerve but same slant in distribution.

the mole pyramidal tract (figure not presented), with axon diameters ranging from less than 0.5 to 4 μ maximum, the scatter of values is similarly along the reference line. In Figure 3, cat optic tract, the ratio values drift at an angle to the reference line, crossing it only in the small fiber range, and sheaths throughout are relatively thinner than in the previous tissues. Figure 4, hedgehog

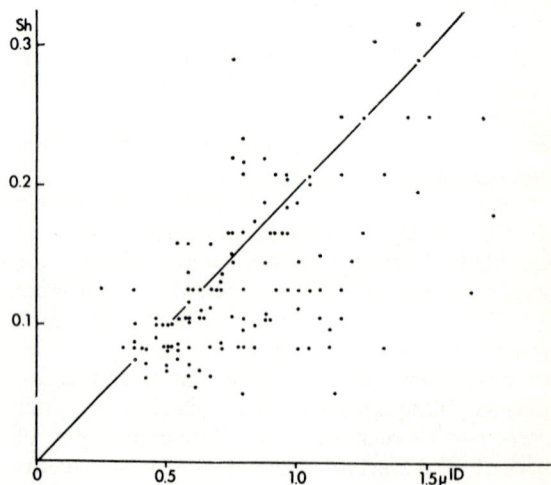

FIGURE 4 Hedgehog pyramidal tract. Perfused with glutaraldehyde cacodylate followed by osmic 2%. E.M. × 24,000. Typical slant of sheath to axon ratios crossing standard line at 0.5 μ.

of the scatter may cross the standard diagonal (Figure 3). Different counts vary from an average scatter that is parallel to or coincides with the standard reference line, to others which cross it at different axon diameter loci, and to still others which fall definitely below it.

In terms of the values for a standard 5 μ axon which we have taken as a reference, the sheath thickness is 1 μ and the outside sheath diameter 7 μ. Thus the areas of axon and total fiber are as the squares of their diameters, or 25 to 49, approximately one to two. The areas of axon and sheath cross sections are then equal. The ratio of sheath thickness to inside sheath diameter is 0.2; the ratio of sheath thickness to outside diameter is one-seventh or 0.14. Larger fibers usually have relatively thinner sheaths; the *g* factor (ID/OD) then increases with increase of fiber diameter, while the ratio of sheath thickness to axon diameter decreases.

In Figures 1 and 2, cat cervical dorsal column areas, the scatter of sheath values is approximately symmetrical with respect to the reference line. In

pyramidal tract, with maximal axon diameters of less than 2 $\mu$, shows a wide scatter but distribution crossing the reference line at about 0.5 $\mu$. Three other preparations, of mole fifth nerve, rat pyramidal tract and squirrel monkey optic nerve showed the same pattern with maximal axon sizes of 6, 5 and 2.5 $\mu$ respectively. In the hedgehog optic nerve with 3 $\mu$ maximal axons (Figure 5) and in the mouse

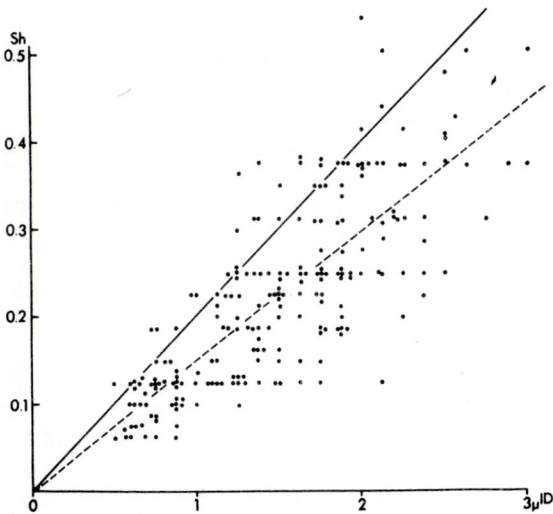

FIGURE 5 Hedgehog optic nerve. Perfused with glutaraldehyde cacodylate buffer followed by osmic. E.M. × 8000. Dash line fails to cross standard ratio line, but distribution shows similar slant to it as in previous figures. Ratio sheath to diameter everywhere less than 1/5. Maximum axons 3 $\mu$.

optic nerve with largest axons 2 $\mu$ (Figures 6, 10) practically all axons, even the smallest, have sheath to axon ratios below one fifth, but these ratios decrease toward the large fiber ends of the distributions as usual.

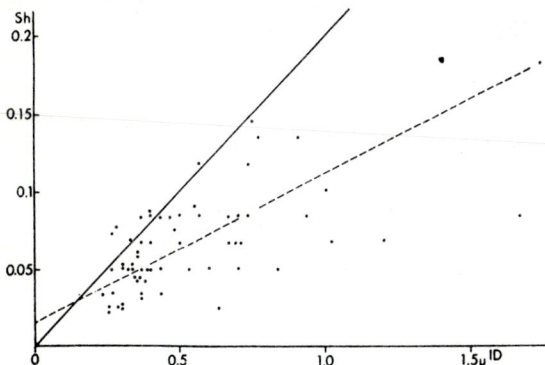

FIGURE 6 Mouse optic nerve, Dalton osmic immersion. E.M. × 30,000. Slant of dash line greater than usual but everywhere below the standard of ratio of 1/5 Largest axons less than 2 $\mu$.

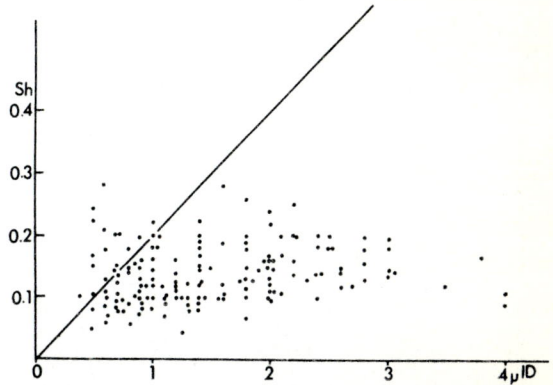

FIGURE 7 Green frog optic nerve. 10% formol followed by Dalton osmic. E.M. × 10,600. Maximum axon diameter 4 $\mu$.

The green frog optic nerve (Figures 7, 9) is at the extreme of this sequence, showing the average sheath thickness to be effectively constant over the whole size range. The ratio of 1 to 5 is still attained for the sheath but only below the 1 $\mu$ size range, and decreases to approximately 1 to 20 for the larger fibers. In the nerve cross section, some areas have only small fibers and others contain most of the large ones. The data for Figure 7 were taken from six photographs in which varying size distributions were typical of the whole cross section.

The cat auditory nerve (Figure 8) presents an extreme found in no other tract examined: a concentration of 80% of the axons have diameters of 2 to 3 $\mu$ with a few larger and smaller ones. Sheath thicknesses for all axon diameters were scattered widely, and change in sheath thickness/axon diameter ratio is equivocal. A similar preparation of cat auditory nerve, dissected out without decalcification, photographs of which were lent to us by Miss Barbara Nelson of our anatomy department, showed an approximate duplication of our findings.

Figures 9 and 10 (see legends) illustrate some of the complications which are met in measuring fibers.

DISCUSSION

Our results can be interpreted only in a statistical sense. The ratio of average Sh/ID is not constant, but is usually higher for smaller fibers than for larger. The larger the axon core diameter, the relatively thinner is the sheath.

Our work is in agreement with the results of Sunderland and Roche (1958) on the structural measurements of peripheral nerve fibers. They note

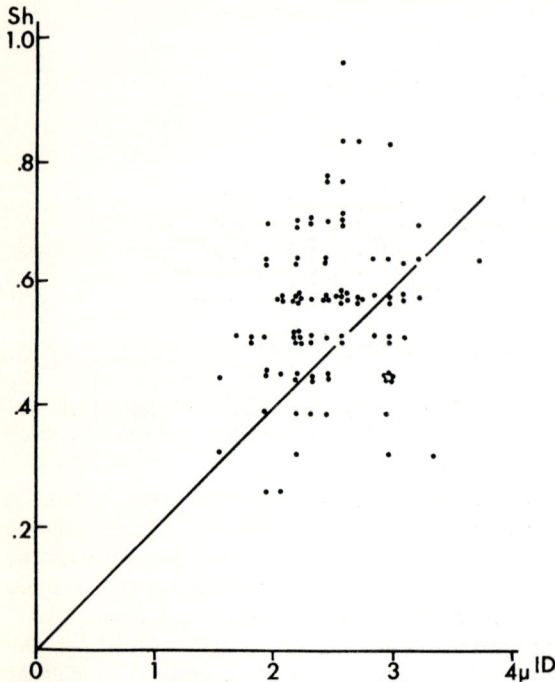

FIGURE 8   Auditory nerve of cat, formalin 10% followed by decalcification in 0.2 M EDTA in 0.1 M PO$_4$ buffer for 3 months. E.M. × 20,000. 80% of axons fell between 2 and 3 $\mu$. Too narrow a range to specify slant of values, with the wide scatter of sheath thicknesses, but average scatter appears to be above the standard 1/5 ratio to axon diameter.

that up to 1954 no account has been taken of sheath thickness, although several workers had reported inside and outside measurements from which sheath thickness values were available. But we cannot accept the drastic inference which Sunderland and Roche draw from their thorough and accurate measurements, that fiber diameters cannot be significantly correlated with their physiological properties. Even with variations in Sh/ID values for cross sections of single fibers, the average of the scatter may be a fair measure that could correlate to conduction rate, for example.

The earlier measurements, anatomical and physiological, were made on peripheral nerve fibers employing the light microscope and not too sophisticated apparatus for recording. Further work has not confirmed the elegant simplicity of the relations between structure and function (Landau, et al.; 1968). Amplitude of potential is not precisely proportional to D$^2$, nor conduction rate to the inverse of the diameter, and threshold relations vary with the form and duration of the stimulating current.

Irregularities in fiber cross sections are not conspicuous in the light microscope. In the electron microscope many fibers are found to deviate widely from circular cross section even when the myelin appears to be properly fixed. The question whether such distortions from a round cross section are due to the processes of fixation and dehydration, or whether they are present in the normal nerve in situ, has not been satisfactorily answered. One may infer, however, that the nerve fiber in a state of nature may be far from an ideal tubular structure, divided by nodal constrictions into equal internodal segments.

We can draw no conclusion as to any rationale of the differences in distributions in the tracts analyzed so far, with the possible exception that there is some correlation between the size of the animal and the maximal size of the fibers of a given tract. The wide scatter of sheath thickness measurements, and the relatively thicker sheaths on smaller fibers, are quite similar, for fibers of both peripheral and central paths, whether the fiber sheaths derive from Schwann cells or oligodendroglia. The sheath thickness thus appears to be a function of the fiber as such, regardless of the physiological action which it mediates.

Our measurement on only one random cross section of a single internodal segment leaves uncertain whether the sheath thickness varies along one internodal length. However, the probability that this single measurement is near that fiber's average dimensions is considerably more than chance. For instance in the 100 sections measured by Hursh (1939) in one fiber, 65% of the 100 values were within ± 10% of the diameters of the maximum in numbers. The measurements of both sheath thickness and axon diameter may vary along the axon's length such that variations in one measurement may compensate for variations in the other; in terms of ohmic resistance and node area the combination may be equivalent to a uniform fiber of approximately the average measured values. It would then be this equivalent fiber that might have the average conduction rate of fibers with a given measured cross section. Further, a region in a fiber of given outside diameter with a sheath thicker than average will have an axon diameter thinner than the average. The former will increase the fiber's insulation, and the latter will decrease its conductance, and the two effects tend to compensate each other.

## SUMMARY

Our measurements of the ratios of sheath thickness

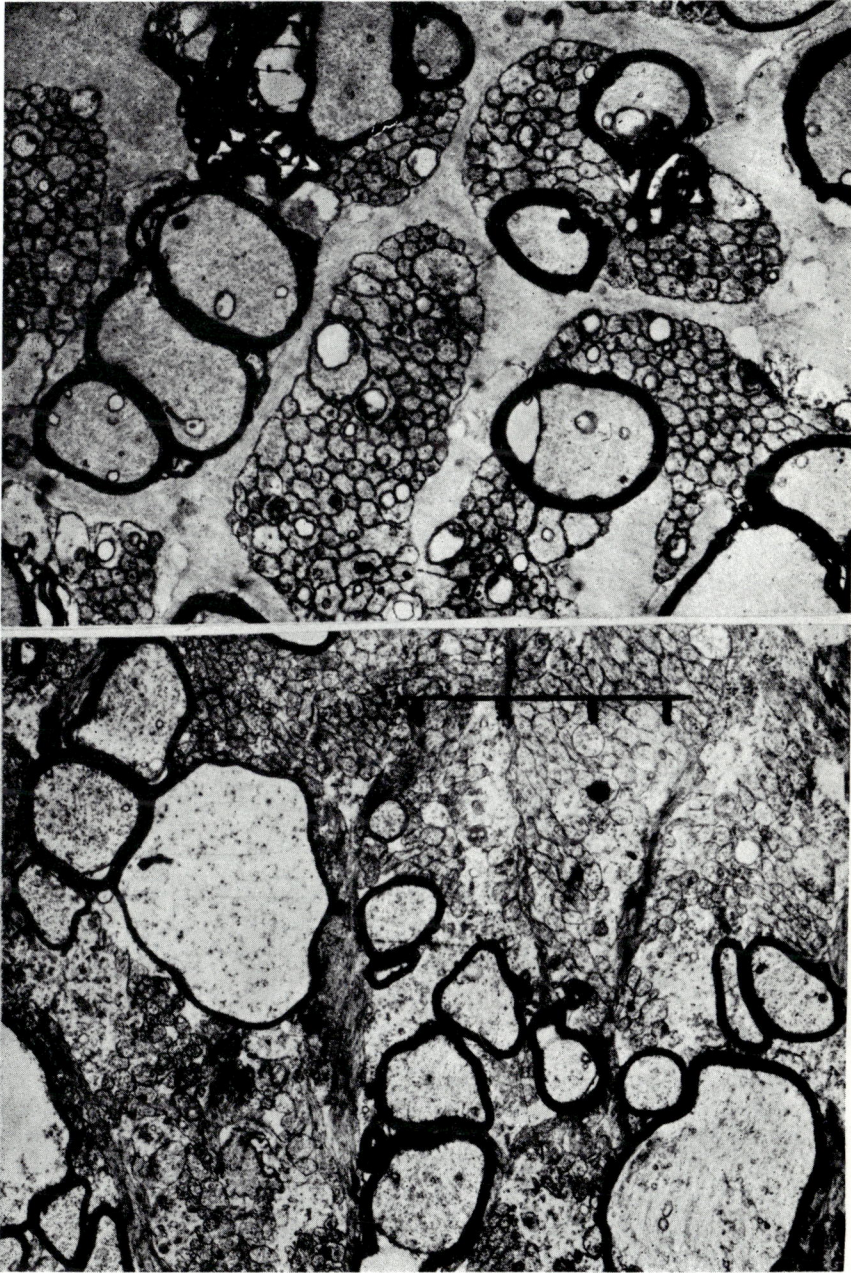

FIGURE 9   Green frog optic nerve, osmic 2% fixation. E.M. × 10,600, horizontal line divided into 1 $\mu$ segments. On background of unmyelinated fibers, the sheaths of myelinated fibers seem to have similar scatter of thicknesses over the whole size range. About 15 unmyelinated fibers per 1 myelinated.

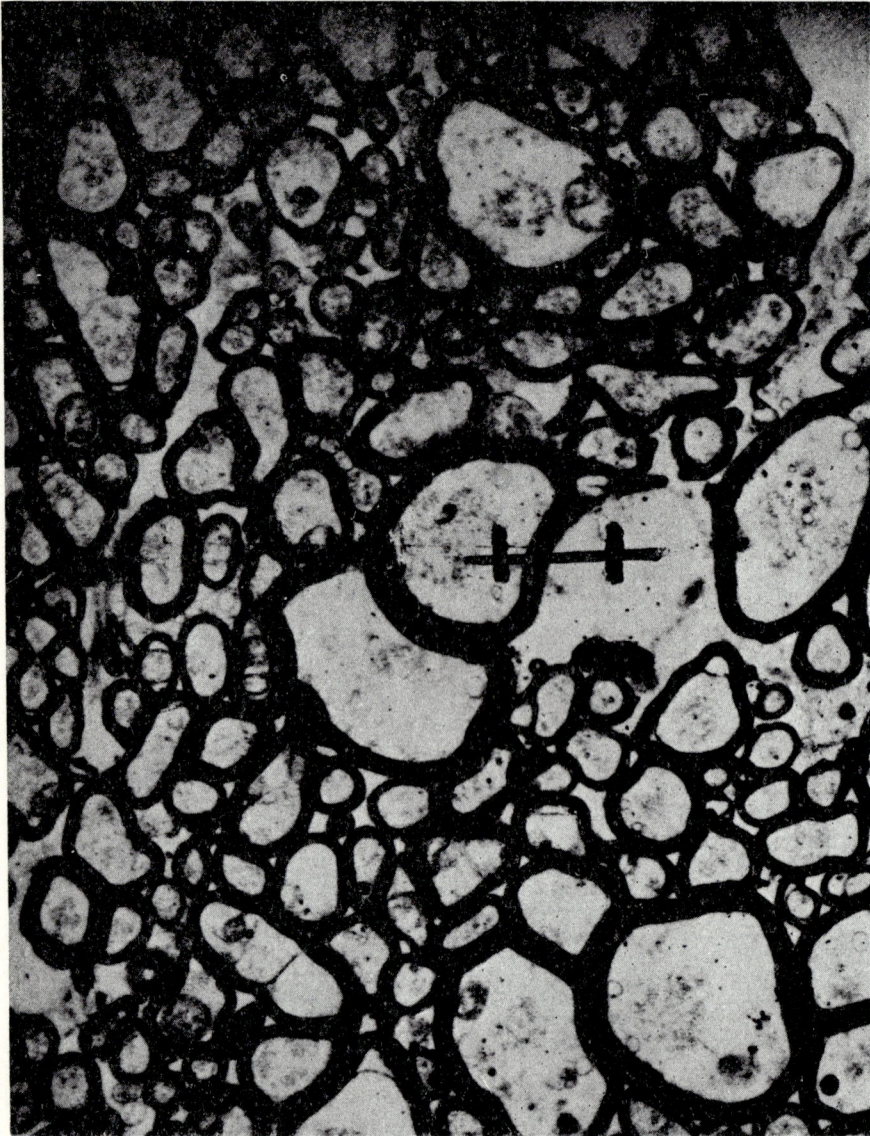

FIGURE 10   Mouse optic nerve, Dalton osmic immersion. E.M. × 10,600. Scale on horizontal line is 1 $\mu$.

to axon diameter in various central tracts show a wide scatter, but with average values of 1/5 or less. Such tracts vary with respect to the maximum fiber sizes, for instance from 10–12 $\mu$ in the cat optic tract to 2 $\mu$ maximum in the mouse optic tract. These tracts also vary with respect to the *change* of sheath to diameter ratio with axon diameter, but even the tracts with smallest maximal diameters approach the 1/5 ratio in the small fiber range of 1 $\mu$ or less. Since only one cross section per fiber is measured, we cannot say what these measurements signify in nerve functioning, except in a statistical sense. Finally, we cannot exclude an error due to distortion during fixation and dehydration, which may be more significant than the probable error in actual measurement on electron microscope photographs.

REFERENCES

Bishop, G. H., and Smith, J. M., 1964, The sizes of nerve fibers supplying cerebral cortex, *Exp. Neurol.* **9**: 483–501.

Bishop, P. O., Jeremy, D., and Lance, J. W., 1953, The optic nerve; properties of a central tract, *J. Physiol.* **121**: 415–432.

Donaldson, H. H., and Hoke, G. W., 1905, On the areas of the axis cylinder and medullary sheath as seen in cross sections of the spinal nerves of vertebrates, *J. Comp. Neurol. Psychol.* **15**: 1–16.

Gasser, H. S., and Erlanger, J., 1927, The role played by the sizes of the constituent fibers of a nerve trunk in determining the form of its action potential wave, *Am. J. Physiol.* **80**: 522–547.

Hursh, J. B., 1939, Conduction velocity and diameter of nerve fibers, *Am. J. Physiol.* **127**: 131–139.

Landau, W. M., Clare, M. H., and Bishop, G. H., 1968, Reconstruction of myelinated nerve tract action potentials: An arithmetic method, *Exp. Neurol.* **22**: 480–490.

Rushton, W. A. H., 1951, A theory of the effects of fibre size in medullated nerve, *J. Physiol.* **115**: 101–122.

Schmitt, F. O., and Bear, R. S., 1937, The optical properties of vertebrate nerve axons as related to fiber size, *J. Cell. Comp. Physiol.* **9**: 261–273.

Sunderland, S., and Roche, A. F., 1958, Axon-myelin relationships in peripheral nerve fibers, *Acta Anat.* **33**: 1–37.

D

# MONOSYNAPTIC TERMINALS ON VENTRAL HORN CELLS OF THE RAT

RAYMOND B. WUERKER

*Department of Anatomy, College of Medicine, University of Arizona, Tucson, Arizona* 85721

Monosynaptic terminals measure approximately 2 x 4$\mu$, have dense post-synaptic thickenings and subsynaptic granules, and are located on the proximal ends of primary dendrites. Following dorsal root section, alpha-glycogen particles accumulate in these synapses.

The distribution of synapses on neurons has become a rallying point for anatomists and physiologists interested in synaptic function. The question is if the size and waveform of synaptic potentials is a simple reflection of the distribution of endings on the cell body, dendrites, and initial segment. Or are there other factors such as variations in synaptic morphology. Using physiological evidence, Rall and his collaborators (1967) suggest that the mono-synaptic connections from the dorsal roots to motoneurons are arrayed along the length of the cell body and dendrites. These synapses have a number of unusual and puzzling characteristics. They give rise to an excitatory post-synaptic potential or *EPSP* that is of short latency, high amplitude, and fast rise time and decay (Coombs, Eccles, and Fatt, 1955b). These properties are compatable with potentials generated close to the cell body and not attenuated by electrotonic spread.

Unlike its muscular counterpart, the endplate potential or *EPP* (Del Castillo and Katz, 1954), the monosynaptic *EPSP* is unaffected by membrane hyperpolarization, and only slightly affected by small depolarizing currents (Eccles, 1955b). Furthermore, two *EPSP*'s add arithmetically rather than non-linearly as is true of *EPP*'s (Burke, 1967). These last two observations reflect the small impedance change occurring during these potentials (Smith, Wuerker, and Frank, 1967). Since the *EPSP* has a reversal potential and thus meets the physiological criterion of a chemical synapse, its small impedance change suggests a chemical synapse with remote terminations. The *EPSP* is

[1]This investigation was supported in part by a General Research Support Grant from the U.S. Public Health Service to the University of Arizona, College of Medicine.

generated by a conductance change on the distal dendrites and spreads electrotonicly into the soma, where it is recorded by an intracellular electrode. The impedance change seen by the electrode is small because it is *in series* with the large internal resistance of the dendritic shaft.

## METHODS

Normal rats (Charles River) and rats with single rhizotomies of the proximal $C_4$, $C_5$, $C_6$ and $C_8$ dorsal roots were fixed by perfusion, and tissue from the cervical ventral horn was examined with the electron microscope. The preparative procedures have been reported elsewhere (Wuerker and Palay, 1969), and the reader is referred to this article for further details. In essence, the fixing solution was a combination of 1% paraformalde-hyde and 1% glutaraldehyde, which was dissolved in 0.12 M phosphate buffer containing 0.02 mM $CaCl_2$. Blocks from the ventral horn were post-fixed in 2% osmium tetraoxide and 7% dextrose in the same buffer. After post-fixation and de-hydration the blocks were embedded in Epon 812 (Shell) and cut on a Porter-Blum MT-2 microtome. Thin sections were mounted on copper grids, coated with both Formvar and carbon, and double stained with alcoholic uranyl acetate and aqueous lead citrate. The sections were examined with an RCA EMU-3G or Hitachi 7S electron microscope.

## RESULTS

In normal rats, large synapses are seen on the proximal ends of the primary dendrites. They are

11

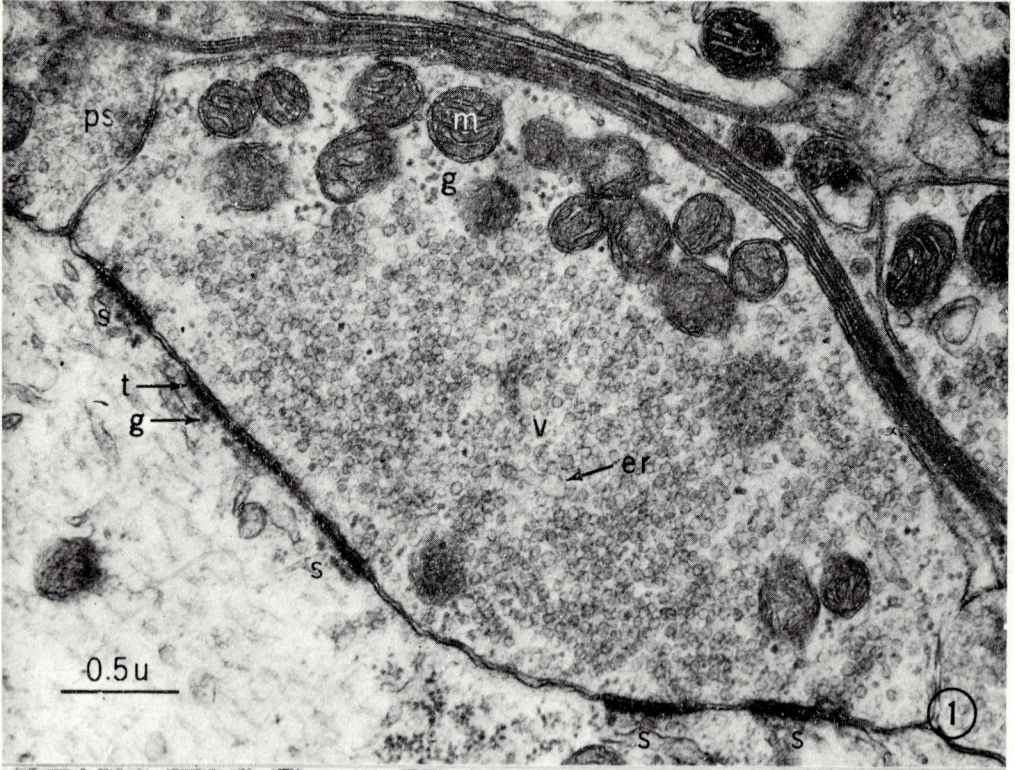

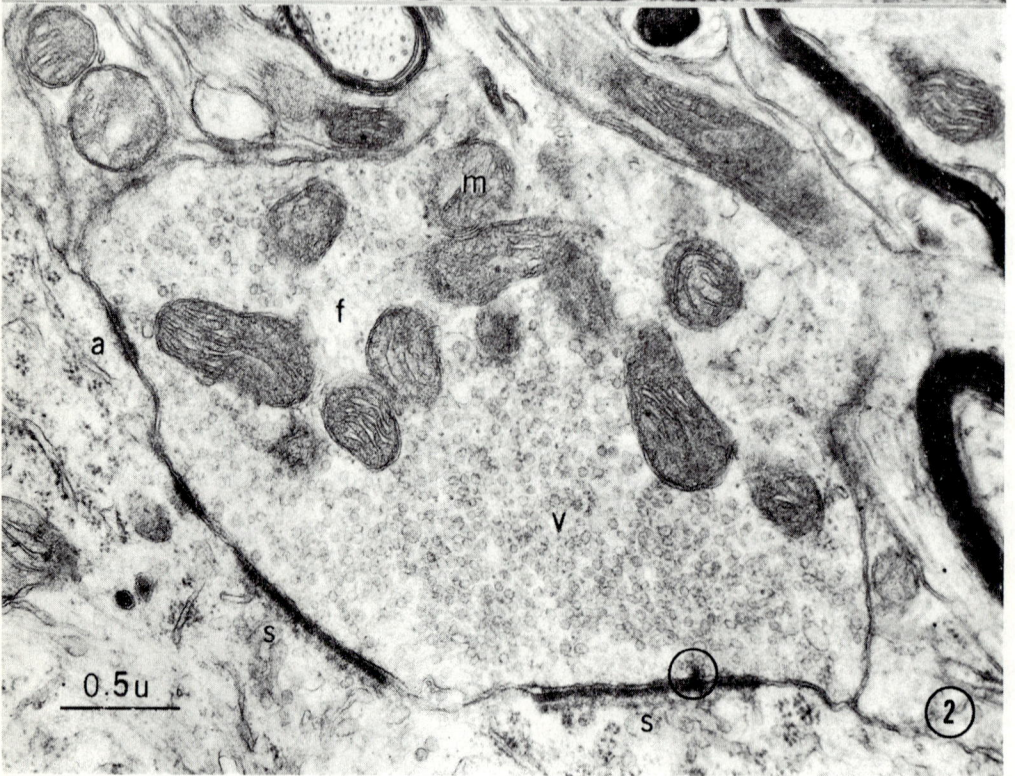

quite remarkable because of their large size, extensive postsynaptic specializations and characteristic proximal position on the dendrite. Their axons end as isolated expansions or *bouton terminaux* (see Figures 1 and 2), which measure approximately $2 \times 4\mu$. They have all the characteristics associated with chemical synapses seen elsewhere in the nervous system. The pre-synaptic terminal contains many mitochondria and synaptic vesicles. The latter are predominately spherical and measure 300–600Å in diameter. There is a synaptic cleft, which is 200–250Å wide and contains an intercellular, dense plaque. And finally, there are post-synaptic thickenings opposite clusters of synaptic vesicles in the pre-synaptic element. The thickenings are unusually wide, filamentous, and dense. They measure 300–350Å in width. The complex of synaptic vesicles, synaptic cleft, and post-synaptic thickening has been called the 'active zone' by Couteaux (1961, 1963) and is the presumed site of transmitter release. Beneath the synaptic thickening, but separate from it, is an accumulation of spherical granules, approximately 400Å in diameter, the 'subsynaptic densities' (Taxi, 1961).

In addition to the active zones, there are nonsynaptic junctions or puncta adhaerentia (Peters and Palay, 1966). Here, the pre- and post-synaptic thickenings are symmetrical and not associated with an accumulation of vesicles. Because of their similarities to the macula adhaerens in various epithelia (Farquhar and Palade, 1963), Palay (1970) considers them points of adhesion.

Intermingled with the vesicles in the pre-synaptic element are single, amorphous granules, approximately 200Å in diameter. Because of their size, shape, and staining, they are identified as beta-particles of glycogen, and are similar to glycogen seen elsewhere in the central nervous system. These particles meet most of the critera for identification of Revel (1960, 1964) and Drochmans (1962) except they do not stain uniformly. Inconsistent staining of glycogen is not unusual with the present method of double staining.

In addition to the glycogen and other organelles already described, occasional profiles of fibers, 100 and 250Å in diameter, indicate that neurofilaments and microtubules enter the terminal but do not penetrate its whole extent. Random profiles of smooth endoplasmic reticulum are seen throughout the terminal but not associated with the glycogen.

An occasional axo-axonic or pre-synaptic ending is seen abutting against the primary synapse.

One week after dorsal root section, changes are seen in these same large synapses (see Figures 3

FIGURE 1 Normal synapse. The axon terminal contains many mitochondria (*m*), synaptic vesicles (*v*), random profiles of smooth endoplasmic reticulum (*er*), and beta-glycogen particles (*g*). The synaptic specializations (*s*) consist of a dense post-synaptic thickening (*t*) and subsynaptic densities (*d*). An axo-axonic or pre-synaptic ending (*ps*) is present.

FIGURE 2 Normal synapse. An 'active zone' is circled in the lower right hand corner and consists of a cluster of synaptic vesicles (*v*) abutting against the pre-synaptic membrane, the synaptic cleft, and the post-synaptic specializations (*s*). In addition to these active areas, there are nonsynaptic junctions or puncta adhaerentia (*a*). The axon terminal contains fascicles of fibers (*f*), which are microtubules and neurofilaments.

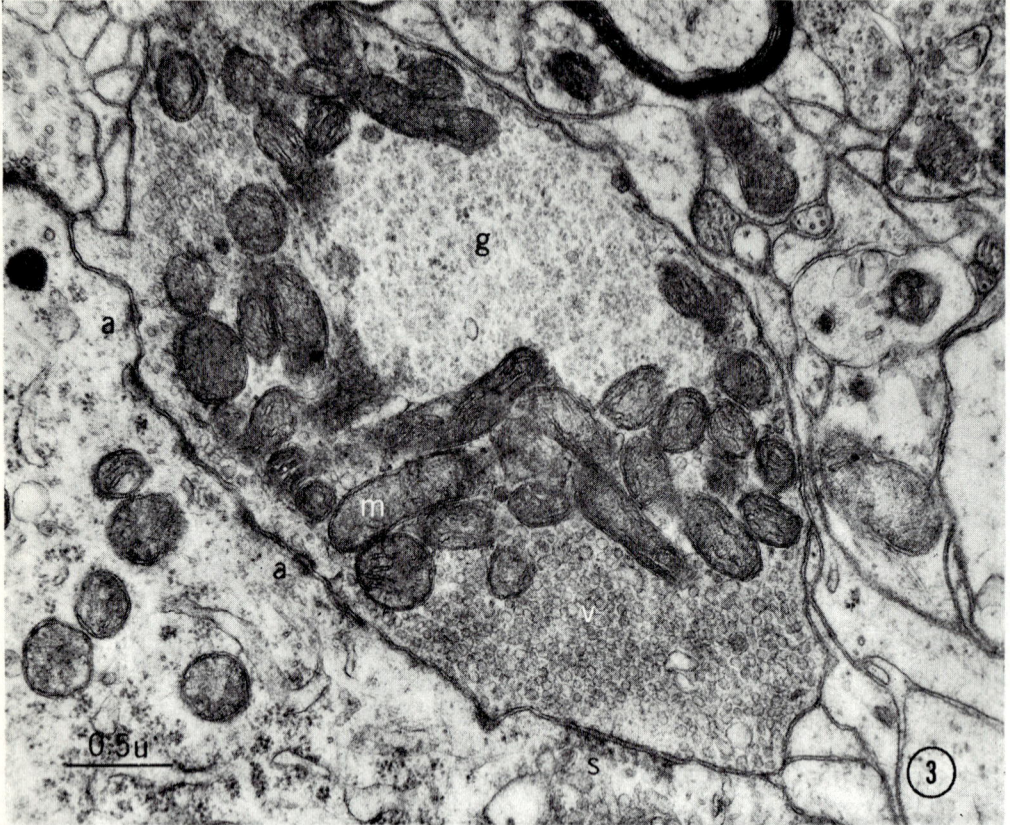

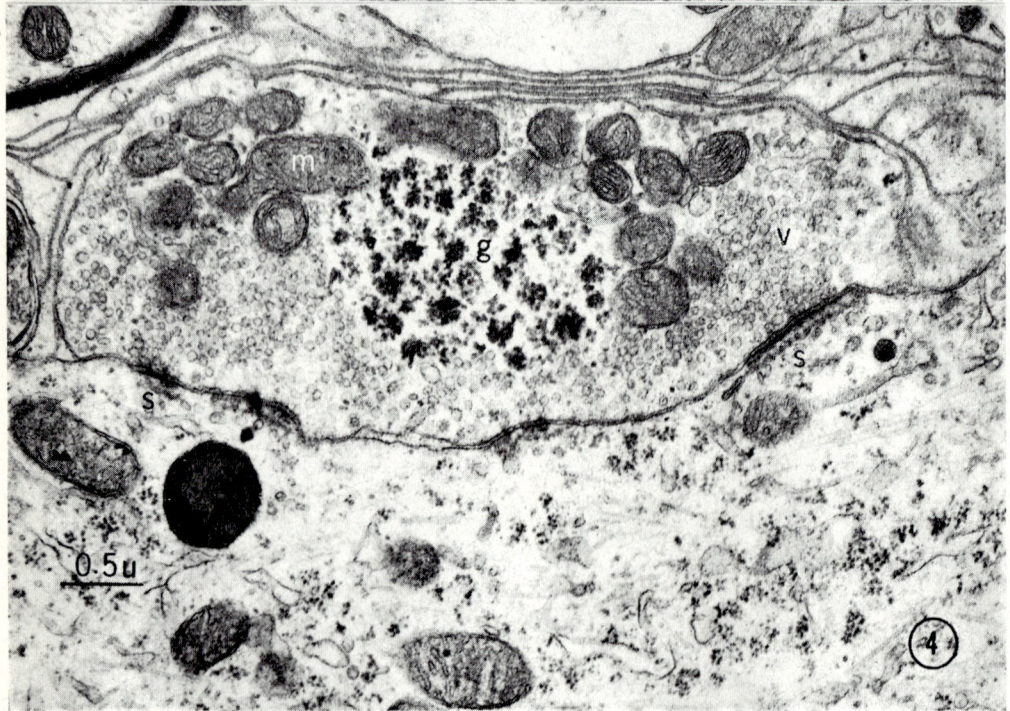

and 4). Again, these synapses could be identified by their size, complex post-synaptic specializations, and characteristic point of termination on the neuron. The most striking change is a central accumulation of granular material that displaces the mitochondria and synaptic vesicles. The terminals still measure $2 \times 4\mu$. The vesicles remain spherical but are slightly larger than those in the normal material, measuring 400–700Å in diameter.

The material found in the degenerating terminals consists of clusters of roughly spherical granules, approximately 200Å in diameter. Their staining is highly variable like the glycogen in normal terminals. Unlike the normal glycogen, it is aggregated in rosettes or alpha-particles, an unusual configuration in nerve terminals.

FIGURE 3 Degenerating synapse. The axon terminal contains an accumulation of lightly staining, alpha-glycogen particles (g), mitochondria (m), and synaptic vesicles (v). The characteristic synaptic specializations (s) are present as are puncta adhaerentia (a).

FIGURE 4 Degenerating synapse. Densely staining, alpha-glycogen particles (g) and the characteristic synaptic specializations (s) are present. All calibration lines represent 0.5$\mu$.

## DISCUSSION

The monosynaptic connections from dorsal roots to ventral horn cells can be identified both in normal and lesioned material by virtue of their characteristic morphology and position on the neuron. These synapses would be classified by Bodian (1966) as *type R synaptic bulbs* and by Conradi (1969a) as *M-endings*. In Gray's classification (1959), they have some of the characteristics of both type I and type II synapses but best represent an intermediate type. In line with Gray's observations, these synapses are not axo-somatic, as are type II, or wholly axo-dendritic, as are type I, but lie at the fringe between these two parts of the neuron. It is not surprising that they are of an intermediate type. Furthermore, their presence on the proximal dendrite suggests a precise localization of synapses on the motoneuron similar to that seen by Cajal (1909) on the Purkinje cell and by Lorrente de No (1934) on the hippocampal pyramidal cell.

In cats with dorsal root lesions, Conradi (1969b) has described the loss of synapses similar in appearance to those described here. No reactive changes were seen, but rather the synapses could not be found in the lesioned material. He postulates that the synapses were phagocytosed by glial cells. The rat differs from the cat in that there is an accumulation of glycogen in these degenerating terminals, although the synapses have similar morphology.

Such an accumulation of glycogen in degenerating terminals has not been described previously. However, glycogen does accumulate in the various elements of the nervous system following injury. Most noticeably, it occurs in astrocytes (Maxwell and Kruger, 1965) but can occur in neurons and their processes (Pick, 1965). The usual forms taken by degenerating synapses are described as (1) shrinkage and condensation of the synapse, (2) an accumulation of neurofilaments, or (3) phagocytosis of the terminal (Gray and Guillery, 1966). Furthermore, the glycogen is of an unusual configuration, being alpha-particles. A number of mechanisms for this accumulation can be suggested: (1) increased accumulation of glycogen associated with the breakdown of organelles, (2) decreased breakdown secondary to a decreased metabolic rate, and (3) accumulation of an abnormal glycogen that cannot be degraded.

The large size of these synapses, or more precisely the large area of their active zones, seems correlated with the large size of its *EPSP*, which is the most prominent excitatory potential seen in the motoneuron. The large size of the active zone means a large surface for ion flow and hence a large current acting on the *whole neuron resistance* (Rall, 1959). The result is a large amplitude potential. This, along with the location of the synapse close to the cell body, explains not only the large size of the potential but its rapid rise and fall times. Theoretically, synapses located more distally on the dendrites should have long rise times and extended falling phases because of the longer time constant imposed by the high internal resistance of the dendrites. Just the reverse is true of the monosynaptic *EPSP*, since it is generated close to the recording site or the cell body. This leaves unexplained the properties of (1) being unaffected by small polarizing currents, (2) having a small impedance change, and (3) arithmetic summation. All these would suggest a remote termination for the synapses unless the active zones were insulated from the cell body. In other words, there is a high resistance between these zones and the intracellular electrode. Structurally, these high resistance components can only be the synaptic thickenings and subsynaptic granules, which are thicker and denser than those of axo-somatic synapses. Axo-somatic junctions, particularly those classified by Gray as type II, characteristically have thin post-synaptic thickenings and should be easily influenced by polarizing currents (Eccles, 1964). If indeed type II synapses are inhibitory, they show marked potential change with polarizing currents (Coombs, Eccles, and Fatt, 1955a). In the monosynaptic excitatory ending, the post-synaptic thickenings, subsynaptic granules, or both might act as barriers to ion diffusion, resulting in a high resistance pathway between the active zone and the cell body. These structures might

insulate the active zone from the shunting effects of surrounding synapses. This is compatible with the physiology of the monosynaptic reflex, since it is relatively insensitive to jamming.

Morphologically, these synapses have all of the characteristics of chemical synapses (Peters, Palay, and Webster, 1970). Their pre-synaptic axon contains many mitochondria and vesicles. There is a synaptic widening or cleft that contains an intercellular plaque. And finally, no tight junctions nor evidence of electrical coupling are present. The vesicles are predominately spherical and compatible with Bodian's thoughts (1966) that such vesicles are filled with excitatory transmitter. Physiologically, the chemical nature of the synapse expresses itself as a reversal potential. The junctional specializations might act as high resistance pathways in order to functionally isolate the action of the transmitter substance. It is not known if these specializations act as simple barriers to diffusion, an ion exchange barrier, or an area of active transport. Any of the mechanisms could account for an increased resistance.

## SUMMARY

Monosynaptic terminals measure approximately $2 \times 4\mu$, have wide post-synaptic thickenings and subsynaptic granules, and are located on the proximal ends of the primary dendrites. They contain spherical synaptic vesicles, 300–600Å in diameter, mitochondria, microtubules, neurofilaments, smooth endoplasmic reticulum and beta-particles of glycogen. Following dorsal root section, alpha-glycogen particles accumulate in these synapses. The synaptic vesicles remain predominately round but are slightly larger, measuring 400–700Å in diameter. A possible function for the usually wide post-synaptic thickenings and subsynaptic granules is discussed.

## REFERENCES

Bodian, D., 1966, Synaptic types on spinal motoneurons: an electron microscopic study, *Bull. Johns Hopkins Hosp.* **119**: 16–45.
Burke, R. E., 1967, Composite nature of the monosynaptic excitatory postsynaptic potential, *J. Neurophysiol.* **30**: 1114–1137.
Cajal, S. Ramon y., 1952, Histologie du System Nerveux, Consejo Superior de Investigaciones Cientifias Instituto Ramon y Cajal, Madrid, 1909–1911, vol. II: 1–71.
Couteaux, R., 1961, Principaux critères morphologiques et cytochimiques utilisables aujourd'hui pour definir les divers types de synapses, *Actualités Neurophysiol.*, 3e série: 145–173.

Couteaux, R., 1963, The differentiation of synaptic areas, *Proc. Roy Soc. (Lond). Ser. B.* **158**: 457–480.

Conradi, S., 1969, Ultrastructure and distribution of neuronal and glial elements on the motoneuron surface in the lumbosacral spinal cord of the cat, *Acta Physiol. Scand. Supp.* **332**: 5–48.

Conradi, S., 1969, Ultrastructure of dorsal root boutons on lumbosacral motoneurons of the adult cat, as revealed by dorsal root section, *Acta Physiol. Scand. Supp.* **332**: 85–115.

Coombs, J. S., Eccles, J. C., and Fatt, P., 1955, The specific ionic conductances and the ionic movements across the motoneuronal membrane that produces the inhibitory post-synaptic potential, *J. Physiol.* **130**: 326–373.

Coombs, J. S., Eccles, J. C., and Fatt, P., 1955, Excitatory synaptic action in motoneuron, *J. Physiol.* **130**: 374–395.

Del Castillo, J., and Katz, B., 1954, The membrane change produced by the neuromuscular transmitter, *J. Physiol.* **125**: 546–565.

Drochmans, P., 1962, Morphologie du glycogène, *J. Ultrastruct. Res.* **6**: 141–163.

Eccles, J. C., 1964, *The Physiology of Synapses*, Academic Press, New York, pp. 11–23 and 152–157.

Farquhar, M. G. and Palade, G. E., 1963, Junctional complexes in various epithelia, *J. Cell Biol.* **17**: 375–412.

Gray, E. G., 1959, Axo-somatic and axo-dentritic synapses of the cerebral cortex: an electron microscope study, *J. Anat. (Lond.)* **93**: 420–433.

Gray, E. G., and Guillery R. W., 1966, Synaptic morphology in the normal and degenerating nervous system, *Intern. Rev. Cytol.* **19**: 111–182.

Lorrente de No, R., 1934, Studies on the cerebral cortex. II. Continuation of the study of the ammonic system, *J. Psychol. u. Neurol.* **46**: 113–177.

Maxwell, D. S., and Kruger, L., 1965, The fine structure of astrocytes in cerebral cortex and their response to focal injury produced by heavy ionizing particles, *J. Cell Biol.* **25**: 141–157.

Peters, A., and Palay, S. L., 1966, The morphology of laminae A and $A_1$ of the dorsal nucleus of the lateral geniculate body of the cat, *J. Anat. (Lond.)* **100**: 451–486.

Peters, A., Palay, S. L., and Webster, H. de F., 1970, *The Fine Structure of the Nervous System. The Cells and their Processes.* (Hoeber Medical Division, Harper and Row, New York), pp. 132–156.

Pick, J., 1965, The fine structure of sympathetic neurons in X-irradiated frogs, *J. Cell Biol.* **26**: 335–351.

Rall, W., 1959, Branching dendritic trees and motoneuron membrane resistivity, *Exptl. Neurol.* **1**: 491–527.

Rall, W., Burke, R. E., Smith, T. G., Nelson, P. G., and Frank, K., 1967, Dendritic location of synapses and possible mechanisms for the monosynaptic EPSP in motoneurons, *J. Neurophysiol.* **30**: 1169–1193.

Revel, J. P., Napolitano, L., and Fawcett, D. W., 1960, Identification of glycogen in electron micrographs of thin sections, *J. Biophys. Biochem. Cytol.* **8**: 575–589.

Revel, J. P., 1964, Electron microscopy of glycogen, *J. Histochem. Cytochem.* **12**: 104–114.

Smith, T. G., Wuerker, R. B., and Frank, K., 1967, Membrane impedance changes during synaptic transmission in cat spinal motoneurons, *J. Neurophysiol.* **30**: 1072–1096.

Taxi, J., 1961, Études de l'ultrastructure des zones synaptiques dans les ganglion sympathiques de la Grenouille, *Comp. Acad. Sci. (Paris)* **252**: 174–176.

Wuerker, R. B., and Palay, S. L., 1969, Neurofilaments and microtubules in anterior horn cells of the rat, *Tissue & Cell* **1**: 387–402.

# BACKGROUND NOISE CONTENT OF EVOKED RESPONSES FROM FACIAL NUCLEUS OF CAT

CHARLES D. WOODY†, EUGENE K. HARRIS‡ and GUSTAV BROZEK

*Institute of Physiology*
*Czechoslovakian Academy of Sciences, Prague*

A correlation dependent technique for determining signal: noise levels in serial evoked responses was used to measure background noise activity in acute and chronic cat preparations. In the former, topographic mapping of an area of the brain stem concerned functionally with performance of an eye blink showed no significant change in background noise over several mm of the signal space. The averaged amplitudes and signal: noise ratios of the evoked potential responses were maximal at the area of the facial nucleus, the motor nucleus through which information related to the movement was transmitted.

Changes in the background noise of evoked electrical responses were observed during performance of a conditioned eye blink. These increases in background noise above levels associated with the unconditioned blink to glabella tap appeared to depend on the nature of the afferent input rather than on learning effects per se. That is, the cortically mediated input was noisier than that from the trigeminal system.

Changes in background activity of less than 10 $\mu$V following the administration of Flaxedil were within the sensitivity of the measurement technique.

Harris and Woody (1969) have demonstrated a technique for determining the signal:noise (S/N) content of serial electrical potentials. Signal is defined as the waveform of highest mutual intercorrelation among the serial responses, and noise is taken as the associated electrical activity about the signal mean. The method has been tested on artificial signals in noise of wide amplitude range and bandwidth and has been found accurate.

The present investigation explores the merit of two possible applications of this technique to neuroelectric data analysis. One question is whether the amplitude of background noise changes significantly over a space of several mm surrounding a brain stem nucleus through which information related to performance of a motor response is transmitted. It is of interest to evaluate changes in the noise amplitude over the space of the brain surrounding a focal generator concerned with the performance of a particular function. Previous studies, Papez (1927), Woody and Brozek (1969 a, b), suggest that the facial nucleus constitutes such a current source during performance of an eye blink.

The second question is whether the amplitude of background noise at a particular locus changes as a function of learning. Since such behavioral modifications will result in altered motor performance, it will also be useful to evaluate changes in the background noise as a function of modification of motor activity as, for example, following the administration of muscle paralyzing drugs.

## METHODS

### Acute preparations

The evoked electrical response to glabella tap was recorded in nine acute cat preparations. Stainless steel electrodes of 0.1 mm diam. and 50–100$\mu$ uninsulated tip were positioned stereotaxically in the plane of the facial nucleus. Monopolar recordings referred to a grounded electrode in neck muscle were taken every one or two mm along the plane of insertion. Anesthesia was with sodium pentobarbital (30 mgm/kg. i.v. or 35 mgm/kg. i.p.) and

† Supported in part by U.S. and Czechoslovakian Academies of Science and by the Harvard Moseley Fellowship (C.D.W.); Dr. Woody's present address is: UCLA Mental Retardation Center, 760 Westwood Plaza, Los Angeles, Calif.

‡ Laboratory of Applied Studies, NIH, Bethesda, Md.

was maintained at such a level that the corneal reflex was preserved. The electrode was inserted stepwise in one direction to minimize problems of tissue drag with electrode movement. The plane of insertion was 30° from the vertical in order to avoid the bony tentorium. In some of the preparations, the electrode also passed through the medial portion of the spinal trigeminal nucleus.

Recordings were made rapidly, usually within the period of one hour. No more than one electrode tract on each side was made. No differences in lateralization of the electrical responses were found, and no distinction was made in this regard in presenting the data.

At the conclusion of the experiments, electrolytic lesions (10 mamp, 5 sec) were made at known electrode positions. The location of the electrode tract in relation to the facial nucleus was confirmed in all animals by histologic examination of serial sections of the brainstem stained with cresyl violet.

Tap to the glabella was delivered once per second by means of a small air hammer attached to a screw implanted in the bone. The adequacy of periosteal receptors for mediation of the glabella reflex is supported by Rushworth's (1962) observations following injection of xylocaine in the skin overlying the glabella in man. A parallel air path to an earphone elicited a trigger signal which was locked in time to the delivery of the tap with less than 0.5 msec variation. A tap of just subthreshold intensity to produce a blink in the anesthetized preparation was used. The same stimulus intensity produced a blink in the awake preparation. Detailed studies of the evoked response at facial nucleus under both intensities of stimulation have been presented elsewhere together with further information regarding the experimental techniques, Woody and Brozek (1969a). Both the responses to sub and suprathreshold stimulation appear to reflect synaptic activity in the facial nucleus related to production of the blink as opposed to the action potentials of the neurons. The use of the subthreshold stimulus avoided possible complication of the results by blink related EMG components.

Anesthesia was used to facilitate rapid recording at multiple loci. Terminally, d-tubocurarine (0.5 mgm/kg.) was injected in the anesthetized preparations to produce muscular paralysis followed by anoxic death. By studying the persistence or disappearance of the evoked response during this sequence of events, the presence of electrostatic, myographic or mechanical movement artefacts was excluded and the neuronal origin of the observed

potential shifts was supported, Erickson et al (1961).

## Chronic preparations

Changes in signal:noise levels of evoked responses recorded chronically from the facial nucleus of cats were studied before and after conditioning of a blink reflex. In five animals a conditioned blink reflex was established by pairing an initially neutral click with suprathreshold glabella tap. The click preceded the tap by 25 or in other cases 400 msec. The reflex was considered established when a 90% level of response to the conditional stimulus was reached. This normally required five or more training sessions of 150 pairings each. Further details on the method and on the characterization of the conditioned evoked response at the facial nucleus are available elsewhere, Woody and Brozek (1969b).

Electrodes were implanted in the same way as in the acute preparations except that aseptic precautions were taken and a Sheatz (1961) pedestal was provided to facilitate electrode connections. Following surgery the animals were observed closely, and appropriate measures were taken to relieve any discomfort. The recording of potentials was performed with the animals comfortably restrained. Some recordings were made in conditioned preparations following the administration of sufficient gallamine triethiodide (Flaxedil) or d-tubocurarine to produce muscular paralysis. Artificial respiration was provided via a previously implanted tracheal cannula. During such immobilization the animals were unrestrained, and no procedures were carried out that would not be tolerated in the normal state.

## Analysis of data

The electrophysiologic data were led through conventional EEG amplifiers to an FM tape recorder. The system was band limited to the range of 10–4000 Hz. Digitization and processing of the recorded data were performed on a standard LINC computer.

Stimulus locked averages of 128 evoked responses were computed as were averages of the same responses, corrected for variation in latency by the adaptive filter method, Woody (1967). Errors in averaged response amplitude can result when uncorrected variations in latency of the individual responses following the initial stimulus are present, Woody (1967), Woody and Brozek (1969 a). Small variations in the latency of activation and discharge of involved neural elements have been demonstrated even along primary afferent pathways, Darian-Smith *et al.* (1965). Use of an adaptive filter employing correlation to correct for shifts in latency, Woody (1967), can reduce such errors. By the latter technique, individual responses in the present study were cross correlated against a template average. The responses were shifted by the amount of time necessary to yield a maximum in the above correlation coefficient and were then

averaged in this new time alignment. This average replaced the old template, and the process was iterated until no further change in the correlation determined average (XCAV) occurred.

Cross correlation of the final average (XCAV) with each of the individual responses, shifted by the appropriate time value as determined above, yielded a series of 128 correlation coefficients, the average of which was termed $\bar{r}$. As shown in Figure 1A for studies of artificially generated signals in noise, $\bar{r}$ varies as a function of the signal:noise

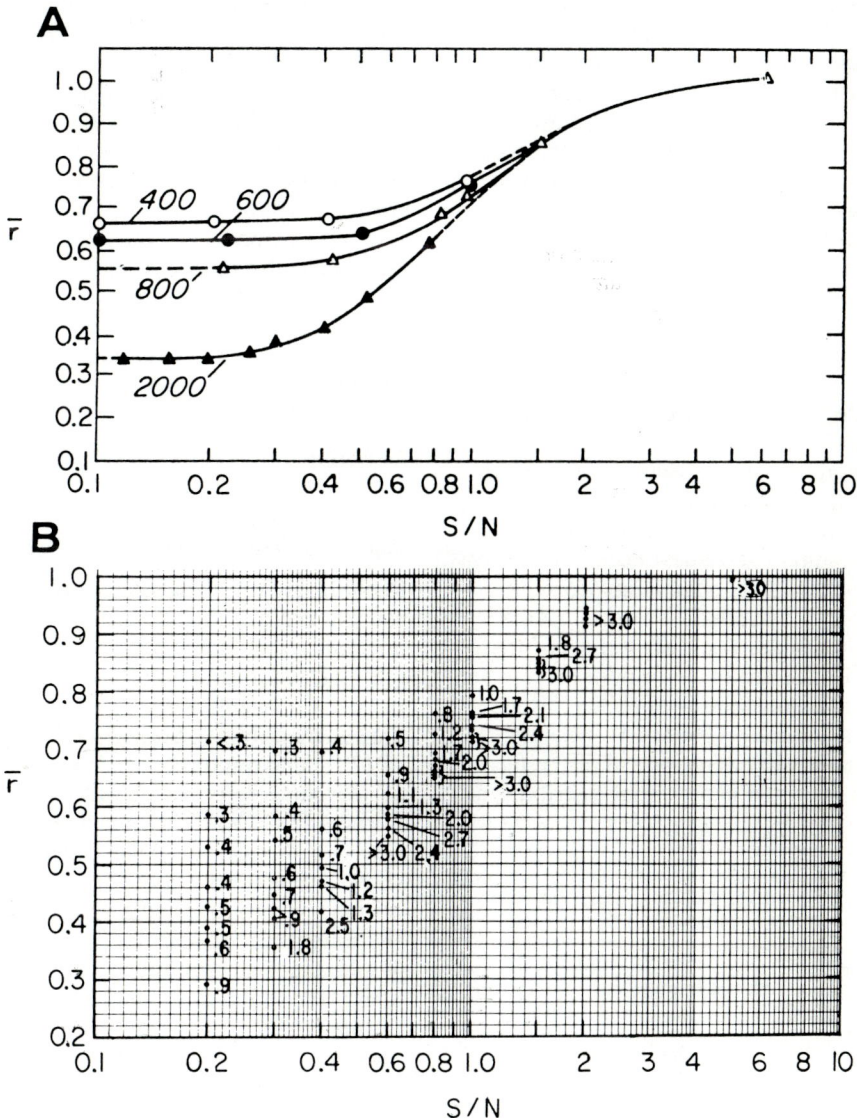

FIGURE 1   (A) Plot of $\bar{r}$ vs S/N for artificial sinusoidal signals (140 Hz) in "white" noise pass filtered between 100 Hz and upper frequency range as shown. (B) Plot of $\lambda$ vs S/N for comparable data to that in A. Numerical values of $\lambda$ are shown as plotted for respective vlaues of $\bar{r}$ in relation to the ordinate. To find S/N one enters the graph at the appropriate value of $\bar{r}$, moves laterally to the associated value of $\lambda$, and then reads directly the corresponding value of S/N.

levels of the responses and the frequency band of the noise relative to that of the signal. The pronounced effect of noise bandwidth relative to the frequency of the signal on the determination of S/N is clearly seen in this figure. As the bandwidth of the noise approaches the frequency of the signal, the covariance reflected in the correlation coefficient, $\bar{r}$, increases. It will be recalled that signal is defined as the waveform of highest mutual inter-correlation among the responses; noise is the variation in associated electrical activity about the signal mean.

The information in Figure 1A was derived from the analysis of sinusoidal waveforms of known R.M.S. amplitude mixed with the noise of known R.M.S. amplitude from a random noise generator. The computed values of $\bar{r}$ from the mixed artificial data were plotted against the known signal:noise (S/N) ratios of the same data. A Krohn-hite 330MR filter was used to reduce the band width of the noise towards that of the 140 Hz signal. R.M.S. voltage measurements of the noise were made after passage through the filter. Further details of more extensive analyses of related data are available elsewhere, Harris and Woody (1969).

As indicated in Figure 1A, a knowledge of correlation alone, $\bar{r}$, is insufficient to provide an accurate estimate of the S/N ratio of the analyzed data. For estimation of the signal:noise ratio, particularly in the mid-range of the graph, an additional variable besides $\bar{r}$ is needed. Examination of the time shifts ($\Delta t$ values) needed to bring individual responses into maximum correlation with the latency corrected average (XCAV) provides such a variable. The distribution of these shifts, taken over the group of responses, can be well fitted by an exponential probability distribution, Harris and Woody (1969). A measure, $\lambda$, (Table I) corresponds to the reciprocal of the mean of the width of the fitted exponential distribution.

### TABLE I

Estimation of $\lambda$ from frequency distribution of $\Delta t$ shifts based on 15 bin histogram for 128 responses. Values of $\lambda$ calculated from the number of responses in three largest bins are shown.

| $\lambda$ | No. of responses | | |
|---|---|---|---|
|  | Bin 1 | Bin 2 | Bin 3 |
| .4 | 42 | 28 | 19 |
| .6 | 57 | 32 | 17 |
| .8 | 70 | 32 | 15 |
| 1.0 | 81 | 30 | 11 |
| 1.5 | 99 | 22 | 5 |
| 2.0 | 111 | 15 | 2 |
| 2.5 | 117 | 10 | 1 |
| 3.0 | 122 | 6 | 0 |

It can be related to the S/N ratio of the analyzed data and is weakly dependent on $\bar{r}$. Figure 1B shows this relationship plotted on the same ordinate as $\bar{r}$ for data comparable to that in Figure 1A. Obviously with a knowledge of $\bar{r}$ and $\lambda$ it is possible to estimate the signal:noise ratio of analyzed data directly from this graph.

The $\lambda$ variable represents the effect on signal alignment by the correlation process of noise which is correlated but out of phase with the signal. Variability in the latency of the signal, if sufficiently great, could also affect the determination of $\lambda$. Since $\lambda$ was calculated from the histogram of the $\Delta t$ shifts, divided into 15 equal time bins over the total record length, T, variation in latency of less than one bin width did not significantly affect the $\lambda$ determination. Variation in latency of the neurophysiologic data presented in this report was normally within one bin width of the period, T, as determined from those preparations in which noise was sufficiently small that only variation in $\lambda$ due to physiologic shifts in latency was observed. Thus variation in $\lambda$ mainly reflected signal associated noise rather than variation in latency of the responses.

## RESULTS

A. *The amplitude of background noise over a signal space of several mm in the brain stem*

Recordings of evoked electrical responses to glabella tap were made in nine cats at 1 to 2 mm intervals along the tract of an electrode penetrating the cerebellum and brainstem in the direction of the facial nucleus. Results in a typical preparation are shown in Fig. 2. Stimulus locked averages (AV) of 128 single responses are compared with averages of the same responses corrected for variation in latency (XCAV). The position of the electrode tracts in this (x) and four other preparations are shown. The corrections of variation in latency resulted in improvements of as much as 100% in the amplitude of some of the response averages. While the variation in latency was sufficient to affect the amplitude of some averages, it was normally less than 1/15th of the response length, T, and thus did not affect the noise determination.

At the level of the medial border of the spinal trigeminal nucleus ($+1$, $+2$), a small, early negative component of the evoked response average was seen corresponding to the N1 wave reported by other investigators, Vyklicky *et al* (1967), Baldissera *et al* (1967). It was associated with a larger, slower

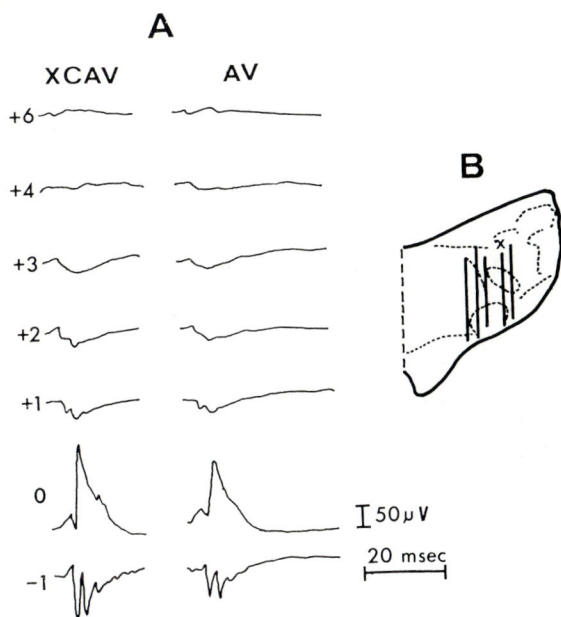

**A**

XCAV   AV

+6
+4
+3
+2
+1
0
−1

I 50 μV
20 msec

**B**

FIGURE 2 (A) Averages of typical evoked responses to glabella tap along path of electrode penetrating brainstem towards facial nucleus. AV is stimulus locked average. XCAV is average of same responses corrected for variation in latency by adaptive filter technique. Numbers are distance in mm. from superior border of facial nucleus. Stimulus delivery synchronous with start of responses. Calibrations as shown; positive up. (B) Relation of electrode tracts in five preparations to facial nucleus (x is preparation described in A).

to peak signal amplitude (S) of the respective response averages were computed for the group of preparations. An example is shown in Table II. On the basis of this data it was possible to assess the degree of correlation between the averaged amplitude (S) measure of response signal and the S/N ratio. Taking maximal values of the respective variables as 100% and expressing values at different recording locations as a percent of the maximum, graphs of S and S/N variation over distance were constructed. Examples are shown for the five preparations in which the electrode tract passed directly through the facial nucleus (Figure 3). There was an obvious correlation between S as determined from peak to peak amplitude response measurements and S/N as determined from measurements of correlation in the individual animals. A regression plot for all measurements in the nine preparations revealed a positive correlation of 0.82 between S and S/N.

The assumption was then made that S, computed from direct measurements of averaged response amplitude, was nearly equivalent to the S term in the S/N computation. This assumption appears to be warranted for areas in which voltage amplitude of the response averages is a reliable indicator of signal content (see method of determining S/N for artificial signals in noise, Harris and Woody (1969). N, the background noise activity, was then calculated directly from S and S/N. Grouped results from the five preparations with electrode tracts through the facial nucleus are shown in Figure 4. Signal (S) and background noise activity (N) were grouped according to the position of the recording electrode, relative to the facial nucleus in the five preparations and plotted over distance. The signal emerged significantly from the background noise at precisely the level of the facial nucleus (0). No significant differences in the background noise at the different locations were found.

negativity (N2 wave) with superimposed, small polyphasic components. At the superior border of the facial nucleus (0), a much larger response with predominantly positive components appeared. At the inferior border of the facial nucleus (−1), the response was again predominantly negative.

The averaged correlation coefficient ($\bar{r}$), the distribution of Δt shifts (λ), and the maximal peak

TABLE II

Changes in signal:noise levels of evoked responses at different recording locations in a single preparation.

| Distance | $\bar{r}$ | λ | S/N | S/N% | SuV | S% | NuV | N% |
|---|---|---|---|---|---|---|---|---|
| +6 | .73 | .4 | .5 | 19 | 10 | 7 | 20 | 14 |
| +5 | .68 | .6 | .6 | 22 | 20 | 14 | 33 | 24 |
| +4 | .73 | .3 | .2 | 74 | 20 | 14 | 100 | 71 |
| +3 | .85 | 2.1 | 1.5 | 56 | 60 | 43 | 40 | 29 |
| +2 | .83 | .7 | 1.0 | 37 | 60 | 43 | 60 | 43 |
| +1 | .93 | 2.4 | 2.0 | 74 | 130 | 93 | 65 | 46 |
| 0 | .97 | 1.3 | 2.7 | 100 | 140 | 100 | 52 | 37 |
| −1 | .84 | 3.0 | 1.5 | 56 | 90 | 64 | 60 | 43 |

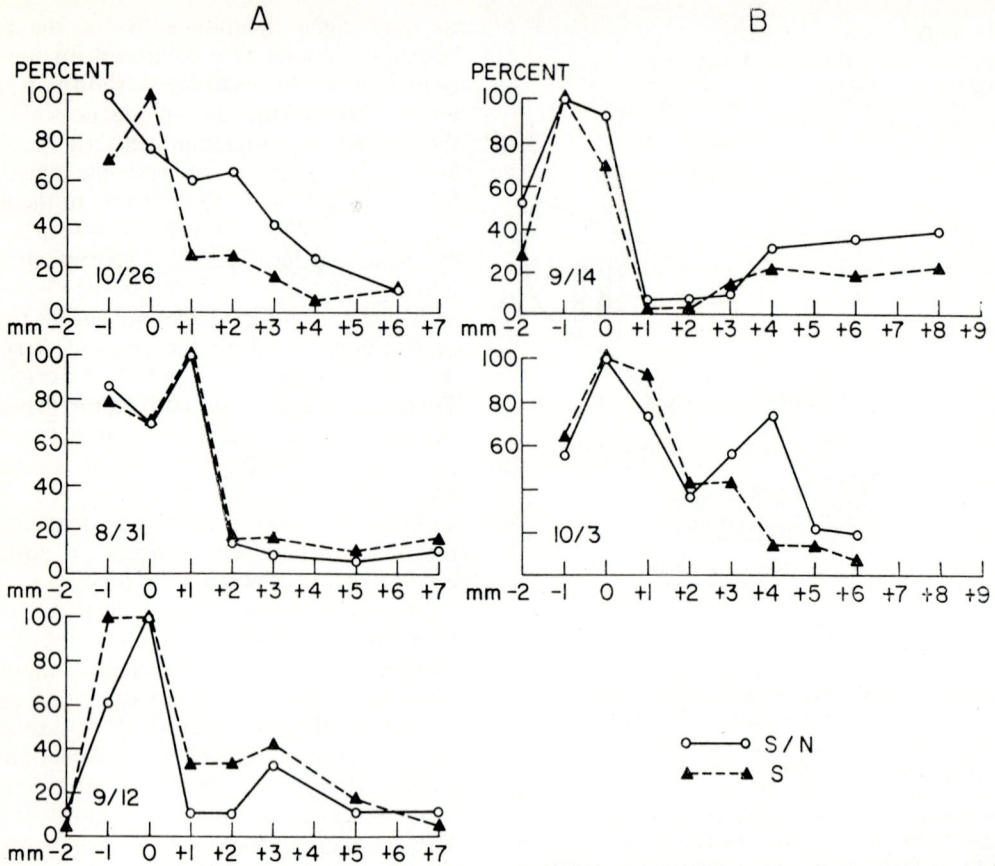

FIGURE 3   Plots of S and S/N vs distance from facial nucleus in five cats. S and S/N expressed as % of respective maxima.

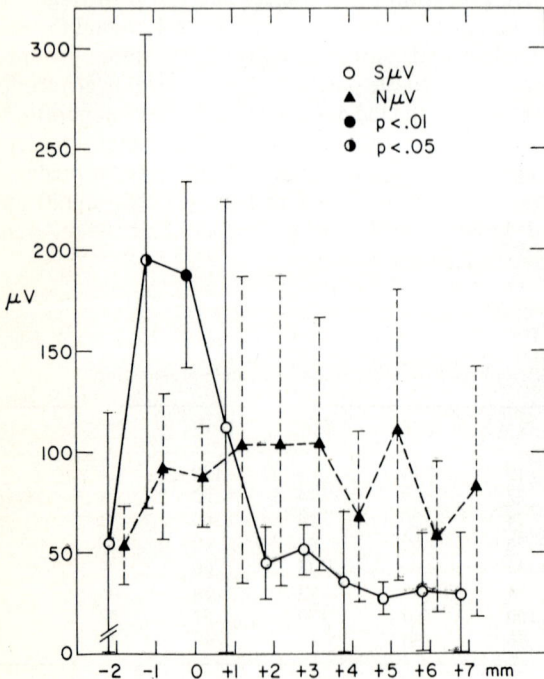

B. *Changes in signal and noise of evoked responses recorded from a single locus in chronic preparations.*
1. *Unconditioned evoked response to glabella tap before and after muscular paralysis.*

In each of three awake cats 128 evoked responses to glabella tap were recorded from the facial nucleus before and after muscular paralysis with Flaxedil. S, S/N and N were calculated as before, assuming equivalence of signal in S and S/N. In the three preparations the level of noise present in the evoked responses was significantly ($p < 0.05$) reduced following Flaxedil administration. The signal amplitude was unaffected or, in two preparations, very slightly increased after paralysis. EMG artefacts of the sort reported by Bickford *et al.* (1964) and Prichard *et al.* (1965) were not present as recognizable components of the averaged evoked

FIGURE 4   Plot of grouped values of signal (S) and background noise activity (N) vs distance from facial nucleus. Amplitude of S and N expressed in $\mu$V. Standard deviations are as shown. The signal is significantly greater than the noise at the level of the facial nucleus.

TABLE III

Changes in signal and noise level in chronic preparations.

| CAT | RESPONSE | $\bar{r}$ | $\lambda$ | S/N | SuV | NuV |
|-----|----------|------|------|------|-----|------|
| CC3 | UR FLAX | .92 | 1.6 | 1.9 | 112 | 63 |
|     | UR | .88 | 1.2 | 1.4 | 100 | 71 |
|     | CR | .80 | .7 | .86 | 94 | 109 |
|     | NAIVE | .49 | .4 | .22 | 33 | 150 |
| CC1 | UR FLAX | .84 | 3.0 | 1.5 | 5 | 3.3 |
|     | UR | .52 | 2.0 | .5 | 5 | 10 |
|     | CR | .45 | .8 | .31 | 5 | 16.2 |
| FLX | UR FLAX | .77 | .6 | .76 | 25 | 33 |
|     | UR | .69 | .5 | .53 | 20 | 38 |
| CC2 | UR | .80 | .9 | 1.0 | 53 | 53 |
|     | CR | .75 | .4 | .5 | 41 | 82 |
|     | NAIVE | .63 | .4 | .33 | 25 | 76 |
| NWT | CR | .79 | 1.1 | 1.0 | 63 | 63 |
|     | NAIVE | .65 | 1.8 | .7 | 28 | 40 |

responses. Without exception the correlation of the individual responses with each other and their signal:noise ratios increased significantly ($p < 0.05$)† after administration of the drug (Table III).

2. *Signal and noise of conditioned vs unconditioned blink evoked response.*

In three preparations evoked responses were recorded from the facial nucleus in relation to a conditioned and unconditioned eye blink. The conditional stimulus was an initially neutral click (no blinking) which acquired significance and elicited a blink after repeated pairing with the unconditional stimulus of glabella tap (for further details see Woody and Brozek (1969b)). Without

† Confidence limits for S/N will reflect statistical sampling errors in the estimates $\bar{r}$ and $\lambda$. The $\lambda$-estimate will have a small percentage error because of the large number of replicate responses. As can be seen in Figure 1B, a small change in $\lambda$ for given $\bar{r}$ will have very little effect on S/N through the range of S/N 0.2–1.0. Therefore, for practical purposes, confidence limits for S/N can be approximated from the sampling errors in $\bar{r}$ alone. As noted by Harris and Woody (1969), $\bar{r}$ can be assumed normally distributed. In the present study, 95% confidence limits for $\bar{r}$ were estimated (conservatively) by extrapolating the sample $\bar{r}$ to the range for the normally distributed population (see chart in Wilks 1951) assuming a "sample size" of 128, that is, assuming that the estimated $\bar{r}$ was at least as reliable as if it had been derived from a set of 128 independent single observations. Using Figure 1B, 95% confidence limits for S/N were then calculated from the confidence range of $\bar{r}$ assuming fixed $\lambda$.

Other tests of significance, comparing responses from the facial nucleus in chronic preparations, were made by computing the mean difference between log $(S/N)_1$ and log $(S/N)_2$ or between $(N)_1$ and $(N)_2$, where subscripts denote the two conditions being tested in each cat, to its observed standard error, based on three animals in each case.

exception higher coefficients of correlation ($\bar{r}$) and higher signal:noise ratios ($p < 0.05$ each animal) were associated with the unconditioned evoked responses than with the conditioned responses (Table III). Noise levels tended to be higher with the conditioned response ($p = 0.10$). However, EMG responses recorded from orbicularis oculi were larger and better defined with the unconditioned blink. Thus the changes in noise level in this case were not attributable to alterations in EMG noise but rather to what is presumed to represent changes in background neural activity.

3. *Signal and noise of naive vs conditioned response.*

Evoked responses to click in naive cats (no blink) were compared to responses in the same preparations after the establishment of a conditioned blink reflex. Correlation coefficients and S/N ratios were significantly higher ($p < 0.05$) for the conditioned responses than for those in the naive animals (Table III). Signal amplitude and S/N ratio clearly increased with learning of the conditioned response. However, noise levels in naive animals were comparable to those in the conditioned animals, both being increased relative to noise levels associated with performance of the unconditioned blink response.

DISCUSSION

The preceding results suggest that the S/N ratio appears to be as useful as the averaged amplitude of serially evoked responses in mapping the topographic representation of a response to a particular stimulus. With repeated glabella tap the signal: noise ratio of the evoked electrical responses recorded over distances of several mm reaches a maximum at the level of the facial nucleus. This is the motor nucleus to which information related to the stimulus is transmitted. A previous study, Woody and Brozek (1969a), has presented evidence for the relatively uncomplicated pattern of current spread about this nucleus. The high correlation between the amplitude of the averaged responses and their S/N ratios suggests that in this experimental situation both measures of signal content are equally powerful.

In situations where current spread between signal generators and the recording electrode is complex, Rall (1962), Rall and Shepherd (1968), and Humphrey (1968), one might expect the correlation determined S/N ratio to be a more reliable indicator of signal activity than the amplitude measure. This is because the correlation between serially evoked responses at a single recording location depends on their mutual linear dependence or in a sense their power spectral densities more than their mean amplitude.

The assumption that recruitment of some elements of a large neuronal pool during signal generation must be accompanied by a significant reduction in the total noise of the system is not supported by our findings. Significant changes in the background noise level were not found over the space of brain stem studied during glabella tap. Yet signal amplitude and signal:noise ratios increased in the region of the facial nucleus. Thus it appears that increases in evoked response amplitude can occur somewhat independently of changes in the associated background activity.

Background noise is significantly greater at the facial nucleus during performance of a conditioned blink than during performance of an unconditioned blink movement. Since EMG activity in facial muscles is greater during performance of the unconditioned than the conditioned blink, Woody and Brozek (1969 a, b), and since reduction in EMG activity is associated with a decrease in background noise level, it is reasonable to assume that the increase in background noise with the conditioned movement reflects an increase in background neural activity rather than in activity from non neural sources such as the EMG.

Previous studies, Woody and Brozek (1969b), suggest that the conditioned blink is mediated cortically, while the unconditioned blink is a brain stem reflex. An obvious interpretation of our results is that the cortical input to the facial nucleus

is noisier than the more direct input from the trigeminal afferent system. This effect appears to be independent of learning since noise levels were found to be comparable in naive and conditioned animal evoked response data.

In the present study the absence of change in background noise level as a specific effect of conditioning (i.e., comparable naive vs conditioned response backgrounds) could conceivably have resulted from insensitivity of the measuring techniques. We feel that this is unlikely since the techniques are sensitive to changes of 5–8 $\mu$V in background noise related to muscle activity and, if our interpretation is valid, to changes in noise depending on the type of afferent input.

Of further support to the idea of an intrinsically noisier cortical input is a comparison of evoked activity from the trigeminal system, Darian-Smith et al. (1965) with that from cortical areas projecting to facial nucleus, Woody, Vassilevsky and Engel (1970). Evoked unit activity in the latter system, even after conditioning, is much noisier; that is, alteration in the discharge rate of individual units such as to mediate the conditioned blink response is of such small degree and is so inconsistent that many units must be involved to achieve a statistically significant change in the discharge rate of the whole population.

## SUMMARY

A correlation dependent technique for determining signal:noise levels in serial evoked responses was used to measure background noise activity in acute and chronic cat preparations. In the former, topographic mapping of an area of the brain stem concerned functionally with performance of an eye blink showed no significant change in background noise over several mm of the signal space. The averaged amplitudes and signal:noise ratios of the evoked potential responses were maximal at the area of the facial nucleus, the motor nucleus through which information related to the movement was transmitted.

Changes in the background noise of evoked electrical responses were observed during performance of a conditioned eye blink. These increases in background noise above levels associated with the unconditioned blink to glabella tap appeared to depend on the nature of the afferent input rather than on learning effects per se. That is, the cortically mediated input was noisier than that from the trigeminal system.

Changes in background activity of less than 10 uV following the administration of Flexedil were within the sensitivity of the measurement technique.

## REFERENCES

Baldissera, F., Broggi, G., and Mancia, M., 1967, Depolarization of trigeminal afferents induced by stimulation of brain-stem and peripheral nerves, *Exp. Brain Res.* **4**: 1–17.

Bickford, R. G., Jacobson, J. L., and Cody, T. T., 1964, Nature of average evoked potentials to sound and other stimuli in man, *Ann. N.Y. Acad. Sci.*, **112**: 204–218.

Communications Biophysics Group and Siebert, W. M., 1959, *Processing Neuroelectric Data*. Cambridge, Massachusetts: Technology Press, pp. 1–121.

Darian-Smith, I., Mutton, P., and Proctor, R., 1965, Functional organization of cutaneous afferents within the semilunar ganglion and trigeminal spinal tract of the cat, *J. Neurophysiol.* **28**: 682–694.

Erickson, R. P., King, R. L., and Pfaffmann, C., 1961, Some characteristics of transmission through spinal trigeminal nucleus of rat, *J. Neurophysiol.* **24**: 621–632.

Harris, E. K., and Woody, C. D., 1969, Use of an adaptive filter to characterize signal-noise relationships, *Comp. Med. Biol.* **2**: 242–273.

Humphrey, D. R., 1968, Re-analysis of the antidromic cortical response. I. Potentials evoked by stimulation of the isolated pyramidal tract, *Electroenceph. clin. Neurophysiol.* **24**: 116–129.

Papez, J. W., 1927, Subdivisions of the facial nucleus, *J. Comp. Neurol.* **43**: 159–191.

Prichard, J. W., Chimienti, J., and Galambos, R., 1965, Evoked response from extracellular sites in the cat, *Electroenceph. clin. Neurophysiol.* **18**: 493–499.

Rall, W., 1962, Electrophysiology of a dendritic neuron model, *J. Biophys.* 2 (No. 2, Part 2): 145–167.

Rall, W., and Shepherd, G. M., 1968, Theoretical reconstruction of field potentials and dendrodendritic synaptic interactions in olfactory bulb, *J. Neurophysiol.* **31**: 884–915.

Sheatz, G. C., 1961, Electrode holders in chronic preparations. A. Multihead techniques for large and small animals. In: *Electrical Stimulation of the Brain.* Sheer, D. E., ed. Austin: University of Texas Press, pp. 45–50.

Vyklicky, L., Maksimova, E. V., and Jirousek, J., 1967, Neurones in the reflex pathway between trigeminal sensory fibres in the cat, *Physiol. bohemoslov.* **16**: 285–296.

Wilks, S. S., 1951, *Elementary Statistical Analysis*. Princeton: Princeton University Press, pp. 257–258.

Woody, C. D., 1967, Characterization of an adaptive filter for the analysis of variable latency neuroelectric signals, *Med. biol. Engng.* **5**: 539–553.

Woody, C. D., and Brozek, G., 1969, Gross potential from facial nucleus of the cat as an index of neural activity in response to glabella tap, *J. Neurophysiol.* **32**: 704–716.

Woody, C. D., and Brozek, G., 1969, Changes in evoked responses from facial nucleus of cat with conditioning and extinction of an eye blink, *J. Neurophysiol.* **32**: 717–726.

Woody, C. D., Vassilevsky, N. N., and Engel, J., Jr., 1970, Conditioned eye blink: unit activity at coronal—precruciate cortex of the cat, *J. Neurophysiol.* **33**: 851–864.

# SEROTONIN AND DOPAMINE: DISTRIBUTION AND ACCUMULATION IN *APLYSIA* NERVOUS AND NON-NERVOUS TISSUES

DAVID CARPENTER, GEORGE BREESE[†], SAUL SCHANBERG,[‡] and IRWIN KOPIN

*Laboratory of Neurophysiology and Laboratory of Clinical Science,*
*National Institute of Mental Health, Bethesda, Maryland 20014*

1. Serotonin is present in several ganglia in *Aplysia* in concentrations of 2–4 $\mu$gm/gm and is also present in the heart (0.6–1.3 $\mu$gm/gm).

2. During incubation of auricles *in vitro* serotonin-[3]H is concentrated tenfold from the medium. This accumulation is inhibited by DMI, ouabain and $Na^+$-free (TRIS) solutions. Ganglia incubated *in vitro* concentrate less serotonin-[3]H and at least a portion appears to be bound in non-nervous tissue.

3. Dopamine is present in high concentration (24 $\mu$gm/gm) in some ganglia but is absent in others. There is a large amount of dopamine in the gill and the walls of the branchial vein, probably in sensory neurons.

4. Although dopamine is not present in the auricle, it can be concentrated there, presumably in serotonergic neurons.

## INTRODUCTION

There is an abundance of evidence for a role of the catecholamines and serotonin in neural synaptic transmission in both vertebrate and invertebrate nervous systems (Garattini and Valzelli, 1965; Gerschenfeld, 1966). Norepinephrine, dopamine and serotonin are each found in specific nerve cells which also have enzymatic systems necessary for synthesis and degradation of these compounds. Furthermore, these nerve cells take up these compounds by an active transport process and will concentrate labelled amines several fold over the concentration in the medium (Gillis and Paton, 1967; Bodanski, Tissari and Brodie, 1968).

In Molluscs there is little or no norepinephrine, but relatively large amounts of dopamine (Sweeney, 1963; Cardot, 1963; Kerkut, Sedden and Walker, 1966) and serotonin (Welsh and Moorhead, 1959) with their synthetic and degradative enzymes

†*Present address:* Departments of Psychiatry and Pharmacology, Child Development Institute, North Carolina School of Medicine, Chapel Hill, North Carolina
‡*Present address:* Department of Physiology and Pharmacology, Duke University School of Medicine, Durham, North Carolina

(Cardot, 1963, 1964). Serotonin is found in the nervous systems of many invertebrates and tends to be present in greater amounts in the more primitive animals (Welsh and Moorhead, 1960). Little is known about the role of dopamine in specific nerve pathways but there is evidence that serotonin may be a cardiac-acceleratory neurotransmitter in several Molluscs (Loveland, 1963; S. Rózsa and Graul, 1964). Serotonin has also been implicated as a muscle-relaxing neurotransmitter in *Mytilus* (Twarog, 1954). By use of the highly specific fluorescent method both dopamine and serotonin have been shown to be contained within a few specific neurons and their processes (Dahl, Falck, Lindquist and von Mecklenburg, 1962).

In this paper we describe studies on content and uptake of serotonin and catecholamines in the nervous system of the marine mollusc, *Aplysia californica*, a preparation which has recently been widely used for electrophysiological investigation because of its very large neuronal cell bodies. Specific cells within this preparation are sensitive to catechols and serotonin. Gerschenfeld and Tauc (1964) found a group of cells ('H' cells) which were intensely excited by dopamine in concentrations as low as $10^{-11}$M, while 'CILDA' and 'DINHI' cells were hyperpolarized by dopamine and excited by serotonin.

Preliminary reports have been given of some of these experiments (Breese and Carpenter, 1967; Chase, Breese, Carpenter, Schanberg and Kopin, 1968).

## METHODS

Animals were obtained from Pacific Biomarine Supply Co., Venice, California, and were kept in artificial sea water at 15°C. Tissues were removed from animals pinned to a wax dissecting tray, blotted and weighed.

Serotonin was determined by a modification of the method of Synder, Axelrod and Zweig (1965). After tissue homogenization in cold 0.4N perchloric acid and centrifugation, the serotonin was extracted with butanol from an aliquot of the supernatant brought to pH 9.5. The organic phase was washed and after addition of two volumes of heptane the serotonin was returned to an aqueous phase by shaking with 0.5N phosphate buffer (pH 7). After incubation with ninhydrin for one hour at 60°C, fluorescence was allowed to develop and was read at 495 m$\mu$.

For determination of the catecholamines an aliquot of supernatant was adjusted to pH 8.6 with NaOH and the catecholamines absorbed onto alumina (Woelhm, neutral) by a modification of the method of Anton and Sayre (1962). No endogenous norepinephrine could be found using the fluorometric method of Haggendahl (1963). Dopamine was assayed according to the method of Anton

and Sayre (1964). Chromatography of alumina eluates showed that the only catechol present had the same mobility as authentic dopamine (butanol: acetic acid:water, 4:1:1).

Accumulation of labelled serotonin or catecholamines was determined by incubation of weighed tissues for 30 min. in artificial sea water containing between 200 and 1,000 m$\mu$ curies/ml of tritiated serotonin, dopamine or norepinephrine. When uptake was tested in the presence of drugs or with a change in ionic medium, the tissue was preincubated in the appropriate solution for 15 min. prior to the addition of the isotope.

After incubation, tissues were blotted and homogenized in 0.4N perchloric acid and the homogenate centrifuged at 4°C. An aliquot of the bathing medium and the supernatant was then added to 15 ml of a toluene-ethanol solution (10:4) containing phosphor to determine total radioactivity. To isolate the radioactive serotonin, aliquots of the supernatant and medium were subsequently passed through a $0.3 \times 2$ cm column of Dowex-50 ($NH_4^+$). After washing with 12 ml of water, the serotonin-$^3$H was eluted with 16 ml of an ethyl alcohol: 3N HCl solution (3:1). Norepinephrine-$^3$H and dopamine-$^3$H were assayed as described above. Internal standards of toluene-$^3$H were used to correct for counting efficiency.

## RESULTS

### Serotonin Content in APLYSIA Tissues

Table I shows the distribution of serotonin in heart and nervous tissues of *Aplysia*. In heart there is a

TABLE I

Serotonin content of Heart and Nervous Tissues of *Aplysia*

|  | Tissue | No. (a) | Content ($\mu$gm/gm) | SEM |
|---|---|---|---|---|
| Heart | Auricle | 4 | 1.33 | 0.36 |
|  | Ventricle | 9 | 0.65 | 0.09 |
|  | Crista Aortae | 2 | trace (b) | — |
| Ganglia | Buccal | 2 | trace (b) | — |
|  | Cerebral | 4 | 2.50 | 0.04 |
|  | Pleural and pedal | 5 | 2.05 | 0.35 |
|  | Visceral | 9 | 3.30 | 0.19 |
|  |    Right half | 2 | 4.36 | — |
|  |    Left half | 2 | 3.21 | — |
|  |    Bag Regions | 2 | trace (b) | — |
| Nerves | Connectives | 7 | 0.50 | 0.10 |
|  | Posterior parapodial | 4 | 2.61 | 0.45 |

(a) Number of determinations. Each sample contained tissues from 2–10 animals, with a pooled weight of at least 50 mg.

(b) Fluorescence spectrum differs from authentic serotonin.

significant amount of serotonin in the auricle. There is much less in the ventricle, and none detectable in the cristae aortae, a non-contractile bag-like appendage interposed between the ventricle and the aorta. The cerebral, pleural-pedal and visceral ganglia all contained between 2–4 $\mu$gm serotonin/gm wet weight, but none was found in the buccal ganglion. Five times as much serotonin was found in the posterior parapodial nerve, which runs from the pedal ganglion to the foot musculature, as in the connective nerves, which connect the pleural and visceral ganglia.

Because of the distribution of various cell types in the visceral ganglion, it is of interest to determine serotonin content in various parts of the ganglion; consequently analysis was done separately on the bag regions and the right and left halves of the remainder. The bag region contains a homogeneous population of small neurosecretory neurons which have relatively few synaptic contacts with other neurons (Frazier, Kandel, Kupfermann, Waziri and Coggeshall, 1967). This region did not contain detectable serotonin. The right half of the ganglion differs from the left in that it contains 13 large neurons containing other neurosecretory granules (Frazier, et al., 1967). These granules are prominent when observed by electron microscopy and impart a white color to the cells in contrast to the orange color of the other neurons. No significant differences in serotonin content were found between right and left halves of the ganglion. It appears very unlikely, therefore, that serotonin is present in the neurosecretory granules.

## Studies on Uptake of Serotonin-$^3$H

Table II shows the relative concentration of serotonin-$^3$H from the media by several nervous

and cardiac tissues during 30 min. incubation with sea water containing the labelled amine. Greatest concentration was consistently observed in the auricle and was about ten times the bath concentration. The ventricle concentrated very much less and the cristae aortae did not concentrate the amine.

Unlike the auricle, none of the ganglia or nerves concentrated serotonin more than about threefold, although these tissues normally contained more serotonin than the auricle. Nerve endings in the auricle may be more efficient than the cell bodies in the ganglia in taking up the amine, or the difference in part may reflect the ease of penetration of the amine into the tissues. The very thin walled auricle contracts regularly during the incubation, while the ganglia are much thicker and are not contractile.

When all NaCl in the incubating sea water was replaced by TRIS$^+$ Cl$^-$ [tris (hydroxy-methyl) aminomethane, (pH 7.8) made isosmotic with normal sea water], the auricle was no longer able to concentrate serotonin (Table III). Ouabain, a specific inhibitor of Na$^+$–K$^+$ activated adenosinetriphosphatase, also dramatically depressed the uptake, as did DMI (desmethylimipramine), a drug which has been shown to inhibit the accumulation of the catecholamines (Axelrod, Whitley and Hertting, 1961).

In contrast to the results in auricle the relatively low concentration of serotonin in visceral ganglion was not at all depressed by incubation in Na$^+$ free solutions, although some depression did result from exposure to ouabain or DMI in two experiments. These results suggest that much of the concentration of serotonin-$^3$H in ganglia is a result of a nonspecific binding rather than uptake into nerve cells and terminals. In agreement Ascher, Glowinski, Tauc and Taxi (1968) have shown that auto-

TABLE II

Serotonin-$^3$H accumulation in Heart and Nervous Tissues of *Aplysia* (a)

|  | Tissue | No. (b) | Accumulation Ratio (c) | SEM |
|---|---|---|---|---|
| Heart | Auricle | 25 | 10.55 | 0.93 |
|  | Ventricle | 9 | 2.58 | 0.39 |
|  | Crista Aortae | 3 | 1.26 | 0.16 |
| Ganglia | Visceral | 12 | 2.29 | 0.22 |
|  | Pleural and pedal | 8 | 1.56 | 0.08 |
| Nerves | Connectives | 7 | 3.02 | 0.09 |
|  | Posterior parapodial | 8 | 3.16 | 0.13 |

(a) Tissues incubated for 30 minutes in artificial sea water containing 200–1,000 m$\mu$ curies/ml at 22–25°C.

(b) Number of determinations. Each sample contained tissues from 2–8 animals with a pooled weight of at least 50 mg.

(c) Accumulation ratio is the ratio of tissue-$^3$H medium-$^3$H.

TABLE III

Effects of Drugs and Na$^+$ free sea water on Serotonin-$^3$H uptake (a)

| | Auricle | | | Visceral Ganglion | | |
|---|---|---|---|---|---|---|
| | Accumulation Ratio (b) | SEM | No. (c) | Accumulation Ratio (b) | SEM | No. (c) |
| Control | 11.22 | 1.33 | 15 | 2.20 | 0.43 | 5 |
| Na$^+$ free (TRIS) | 1.26 | 0.46 | 3 | 2.41 | 0.56 | 4 |
| Ouabain 10$^{-4}$M | 3.21 | — | 2 | 1.12 | — | 2 |
| DMI 10$^{-4}$M | 1.68 | 0.24 | 10 | 1.58 | — | 2 |
| LSD 10$^{-4}$M | 16.37 | 3.69 | 6 | 5.35 | — | 2 |
| BOL 10$^{-4}$M | 8.35 | 0.35 | 3 | 1.30 | — | 2 |

(a) Tissues incubated in sea water containing drug for 15 minutes prior to addition of serotonin-$^3$H, then for 30 minutes in presence of isotope.

(b) Accumulation ratio is the tissue/medium ratio. Serotonin-$^3$H concentration was between 200–1,000 m$\mu$ curies/ml.

(c) Number of determinations. Each sample contained tissues from 2–6 animals with a pooled weight of at least 50 mg.

radiography of serotonin-$^3$H in visceral ganglion shows deposits over non-nervous tissues.

Lysergic acid diethylamide (LSD) and bromolysergic acid diethylamide (BOL) are known to block the action of serotonin (Giarman and Freedman, 1965). These agents have not been known to effect uptake, but were tested because of several surprising effects of both drugs on the *Aplysia* heart. As previously reported (Chase, *et al.*, 1968, Figure 2), serotonin-$^3$H taken up by an intact canulated and perfused heart is released on exposure to LSD, with an accompanying dramatic increase in heart rate and strength of contraction and a large spontaneous efflux of radioactivity. BOL causes similar but less marked effects. These results suggested that LSD and BOL stimulated spontaneous release of serotonin so markedly that the ability to concentrate the chemicals might be impaired. The results, however, showed that LSD, but not BOL, appeared to stimulate accumulation to about 150% of control values. The increased uptake is probably not a result only of the stimulation of heart rate (which is dramatic even in the isolated auricle) since there is an increase in uptake in the visceral ganglion as well.

*Catecholamines in Nervous and Non-nervous Tissues*

The distribution of dopamine in nervous tissues is shown in Table IV. The most striking finding is the relatively large amount of dopamine present in the pedal ganglion and the posterior parapodial nerve which runs distally from the pedal ganglion in contrast to the almost complete absence of dopamine in the pleural ganglion and the pleural visceral connective nerve. Smaller but significant

TABLE IV

Dopamine content of nervous tissues of *Aplysia*

| Tissue | No. (a) | Content ($\mu$gm/gm) | SEM |
|---|---|---|---|
| Cerebral ganglion | 4 | 2.61 | 1.05 |
| Buccal ganglion | 4 | 0.80 | 0.35 |
| Pleural ganglion | 4 | trace | — |
| Pedal ganglion | 6 | 24.32 | 7.83 |
| Visceral ganglion | 5 | 1.85 | 0.31 |
| Posterior parapodial nerve | 4 | 5.99 | 2.41 |
| Connective nerve | 4 | trace | — |

(a) Number of determinations. Each sample contained tissues from 4–10 animals with a pooled weight of at least 50 mg.

amounts of dopamine are present in the cerebral, buccal and visceral ganglia.

Norepinephrine was not detected in significant amounts in the auricle, ventrical or the visceral and pedal ganglia.

The ctenidium, which includes the gill and efferent branchial vein and which receives its innervation through nerves from the visceral ganglion (Eales, 1921), had the highest concentration of dopamine (2.9 $\mu$gm/gm) of the non-nervous tissue studied (Table V). The auricle and ventricle, however, which also receive nerves from the visceral ganglion, contain no measurable dopamine. Dopamine was present in amounts between 0.1–0.5 $\mu$gm/gm in the tentacles, digestive gland, penis, kidney and skeletal muscle while dopamine was present in amounts between 0.6–1.0 $\mu$gm/gm in crop and the accessory genital mass. Only trace amounts were detected in salivary gland and the buccal muscle mass.

## TABLE V

Dopamine content of non-nervous tissues of *Aplysia*

| Tissue | No. (a) | Content ($\mu$gm/gm) |
|---|---|---|
| Tentacles | 2 | 0.24 |
| Salivary gland | 2 | trace |
| Buccal muscle | 2 | trace |
| Digestive gland | 2 | 0.47 |
| Crop | 2 | 0.72 |
| Auricle | 2 | none detected |
| Ventricle | 4 | none detected |
| Accessory genital mass | 2 | 0.91 |
| Penis | 2 | 0.49 |
| Kidney | 2 | 0.11 |
| Ctenidium | 2 | 2.92 |
| Skeletal muscle | 2 | 0.22 |

(a) Number of determinations. Each sample contained pooled tissues from 2–10 animals with a total weight of 200–1,000 mg.

## UPTAKE OF CATECHOLAMINES

Table VI shows results of experiments on uptake of dopamine-$^3$H and norepinephrine-$^3$H in the auricle and the visceral ganglion. Although dopamine could not be detected at all in either ventricle or auricle, the auricle did concentrate this compound. This uptake was blocked by DMI suggesting that the uptake is into nerve terminals.

The visceral ganglion, which does contain dopamine, did concentrate dopamine-$^3$H. At least a portion of the uptake of this amine was inhibited by DMI. Norepinephrine was not concentrated in the visceral ganglion or the auricle.

## TABLE VI

Uptake of Dopamine-$^3$H and Norepinephrine-$^3$H by *Aplysia* Auricle and Visceral Ganglion (a)

| | No. (b) | Accumulation Ratio (c) |
|---|---|---|
| *Auricle* | | |
| Control—DA$^3$H | 3 | 2.54 |
| DA-$^3$H plus DMI, $10^{-4}$M | 1 | 0.50 |
| Control—NE-$^3$H | 3 | 0.52 |
| NE-$^3$H plus DMI, $10^{-4}$M | 1 | 0.46 |
| *Visceral Ganglion* | | |
| Control—DA-$^3$H | 2 | 2.42 |
| DA-$^3$H plus DMI, $10^{-4}$M | 1 | 1.71 |
| Control—NE-$^3$H | 2 | 0.41 |
| NE-$^3$H plus DMI, $10^{-4}$M | 1 | 0.56 |

(a) Tissues incubated for 30 minutes in artificial sea water containing either 500 m$\mu$ curies DA-$^3$H at 23°C. or 400 m$\mu$ curies NE-$^3$H at 23°C.
(b) Number of determinations. Each sample consisted of tissues from four animals.
(c) Accumulation ratio is ratio of tissue-$^3$H to medium-$^3$H.

## DISCUSSION

These experiments add to the expanding body of experimental evidence that serotonin and dopamine are neurotransmitters in Molluscs. The existence of serotoninergic fibers to the heart is best documented. These experiments have shown that serotonin is normally present and that serotonin-$^3$H can be concentrated in the heart by an active uptake process. In addition, Taxi and Gautron (1969) have demonstrated serotonin in nerve fibers of the heart of *Aplysia* by fluorescence microscopy and have shown that the uptake of serotonin-$^3$H is into nerve terminals of the heart by autoradiography. In *Helix*, S. Rózsa and Perenyi (1966) have been able to identify serotonin in the perfusate of the heart after stimulation of the cardiac nerves. After loading of the heart with serotonin-$^3$H, radioactive serotonin has also been shown to be released from the perfused *Aplysia* heart upon stimulation of the connective nerves to the visceral ganglion (Breese and Carpenter, 1967; Chase *et al.*, 1968).

In the heart there is a good relationship between the content of endogenous serotonin and the ability of the tissue to concentrate serotonin-$^3$H. This fact, plus the inhibition of uptake by Na$^+$-free solutions and ouabain all strongly support the proposition that the accumulation of serotonin-$^3$H in auricle is an active transport process in the nerve terminals.

Using platelets, Marshall, Stirling, Tait and Todrick (1960) were first to demonstrate that imipramine could alter the uptake of serotonin. Extending this observation to mammalian brain, Carlsson (1970) has reported that DMI as well as other thymoleptics can inhibit the uptake system in serotonergic fibers. The inhibition of uptake of serotonin-$^3$H by DMI in the *Aplysia* auricle thus adds further support to the view that serotonergic fibers are influenced by imipramine-like compounds (Table III).

It is of interest that while the *Aplysia* heart contains no dopamine, it will concentrate dopamine several times over the bath concentration. Since this accumulation is DMI sensitive, it is probably into nerve terminals and may, in fact, be into the serotonin terminals. The formation of dopamine from DOPA in serotonergic neurons results in release of serotonin from mammalian brain slices (Ng, Chase, Colburn and Kopin, 1970) and the dopamine so formed can be released by electrical stimulation (I. J. Kopin, unpublished).

In ganglia there is a sizable accumulation of serotonin-$^3$H which is not sensitive to Na$^+$-free solution, and which is probably a result of a nonspecific binding to connective tissue elements. A similar conclusion was reached by Ascher et al. (1968), who report autoradiographic evidence for localization of serotonin-$^3$H in non-nervous parts of the ganglion. Furthermore, we (Breese and Carpenter, unpublished) have obtained significant release of substances which are not concentrated, such as norepinephrine-$^3$H, and substances which are inert, such as urea-$^3$H, on electrical stimulation of the nerves to the ganglion. Inhibition of accumulation by absence of Na$^+$, treatment with DMI or ouabain, etc., should be demonstrated before the accumulation of labelled amines is assumed to be due to active uptake into nerve terminals. Caution must also be taken in experiments on release of labelled amines, since when only a portion of the label is in nerve terminals there may be the appearance of isotope on stimulation from non-neuronal sites.

The distribution of dopamine in the various nervous and non-nervous tissues suggest that this compound may have several very specific functions as a neurotransmitter in *Aplysia*. The pedal ganglion is concerned primarily with control of motor functions of the foot (Eales, 1921). The very much higher concentration of dopamine in this ganglion suggests that dopamine may be of special importance in motor integration, although no evidence for dopamine motor neurons has been found in some other Molluscs (Sweeney, 1968).

The very high concentration of dopamine in the ctenidium is of particular interest because of the lack of significant dopamine in the pleural ganglion or the pleural-visceral connective and the relatively small amount in the visceral ganglion, from which the ctenidium receives its innervation. Sensory neurons containing dopamine are common in a variety of invertebrates (Dahl, Falck, von Mecklenburg and Myhrberg, 1963). The very large amount of dopamine in the ctenidium as compared to the total amount present in the visceral ganglion (the ctenidium weighs at least 10–20 times as much as the visceral ganglion) suggests that dopamine in the ctenidium is present in sensory rather than motor cells.

## REFERENCES

Anton, A. H. and Sayre, D. F., 1962, A study of the factors affecting the aluminum oxide—trihydroxyindole procedure for the analysis of catecholamines, *J. Pharmacol. Exp. Ther.* **138**: 360–372.

Anton, A. H. and Sayre, D. F., 1964, The distribution of dopamine and DOPA in various animals and a method for their determination in diverse biological material, *J. Pharmacol. Exp. Ther.* **145**: 326–336.

Ascher, P., Glowinski, J., Tauc, L. and Taxi, J., 1968, Discussion of stimulation-induced release of serotonin, *Adv. Pharmacol.* **6A**: 365–368.

Axelrod, J. A., Whitby, T. G., and Hertting, G., 1965, Effect of psychotropic drugs on uptake of H$^3$-Norepinephrine by tissues, *Science* **133**: 383–384.

Bogdanski, D. F., Tissari, A. and Brodie, B. B., 1968, Role of sodium, potassium, ouabain and reserpine in uptake, storage and metabolism of biogenic amines in synaptosomes, *Life Science* **7**: 419–426.

Breese, G. R. and Carpenter, D. O., 1967, Evidence for serotonin (5-HT) as a neurotransmitter in *Aplysia*, *Fed. Proc.* **26**(2): 2218.

Carlsson, A., 1970, Effect of drugs on amine uptake mechanisms in the brain. *Bayer—Symposium II*, Springer-Verlag, pp. 223–233.

Chase, T. N., Breese, G. R., Carpenter, D. O., Schanberg, S. M. and Kopin, I. J., 1968, Stimulation-induced release of serotonin, *Adv. Pharmacol.* **6A**: 351–364.

Cardot, J., 1963, Sur la présence de dopamine dans le systéme nerveux et ses relations avec la décarboxylation de la dioxyphénylalanine chez le Mollusque *Helix pomatia*, *C. R. Acad. Sci.* (Paris) **257**: 1364–1366.

Cardot, J., 1964, Considerations sur le metabolisme de la 5-hydroxytryptamine et de la tryptamine chez le Mollusque *Helix pomatia*, *C. R. Acad. Sci.* (Paris) **258**: 1103–1195.

Dahl, E., Falck, B., Linquist, M. and von Mecklenburg, C., 1962, Monoamines in mollusc neurons, *Kungl. Fysiograf. Sallskapets i Lund Forhandl.* **32**: 88–92.

Dahl, E., Falck, B., von Mecklenburg, C. and Myhrberg, H., 1963, Adrenergic Sensory neurons in Invertebrates, *Gen. Comp. Endocrinol.* **3**: 693.

Eales, N. B., 1921, *Aplysia*, *Trans. Liverpool Biol. Soc.* **35**: 183–266.

Frazier, W. T., Kandel, E. R., Kupfermann, I., Waziri, R. and Coggeshall, R. E., 1967, Morphological and functional properties of identified neurons in the abdominal ganglion of *Aplysia californica*, *J. Neurophysiol.* **30**: 1288–1351.

Garattini, S. and Valzelli, L., 1965, *Serotonin*, Amsterdam: Elsevier.

Giarman, N. J. and Freedman, D. X., 1965, Biochemical aspects of the actions of psychotomimetic drugs, *Pharmacol. Rev.* **17**: 1–25.

Gillis, C. N. and Paton, D. M., 1967, Cation dependence of sympathetic transmitter retention by slices of rat ventricle, *Br. J. Pharmac. Chemother.* **29**: 309–318.

Gerschenfeld, H. M., 1966, Chemical transmitters in invertebrate nervous systems, *Soc. Exper. Biol. Sympos.* **20**: 299–323.

Gerschenfeld, H. M. and Tauc, L., 1964, Différents aspects de la pharmacologie des synapses dans le systéme nerveux central des Mollusques, *C. R. Soc. Biol.* (Paris) **56**: 360–361.

Haggendal, J., 1963, An improved method for fluorimetric determination of small amounts of adrenaline and nor-adrenaline in plasma and tissues, *Acta Physiol. Scand.* **59**: 242–254.

Kerkut, G. A., Sedden, C. B., and Walker, R. J., 1966, The effect of DOPA, α methyldopa and reserpine on the dopamine content of the brain of the snail, *Helix aspersa*, *Comp. Biochem. Physiol.* **18**: 921–930.

Loveland, R. E., 1963, 5-Hydroxy tryptamine, the probable mediator of excitation in the heart of Mercenaria (Venus) mercenaria, *Comp. Biochem. Physiol.* **9**: 95–104.

Marshall, E., Stirling, G. S., Tait, A. C. and Todrick, A., 1960, The effect of iproniazid and imipramine on the blood platelet serotonin level in man, *Brit. J. Pharmacol.* **15**: 35–51.

Ng, K. Y., Chase, T. N., Colburn, R. W. and Kopin, I. J., 1970, L-Dopa-induced release of cerebral monoamines. *Science*, **170**: 76–77.

Snyder, S. H., Axelrod, J. and Zweig, M., 1965, A sensitive and specific fluorescence assay for tissue serotonin, *Biochem. Pharmacol.* **14**: 831–835.

S.-Rózsa K., and Graul, C., 1964, Is serotonin responsible for the stimulative effect of the extracardiac nerve in *Helix pomatia?*, *Annal. Biol. Tihany* **31**: 85–96.

S.-Rózsa, K. and Perenyi, L., 1966, Chemical identification of the excitatory substance released in *Helix* heart during stimulation of the extracardial nerve, *Comp. Biochem. Physiol.* **19**: 105–113.

Sweeney, D., 1963, Dopamine: Its occurrence in Molluscan ganglia, *Science* **139**: 1051.

Sweeney, D. C., 1968, The anatomical distribution of monoamines in a fresh-water bivalve Mollusc, *Sphaerium Sulcatum* (L.), *Comp. Biochem. Physiol.* **25**: 601–613.

Taxi, J. and Gautron, J., 1969, Données cytochimiques en faveur de l'existence de fibre nerveuses sérotoninergiques dans le coeur de l'Aplysie, *Aplysia californica*, *J. Microscopie* **8**: 627–636.

Twarog, B. M., 1954, Responses of a molluscan smooth muscle to acetylcholine and 5-hydroxytryptamine, *J. Cell. Comp. Physiol.* **44**: 141–163.

Welsh, J. H. and Moorhead, M., 1959, The In Vitro synthesis of 5 hydroxytryptamine from 5 hydroxytryptophan by nervous tissues of two species of molluscs, *Gumma J. Med. Sci.* **8**: 211–218.

Welsh, J. H. and Moorhead, M., 1960, The quantitative distribution of 5-hydroxytryptamine in the invertebrates, especially in their nervous systems, *J. Neurochem.* **6**: 146–169.

# THE CORTICAL ACIDIC RESPONSE TO INTRAVENOUS NaHCO₃ AND THE NATURE OF BLOOD BRAIN BARRIER DAMAGE

STANLEY I. RAPOPORT

*Laboratory of Neurophysiology, National Institute of Mental Health, Bethesda, Md.* 20014

The cortical adicic response to intravenous NaHCO₃ in the cat was analyzed critically, and its relation to the blood brain barrier to $HCO_3^-$ was considered. Correct measurement of the response requires intact cortical and pia-arachnoid blood vessels, accurate identification and recording of DC and pH potential differences at the brain surface, and control of ventilation. Changes in the response could be quantified and used to demonstrate damage of the blood brain barrier to $HCO_3^-$. With progressive damage, $HCO_3^-$ probably enters the brain more readily and the acidic response changes until, with severe damage, the response is only alkaline and resembles the alkaline change of the blood following i.v. NaHCO₃. Reduction of the response is graded and may depend on the extent and site of vessel damage. The response is resistant to anoxia and metabolic inhibitors, but is reduced by topically applied alcohols and surface active compounds. Analysis of the changes following increasing quantities of concentrated solutions suggests that some agents may damage the barrier osmotically by shrinking barrier cells, possibly at the vascular endothelium. In its resistance to change, the acidic response resembles the blood brain barrier to trypan blue.

## PREFACE

In 1960, when I came to the Laboratory of Neurophysiology, Dr. Marshall and I began to work on the pH change on the cat cortex in spreading cortical depression, with the technical collaboration of Mr. Anthony Bak and Mr. Alvin Ziminsky. In the course of the investigation, it was found that the cortical pH became acidic within 10 seconds following an injection of NaHCO₃ into the femoral vein, while the blood became alkaline. After interpreting this phenomenon as due to the blood brain barrier to $HCO_3^-$, I proceeded to study the effect of a variety of agents and metabolic states on the cortical acidic response. It was concluded that the response represented a barrier very like the blood brain barrier to trypan blue. In addition, since changes in the response could be quantified and compared with respect to the action of different agents, these changes could be interpreted in terms of models for blood brain barrier damage. In particular, it was suggested that some concentrated solutions could act osmotically on the barrier and damage it by shrinking barrier cells. Since these results have never been reviewed, it was thought worthwhile to consider them at length in this paper.

## THE NATURE OF THE CORTICAL ACIDIC RESPONSE TO INTRAVENOUS NaHCO₃

One function of the blood brain barrier is to regulate the pH of the brain and cerebrospinal fluid (c.s.f.) under a variety of physiological conditions. In order for respiration to continue within non-lethal limits, the pH of the c.s.f. must be kept between 7.18 and 7.38 pH units (Pappenheimer, 1967). This pH regulation operates at least at two levels with respect to the blood brain barrier. First, it is probable that $HCO_3^-$ (Pappenheimer, 1967) and possible that $H^+$ (Siesjö and Kjällquist, 1969) are transported actively at the barrier. Second, this transport is associated with a relative impermeability to $HCO_3^-$ across the blood brain barrier, so as to permit the active transport mechanisms to maintain significant concentration differences in the steady state.

As early as 1926, Gesell and Hertzman demonstrated in dogs that an intravenous injection of NaHCO₃, which alkalinized the blood, produced an initial acidification of the c.s.f. in the cisterna magna. They ascribed this to the ready passage of $CO_2$, liberated in the blood, into the c.s.f. (equation 1), and the slower equilibration of $HCO_3$ between blood and c.s.f. because of the 'relative impermeability of cerebral membranes to base',

$$HCO_3^- + H^+ \rightarrow H_2CO_3 \rightarrow H_2O + CO_2\uparrow \qquad (1)$$

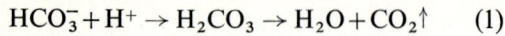

Their observations, however, were disputed by later workers, who recorded pH at the surface of the cerebral cortex (Dusser de Barenne, McCulloch and Nims, 1937; Jasper and Erickson, 1941; Meyer and Gotoh, 1961), until Rapoport and Marshall (1961) and Rapoport (1964a) confirmed the validity of their results. These latter authors measured pH on the pia-arachnoid membrane of the cat cortex with a 1 mm² flat-surfaced pH electrode (MacInnes and Dole, 1929) juxtaposed with a pore reference electrode developed by Marshall (1959). The pore electrode is a glass capillary with an inner diameter of 1 mm, filled with saline and into which a Ag/AgCl electrode is inserted more than one centimeter from its mouth (pore). It records the DC potential at a localized cortical region with reference to earth. The pH electrode records the DC potential + pH potential with reference to earth. Subtraction of the pore recording from the pH recording gives the pH potential, −55 mV/pH unit (Figure 1).

It was shown that three criteria should be met before an accurate cortical pH response to NaHCO₃ could be obtained: (1) the cortical and pia-arachnoid blood vessels should be intact, (2) the cortical DC potential and its changes must be measured and subtracted from the net potential change recorded by the pH electrode, which is the sum of the DC potential and a potential due to a change in [H⁺], and (3) the animal should be under controlled ventilation in order to eliminate ventilatory changes following i.v. NaHCO₃ injection. NaHCO₃ will produce hyperventilation (Gesell and Hertzman, 1926) and CO₂ may be blown off to make the pH of the c.s.f. alkaline (De Bersaques, 1955).

## EFFECT OF CO₂, HYPOXIA AND DAMAGE ON THE CORTICAL ACIDIC RESPONSE

In the course of investigating the nature of the acidic response (Figure 1), Rapoport (1964a) showed that it existed in kittens less than one week old and that it was resistant to 10%(v/v) but not to 20%(v/v) CO₂ and to hypoxia (2.8%(v/v) O₂ in N₂). The cortical acidic response was present after removal, without damage to blood vessels, of the arachnoid membrane.

Similarly, the barrier to trypan blue is unchanged with 10%(v/v) but not with 20%(v/v) CO₂ (which produces petechiae) (Clemedson, Hartelius and Holmberg, 1958), while the barrier to albumin (which binds trypan blue) also is intact at 10%(v/v) CO₂ in the adult dog (Lending, Slobody and Mestern, 1961). The barrier to trypan blue is found in the fetal rat (Grazer and Clemente, 1957) and, like the acidic response, is resistant to anoxia in several animals (Broman, 1949).

Thus, the barriers to both trypan blue and HCO₃⁻ have some similar properties and may both arise at

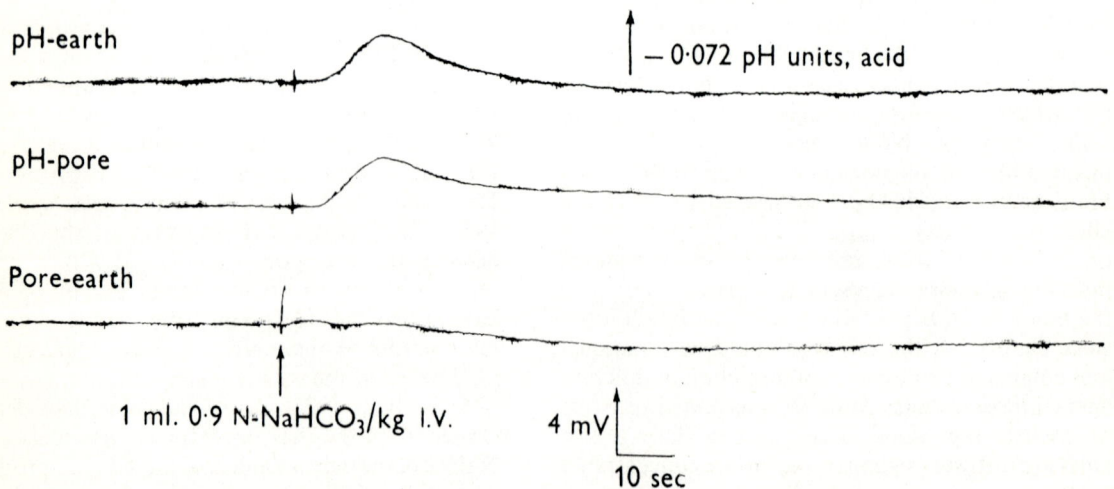

FIGURE 1   *The cortical pH and DC responses to intravenous NaHCO₃.* The pH and pore electrodes are on the cortical surface, and the earth electrode is in the neck muscles of the cat. Records represent potential differences between electrodes. The pH-earth record gives the pH + DC p.d. (The pore-earth electrode gives the DC p.d.) The pH-pore gives the pH p.d., 55 mV/pH unit (change in upward direction represents acidic response.) After injection of NaHCO₃, there is an acidic pH change and a slow negative DC shift. (Reprinted with permission of the *Journal of Physiology* [London]).

the level of the endothelial cells of the brain vascular system, which have been shown to have apposed or fused intercellular clefts (Reese and Karnovsky, 1967). The work of Broman and Lindberg-Broman (1945) on the effect of physico-chemical agents applied to the surface of the pia-arachnoid of the mammalian cerebral cortex, or injected into the carotid artery, suggested earlier that the blood brain barrier to trypan blue lay at the endothelial cell level. Their method of topical application demonstrated also what appeared to be an all-or-none staining of the brain surface above a threshold concentration of a damaging agent.

## EFFECT OF TOPICAL AGENTS ON CORTICAL ACIDIC RESPONSE

In order to obtain more information about the cortical acidic response and to investigate its usefulness as a means to understand the blood brain barrier to $HCO_3^-$, Rapoport (1964b) studied the effect of topically applied substances on the response, using the method of Broman and Lindberg-Broman (1945) (Figure 2). A fixed dose of NaHCO₃ (1 ml. 0.9 N NaHCO₃/kg) was injected into the femoral vein of the cat. A control response first was recorded at a cortical region by the electrode pair (pore and pH electrodes), the electrodes were removed and a filter paper pledget soaked with a test substance applied to the region for 10 minutes, the region washed with Ringer, and the response to the same dose of NaHCO₃ was again recorded after the electrode pair was replaced on the treated region.

The test response was related to the control in three ways (Figure 2): (1) by comparing the relative heights of the maximum acidic change (Y) (maximum acidic response ratio, test/control), (2) by comparing the cortical pH, 60 seconds after i.v. injection of NaHCO₃, with the pH of the maximum acidic response, the difference being $\Delta pH_{60}$, and (3) by comparing the rate of alkalinization following the maximum acidic response (d pH/d time, or short term slope effect).

It was observed that different parts of the cortical acidic response were changed as the barrier progressively was disturbed. Some agents changed only the short term slope effect (rate of alkalinization) and the pH one minute following injection ($\Delta pH_{60}$), which showed that the cortical pH had become more alkaline presumably because of increasing entry of $HCO_3^-$ into the brain. With other agents, the changes of the short term slope

effect and of $\Delta pH_{60}$ were exaggerated, and at the same time the height of the maximum acidic response (Y) was reduced.

The agents that changed only the short term slope effect and $\Delta pH_{60}$ were 0.02 M p-chloromercuribenzoate and 0.1 M sodium glycocholate. Lower chain alcohols (methanol, ethanol, n-propanol, n-butanol) and some surface active compounds changed all parameters, including the maximum acidic response (Y). In fact, n-butanol was so effective that it usually produced a monotonic alkaline response to i.v. NaHCO₃ without any acidic response whatsoever. Many agents, however, such as metabolic inhibitors (0.1 M Na azide, 0.1 M 2,4-dinitrophenol, 0.02 M NaF, 0.1 M NaCN), distilled water and isotonic NaCl, did not change any of the parameters of the cortical acidic response. Whether or not the acidic response was changed, the DC change following i.v. NaHCO₃ was unaffected, which indicated that it probably did not arise at the blood brain barrier.

The effectiveness of the topical agents on the acidic response was similar to their effectiveness in letting trypan blue enter the brain (Broman and Lindberg-Broman, 1945), which supported the earlier suggestion that trypan blue exclusion and the acidic response represented a barrier at the same structural or functional level. The relative inability of the metabolic inhibitors to change the response, although due perhaps to the short period of application of 10 minutes, suggested that this

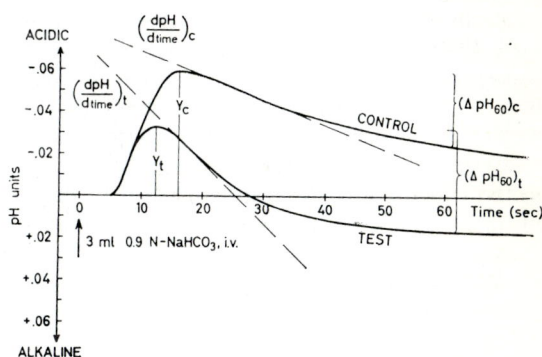

FIGURE 2 *Comparison of representative control and test cortical pH responses to intravenous NaHCO₃ (pH-pore electrode records of Figure 1). Subscripts 'c' and 't' refer to control and test curves, respectively. The control curve represents a normal acidic response. The test curve is the pH response at the same cortical region after 10 minutes of topical application of a test substance. $Y_c$ and $Y_t$ are maximum acidic responses. (dpH/dtime) is the maximum slope after the peak acidic response, when the pH is becoming more alkaline (short term slope effect). $\Delta pH_{60}$ is the difference (Y—pH at 60 sec) for a specific response. (Reprinted with permission of the Journal of Pharmacology and Experimental Therapeutics).*

barrier did not depend on intact metabolic function during this period (the acidic response is resistant also to hypoxia), but acted rather like a structural hinderance to diffusion. Thus, it would be expected that agents that could destroy cell membranes, such as the surface active compounds and the lower chain alcohols, would also reduce the response.

The above observations suggested the following questions: (1) Could one agent at one concentration affect the short term slope effect and $\Delta pH_{60}$ and then, at a higher concentration, change the maximum acidic response (Y) as well? In other words, is blood brain barrier damage all-or-none or graded and, if graded, is it concentration dependent? (2) In what way do the agents damage the barrier? (3) What is the site (or sites) of their action? (4) Why did the short term slope effect and $\Delta pH_{60}$ change in some cases when the maximum acidic response did not? (5) Were some kinds of damage and reduction of the cortical acidic response reversible?

## EFFECT OF CONCENTRATED SOLUTIONS ON CORTICAL ACIDIC RESPONSE

These questions were tackled by two sets of experiments, the first by Rapoport (1970) on the effect of concentrated topical solutions on the cortical acidic response, the second, which is in progress, by Rapoport in collaboration with Dr. Masararu Hori and Dr. Igor Klatzo, on the effect of the concentrated solutions on the blood brain barrier to Evans blue.

In 1970, Rapoport showed that for increasing concentrations of $NaNO_3$, NaCl, urea or ethanol, there was first a change in $\Delta pH_{60}$ (which is correlated with the long term slope effect (Rapoport, 1964b)), followed, at higher concentrations, by a decrease in the maximum acidic response (Y). The changes in $\Delta pH_{60}$ and in the maximum acidic response (Y) increased with increasing concentrations of the topically applied agents. These observations were interpreted to mean that barrier damage was a graded, rather than an all-or-none process.

The reasons remained obscure for the lower threshold of $\Delta pH_{60}$ than that of the maximum acidic response (Y) and for the concentration dependent effects on both parameters, because although it was probable that the concentrated solutions worked at the level of the vascular endothelium, the distribution and mode of their effect were unknown. However, Hori, Rapoport and Klatzo (unpublished) have shown recently that the concentrated solutions first obviously damage the small veins of the pia-arachnoid in a sporadic distribution,

so as to let Evans blue enter the brain, and then, at higher concentrations, damage as well the larger blood vessels over the entire cortical surface to which they have been applied. Concentration-dependent damage and area of distribution of damage suggest that the lower threshold changes in $\Delta pH_{60}$ and in the long term slope effect represent sporadic small vein damage with somewhat increased entry of $HCO_3^-$ over the region as a whole. A change in the maximum acidic response (Y) requires larger vessel damage over a wider area of distribution.

In considering the mode of barrier damage, it should be recognized that interpretation of results using the topical method must be made cautiously, since the concentration at the site of action is unknown and probably depends on the diffusion coefficient of the topical agent through the extracellular space and through cells of the pia-arachnoid and cortex. In addition, the barrier functions in different ways. It is probable that glucose (Crone, 1965) and $HCO_3^-$ (Pappenheimer, 1967), for example, are actively transported at a level at or near the brain vascular endothelium, and agents which inhibit such transport will therefore damage the barrier at the metabolic stage (Eidelberg, Fishman and Hams, 1967; Steinwall, 1968). Lipid soluble substances, such as $CO_2$, $O_2$ and various compounds undissociated at physiological pH (Rall, Stabenau and Zubrod, 1959) are little interfered with by the barrier because they can pass through the cell membrane of the endothelial cells (Davson, 1967). On the other hand, ionic compounds and non-charged compounds which cannot pass through an effective pore radius of 7–9 A, the equivalent pore radius of the barrier endothelium (Fenstermacher and Johnson, 1966), will have relatively lower permeabilities as compared to their permeabilities across blood vessels elsewhere in the body. The blood brain barrier to these substances probably arises because the endothelial cells have apposed or fused intercellular clefts (Reese and Karnovsky, 1967).

The cortical acidic response probably represents the ability of the blood brain barrier to restrict $HCO_3^-$ movement. Rapoport (1970) found that the concentration required for urea to exert its effect on either the maximum acidic change (Y) or on $\Delta pH_{60}$ was somewhat more than twice, on a mole for mole basis, than the concentration of NaCl required to exert a like effect. Since NaCl dissociates into two colligatively active solutes, $Na^+$ and $Cl^+$, he suggested that the concentrated agents shrink cells, possibly the vascular endothelium, because of their osmotic activity, and thereby open up intercellular spaces so as to permit increased exchange of formerly relatively impermeant substances across the barrier. The general formulation for osmotic cell shrinkage is derived from the observations of Overton (1895). Three criteria should be satisfied according to his treatment: (1) the amount of shrinkage should increase with increasing concentration, (2) at a given concentration, shrinkage is inversely related to the ability of the substance to penetrate the cell and (3) within certain concentration limits of the osmotic agent, shrinkage should be reversible. Very high concentrations may denature proteins or disrupt cell membranes (Rapoport, 1970).

The effects of some concentrated solutions on the acidic response do not apparently violate any of these criteria. As has been pointed out, the change in $\Delta pH_{60}$ and in the maximum acidic response (Y) increases with increasing concentration, but this may be ascribed in part to small vein damage and an incomplete area of distribution of damage at the lower concentrations. The order of effectiveness of the four agents tested by Rapoport (1970) was: $NaNO_3 > NaCl \geq$ urea > ethanol, which could be interpreted tentatively as due to the relative permeances of these agents across the blood brain barrier cells. Ethanol, which is very lipid soluble, would be more permeant and much less effective than NaCl, as has also been shown by Overton (1895) for plant cell shrinkage.

Until recently, there has been a suggestion that intracisternal injection of a hypertonic solution (cf. Weir, 1945) produces barrier damage which is reversible after 24 hours (Streicher, Rall and Gaskins, 1964). Hori, Rapoport and Klatzo (unpublished) have now demonstrated that the damage caused by urea to the barrier to Evans blue is reversible within one half hour under the conditions of these experiments. The damage caused by ethanol, however, is not reversible, which means that it doesn't satisfy the third criterion of the osmotic hypothesis, and may disrupt cell membranes.

These observations support the suggestion (Rapoport, 1970) that some concentrated solutions may damage the barrier by shrinking barrier cells, in accord with the considerations of cell shrinkage by Overton (1895). Additional support for this hypothesis comes from a recent demonstration that brain perfusion with hypertonic solutions in rats produces an increased entry of labelled solutes into the brain, but reversibility and concentration dependence of this entry were not tested (Thompson,

1970). The osmotic hypothesis must be confirmed by use of more substances with different lipid and water solubilities, and different techniques, to test the three criteria. It is expected that direct observations of endothelial cells with light and electron microscopy will help to localize the sites of action of topical agents.

ACKNOWLEDGEMENTS

I thank Dr. Wade Marshall for his encouragement and support to do this work, and Dr. Eugene Streicher for the many fruitful and stimulating discussions over the years. I also thank Drs. Joseph Fenstermacher, Igor Klatzo and Milton Brightman for their critical comments.

REFERENCES

Broman, T., 1949, The Permeability of the Cerebrospinal Vessels in Normal and Pathological Conditions. Copenhagen: Munksgaard.
Broman, T. and Lindberg-Broman, A. M., 1945, An experimental study of disorders in the permeability of the cerebral vessels ('the blood-brain barrier') produced by chemical and physico-chemical agents, Acta Physiol. Scand. 10: 102–123.
Clemedson, C. J., Hartelius, H. and Holmberg, G., 1958, The influence of carbon dioxide inhalation on the cerebral vascular permeability to trypan blue (the blood-brain barrier), Acta Path. Microbiol. Scand. 42: 137–152.
Crone, C., 1965, Facilitated transfer of glucose from blood into brain tissue, J. Physiol. (London) 181: 103–113.
Davson, H., 1967, Physiology of the Cerebrospinal Fluid, Boston: Little, Brown, and Co.
De Bersaques, J., 1966, pH et CO₂ du sang et du liquide céphalorachidien dans l'acidose et l'alcalose métaboliques, Arch. Int. Physiol. 63: 1–6.
Dusser de Barenne, J. G., McCulloch, W. S. and Nims, L. F., 1937, Functional activity and pH of the cerebral cortex, J. Cell. Comp. Physiol. 10: 277–289.
Eidelberg, E., Fishman, J. and Hams, M. L., 1967, Penetration of sugars across the blood-brain barrier, J. Physiol. (London) 191: 47–57.
Fenstermacher, J. D. and Johnson, J. A., 1966, Filtration and reflection coefficients of the rabbit blood-brain barrier, Amer. J. Physiol. 211: 341–346.
Gesell, R. and Hertzman, A. B., 1926, The regulation of of respiration. IV. Amer. J. Physiol. 78: 610–629.
Grazer, F. M. and Clemente, C. D., 1957, Developing blood-brain barrier to trypan blue, Proc. Soc. Exp. Biol. N.Y., 94: 758–760.
Jasper, H. and Erickson, T. C., 1941, Cerebral blood flow and pH in excessive cortical discharge induced by metrazol and electrical stimulation, J. Neurophysiol. 4: 333–347.
Lending, M., Slobody, L. B. and Mestern, J., 1961, Effect of hyperoxia, hypercapnia, and hypoxia on the blood-cerebrospinal fluid barrier, Amer. J. Physiol. 200: 959–962.

MacInnes, D. A. and Dole, M., 1929, Tests of a new type of glass electrode, *Ind. Eng. Chem. Anal. Ed.* **1**: 57–59.

Marshall, W. H., 1959, Spreading cortical depression of Leão, *Physiol. Rev.* **39**: 239–279.

Meyer, J. and Gotoh, F., 1961, Interaction of cerebral hemodynamics and metabolism, *Neurology* **11**: Part 2, 46–65.

Overton, E., 1895, Über die osmotischen Eigenschaften der lebenden Pflanzen und Tierzelle, *Vierteljahresschr. Naturforsch. Ges. Zurich* **40**: 159–201.

Pappenheimer, J. R., 1967, The ionic composition of cerebral extracellular fluid and its relation to control of breathing, *Harvey Lectures* **61**: 71–94.

Rall, D. P., Stabenau, J. R. and Zubrod, C. G., 1959, Distribution of drugs between blood and cerebrospinal fluid, *J. Pharm. Exptl. Therap.* **125**: 185–193.

Rapoport, S. I., 1964a, Cortical pH and the blood-brain barrier, *J. Physiol. (London)* **170**: 238–249.

Rapoport, S. I., 1964b, The effect of topically applied substances on the blood-brain barrier, *J. Pharmacol. Exptl. Therap.* **144**: 310–315.

Rapoport, S. I., 1970, Effect of concentrated solutions on blood-brain barrier, *Amer. J. Physiol.* **219**: 270–274.

Rapoport, S. I. and Marshall, W. H., 1961, The 'Gesell Phenomenon' and blood-brain barrier function, *Physiologist* **4**: 91.

Reese, T. S. and Karnovsky, M. J., 1967, Fine structural localization of a blood-brain barrier to exogenous peroxidase, *J. Cell. Biol.* **34**: 207–217.

Siesjö, B. K. and Kjällquist, A., 1969, A new theory for the regulation of the extracellular pH in the brain, *Scand. J. Clin. Lab. Invest.* **24**: 1–9.

Steinwall, O., 1968, Transport inhibition phenomena in unilateral chemical injury of blood-brain barrier, *Progr. Brain Res.* **29**: 357–365.

Streicher, E., Rall, D. P. and Gaskins, J. R., 1964, Distribution of thiocyanate between plasma and cerebrospinal fluid, *Amer. J. Physiol.* **206**: 251–254.

Thompson, A. M., 1970, Hyperosmotic effects on brain uptake of non-electrolytes. In: *Capillary Permeability*. Crone, C. and Lassen, N. A., eds. Copenhagen: Munksgaard.

Weir, E. G., 1945, The effects of intracisternal injection of sodium bromide on the blood-spinal fluid barrier, *Amer. J. Physiol.* **143**: 83–88.

# ACCOMMODATION TO CURRENT RAMPS IN MOTONEURONS OF FAST AND SLOW TWITCH MOTOR UNITS

R. E. BURKE† and P. G. NELSON‡

*Spinal Cord Section, Laboratory of Neurophysiology*
*National Institute of Neurological Diseases and Stroke National Institutes of Health*
*Bethesda, Maryland* 20014

1. The response of triceps surae motoneurons to linearly rising ramps of depolarizing current injected through intracellular micropipette electrodes has been studied in cats. Most of the data were obtained in animals anesthetized with pentobarbital but some units were studied in spinal unanesthetized preparations with anemic destruction of the brain. There were no essential differences in results from the two types of preparations.

2. The speed of contraction of the muscle fibers innervated by the motoneurons studied was measured following stimulation of the cell through the intracellular micropipette. Motor units could then be classified into two groups, fast twitch or type F and slow twitch or type S, on the basis of the speed of contraction of the muscle fibers innervated.

3. The presence or absence of accommodation in each motoneuron was determined based on the shape of the current-latency curve obtained during passage of linearly rising depolarizing currents of widely varying slope. About one half of the total sample of 22 type F units studied exhibited accommodation to linearly rising current ramps according to the criteria adopted in this study, and the other one half did not. Only 2 of the 18 type S units similarly examined showed accommodation to current ramps. The difference in accommodative behavior between types F and S motoneurons was statistically significant and did not appear to be due to any systematic difference in cell injury caused by microelectrode penetration.

## INTRODUCTION

Changes in the excitability of cat spinal motoneurons during passage of prolonged transmembrane depolarizing currents, and the process of accommodation in these cells, have been studied by a number of investigators (Bradley and Somjen, 1961; Frank and Fuortes, 1960; Sasaki and Oka, 1963; Sasaki and Otani, 1961, 1962; Ushiyama, Koizumi and Brooks, 1966; Yamashita, 1966). The variety of methods and results in these studies has made it difficult to arrive at a unified view as to what constitutes the mechanism of accommodation in motoneurons (see Sasaki and Otani, 1962; Stoney and Machne, 1969; Yamashita, 1966). However, it is clear that alpha motoneurons exhibit

a rather broad range of response to steady or changing depolarizing currents of long duration.

Several authors have reported a correlation between the presence or absence of accommodation and the duration of post-spike hyperpolarization in motoneurons, suggesting that cells exhibiting accommodative reactions may be 'phasic' motoneurons with short post-spike hyperpolarizations, which presumably innervate rapidly contracting muscle (Sasaki and Otani, 1961; Ushiyama *et al.*, 1966). It has been demonstrated that the duration of post-spike hyperpolarization in triceps surae motoneurons, in addition to other intrinsic properties, is correlated with the contraction speed of the muscle fibers innervated by the motoneurons (Burke, 1967). The present work was designed to examine the hypothesis that the response of triceps surae motoneurons to long-lasting linear current ramps, and the presence or absence of accommodation, is correlated with the speed of contraction of the muscle fibers innervated by the particular cell under study. A preliminary account of this work has appeared (Nelson and Burke, 1968).

† Present address: Laboratory of Neural Control NINDS, NIH, Bethesda, Md. 20014.

‡ Present address: Behavioral Biology Branch, National Institute of Child Health and Human Development Bethesda, Md. 20014.

## METHOD

Adult cats (2.5–3.5 kg) were used, either anesthetized with pentobarbital (20–25 mg/kg) after surgical preparation under ether or, in several experiments, spinalized and unanesthetized after anemic destruction of the brain under transient halothane anesthesia. The details of surgical preparation and mounting of the animals, permitting intracellular recording and stimulation of motoneurons plus recording of the twitch responses of the muscle fibers innervated by the cell under study (the muscle unit portion of the motor unit), have been described previously (Burke, 1967). Blood pressure and body temperature were monitored continuously, and the temperature of the body and of the oil pools covering the spinal cord and muscle bellies in the leg were kept between 36°–38° C by heating coils.

Intracellular recording and stimulation of triceps surae motoneurons were done using glass micropipette electrodes filled with 3M KCl or 2M potassium citrate solution. Motoneurons were identified by antidromic invasion from either the medial gastrocnemius (MG) nerve or from the lateral gastrocnemius and soleus (LG-Sol) nerve, and by the location of the muscle unit innervated by the cell under study (see Burke, 1967). Wideband cathode follower amplifiers were used which permitted passage of current pulses or ramps of varying shape through the micropipette, either with an external bridge circuit configuration (Frank and Fuortes, 1956) or through an internal operational amplifier circuit (Eide, 1968).

Following penetration of a triceps surae motoneuron and recording of the antidromic action potential, the response of the cell to linearly rising depolarizing currents was examined. Current ramps lasting at least 300 to 400 msec were used. In some cases these were triangular in shape and in other cases trapeziodal, with current rising linearly to a plateau which was held until the end of the oscilloscope sweep, at which time the current returned to zero (see Figures 1–4). The latency from the start of the current ramp to the point of initiation of the first action potential evoked by the current ramp was determined. The relation between this latency and the current strength at the initiation of the first spike response was determined for many different slopes of depolarizing current ramps and this was plotted as the current-latency relation (see Figures 1–4; Bradley and Somjen, 1961; Frank and Fuortes, 1960). During determination of the current-latency relation, the cell membrane potential was recorded on a DC penwriter and the amplitude of the antidromic action potential was recorded before and after current ramp testing. In a number of cells the response of the membrane to long square-wave current pulses of small amplitude were also recorded in order to assess the presence or absence of nonlinear membrane responses during current passage (see Ito and Oshima, 1965).

Following completion of the above procedure, the appropriate head of the triceps surae muscle was connected mechanically to a strain gauge (Statham G1–16 or Grass FT 03) via a stout suture attached to the tendon. The tendons of the three heads of the triceps surae (MG, LG and Sol) had been separated and arranged for independent attachment to the strain gauge. With an initial tension between 50 and 100 gm on the muscle, the mechanical response of the muscle unit innervated by the cell under study was recorded during stimulation of the motoneuron with short depolarizing pulses delivered through the intracellular micropipette (see Burke, 1967).

## RESULTS

The objective of the present work was to apply a more or less standardized test of the response to linearly rising current ramps of as many triceps surae motoneurons as possible in order to determine whether the degree of accommodative response in different cells correlated in any way with the properties of the muscle unit innervated by the tested cell. In practice, this objective required considerable data selection, primarily on the basis of the quality of the microelectrode penetration and the condition of the cell under study. It seems clear that poor penetrations with cell injury and membrane depolarization cause significant changes in the current-latency relation on which our assessment of accommodation is based (Bradley and Somjen, 1961; Ushiyama et al., 1966). Therefore, this factor was carefully considered in each motoneuron tested. The present report is based on a total sample of 40 triceps surae motoneurons, selected from a larger number of cells, which fulfilled criteria of stability of microelectrode penetration and absence of apparent cell injury as judged by stable resting potential, stable action potential amplitude and absence of apparent inactivation of action potential production. All units included in the present report had membrane potentials in excess of $-50$ mV and action potential amplitudes greater than 60 mV.

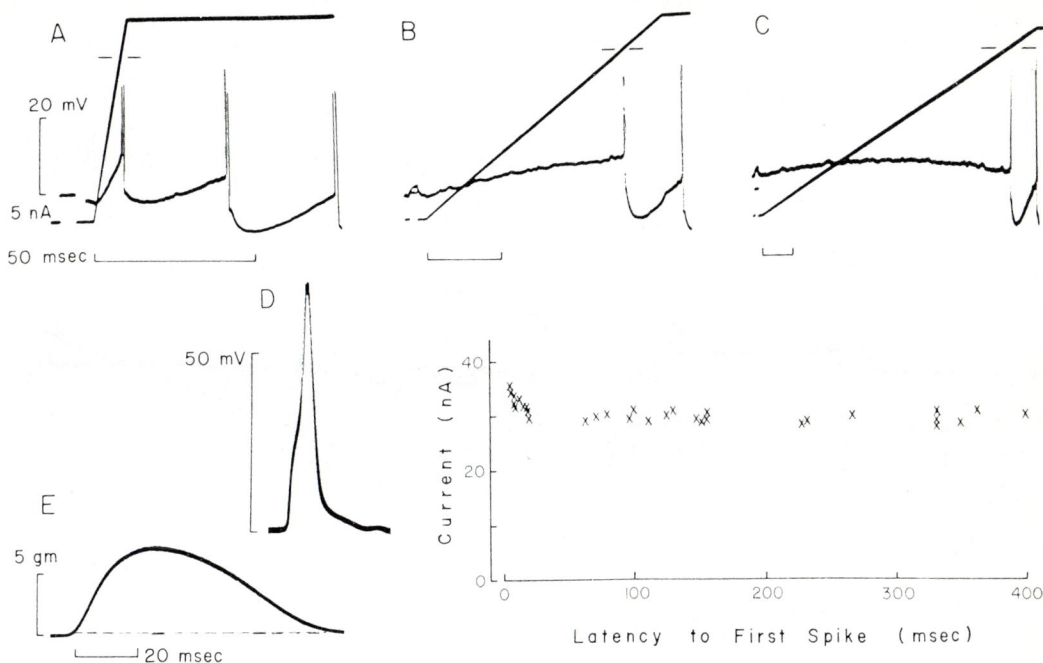

FIGURE 1   Records from a type F MG motor unit with action potential amplitude of 72 mV. A - C: Intracellular potential (upper trace) showing spike responses to depolarizing current injected through the microelectrode. Spikes truncated due to relatively high gain used. Lower trace shows the time course of the current injection, with linearly rising leading edge terminating in a plateau which was held until the end of the sweep. Current strength at the time of initiation of the first spike response is denoted by the short lines beside the initial ramp. Calibration pulse in the current trace is 5 nA. The slope of the initial current ramp was made progressively less from A to C (each time calibration bar denotes 50 msec) but the threshold current at initiation of the first spike was constant. The graph at lower right shows the threshold current level at initiation of the first spike for many different slopes of current ramps (current-latency curve). D: antidromic action potential recorded just before testing with current ramps; action potential after current ramp testing was similar in amplitude. E: Mechanical twitch response of the muscle unit innervated by the motoneuron studied following stimulation with a single short depolarizing current pulse delivered through the microelectrode.

The records shown in Figure 1 are typical of a majority of the cells tested. In this case, the motoneuron had a stable membrane potential of about − 60 mV and antidromic action potential amplitude of 72 mV (Figure 1 D). The traces in A through C show spike responses, recorded intracellularly, produced by linearly rising current ramps of varying slope, with slope decreasing from A to C. Cell firing was maintained during the plateau of current (in the cells illustrated in Figures 1–4, trapezoidal currents were used; see Method). The relation between the current strength at initiation of the first spike (threshold current) and the latency from the start of current to the first spike is shown in the graph at the lower right of Figure 1. The threshold current declined somewhat as very steeply rising ramps were made somewhat slower and then threshold current stabilized at about 30 × 10$^{-9}$ A (30 nA). Threshold currents were then constant even with the most slowly rising ramps tested,

which reached maximum in about 500 msec. The record in Figure 1 E illustrates the mechanical twitch response of the muscle unit innervated by this motoneuron. The twitch time to peak, measured from the initial deflection point in the mechanical record, was about 24 msec and the twitch tension was about 7 gm. The unit was classified as a fast twitch, or type F, motor unit. In this report, motor units will be designated type F if the twitch time to peak values were less than 35 msec and will be termed type S if the time to peak values were greater than 35 msec (see Burke, 1967, 1968).

An example of a slow twitch, or type S, motor unit exhibiting a current-latency relation similar to the above is shown in Figure 2. Again, after an initial decline in threshold current with very steep ramps, the current-latency relation curve became flat for more slowly rising ramps. The record in Figure 2 E illustrates the mechanical response of the muscle unit to a single intracellular stimulus. The twitch time to peak was in this case about 75 msec. Both this unit and the one illustrated in Figure 1 were MG motor units, although not from the same animal.

The type of current-latency curve illustrated in

Figures 1 and 2, with constant current threshold for ramps of long duration, was found in about 2/3 of the total sample of cells considered acceptable for inclusion in the present results. A less frequently encountered response to linearly rising current

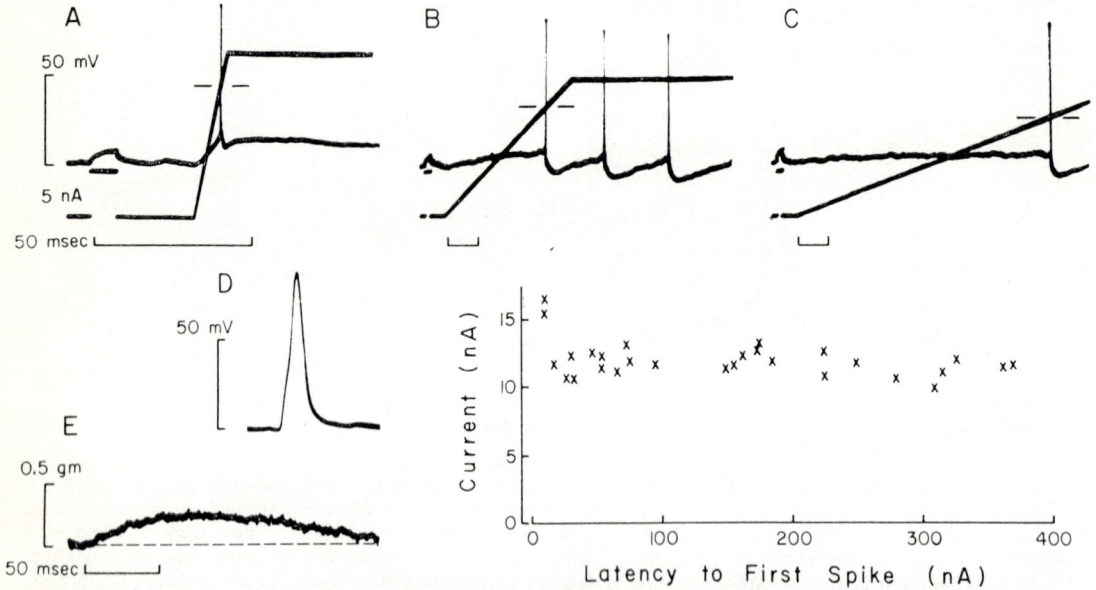

FIGURE 2 Records obtained from a MG type S motor unit. Format and explanation of figure as in Fig. 1. Gain in intracellular records (A - C) relatively low and full spikes are seen. Note constant current threshold level for initiation of the first spike despite widely varying slope of current ramp. The mechanical twitch response of the muscle fibers innervated by the cell under study (E) had a time to peak of about 75 msec.

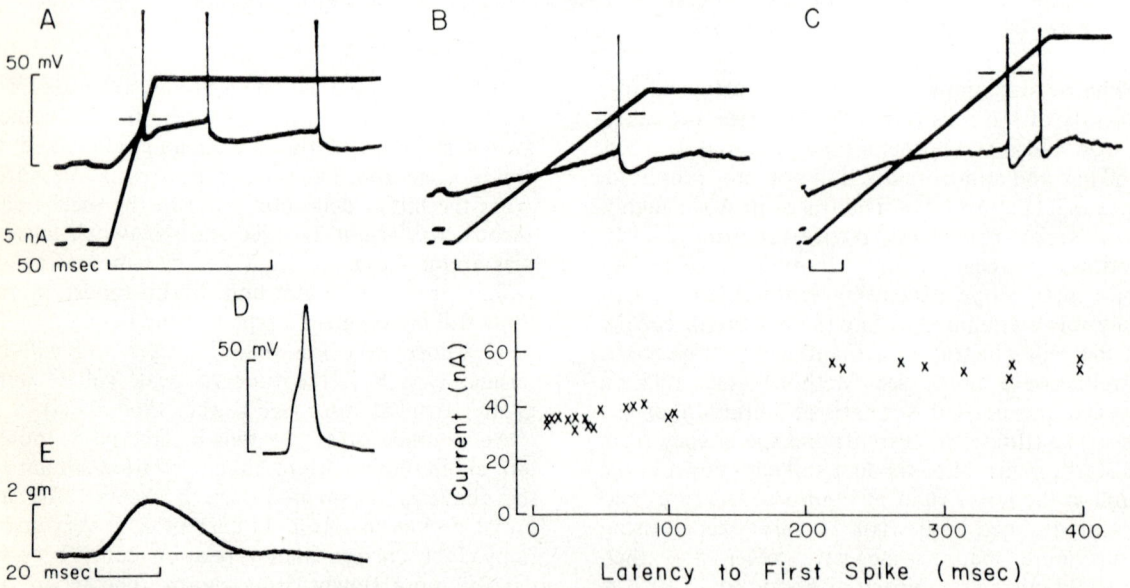

FIGURE 3 Records from a MG type F motor unit. Format as in the previous figures. Note increase in current threshold with the more prolonged current ramps (B. C and graph) and lack of repetitive firing after the first two responses with the longest current ramp shown (C).

ramps is shown in Figure 3. The motoneuron, in this case a MG type F cell, required increasing current for the production of the first spike response as current ramps were prolonged from 10 msec to 400 msec duration. Also, repetitive firing was not sustained following long slow current ramps (Figure 3 C). The maximum current ramp durations used in these experiments were insufficiently long to answer the question whether cells such as that shown in Figure 3 had a minimal current gradient requirement for excitation, with a long time constant of accommodation (Hill, 1936; Ushiyama et al., 1966) or alternatively, whether the current threshold with more prolonged ramps may have stabilized at some constant value. Current-latency curves similar to that illustrated in Figure 3 were found in 6 (4 type F and 2 type S) of the 40 motoneurons in the total sample.

The records in Figure 4, obtained from a LG type F motor unit, illustrate a case in which the current-latency curve shows a clear minimal current gradient requirement for excitation, with a very short time constant of accommodation. Small

decreases in the slope of rapidly rising current ramps led to rather large increases in the threshold current needed to elicit the first spike (Figure 4, graph, crosses) and with ramps rising somewhat more slowly, there was frequent failure of spike initiation up to the maximum current applied (Figure 4 C and graph, open circles). During current ramp testing, both the membrane potential and the antidromic spike amplitude remained relatively constant. The type of current-latency curve illustrated in Figure 4 was found in 7 of the 40 cells in the present sample, all type F.

In order to deal with the current-latency curves found in the present experiments in a simple manner, we have adopted an arbitrary scheme of description. Accommodation will be considered as 'absent' if the mean threshold current for first spike latencies around 300 msec was no more than 20% greater than the mean minimum current needed to excite the cell at shorter latencies. This describes the current-latency curves shown in Figures 1 and 2. Conversely, if the mean threshold current level around 300 msec latency was more than 20%

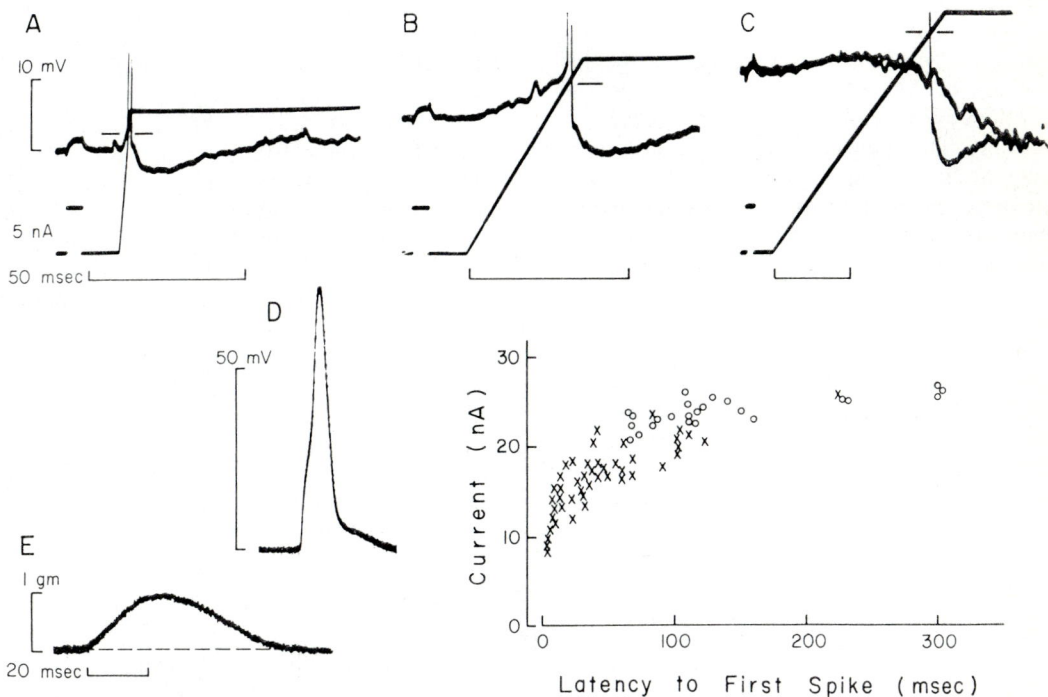

FIGURE 4 Records from a LG type F motor unit. Format as other figures. Spikes in intracellular records are truncated as in Fig. 1. Note rapid increase in current threshold for first spike response as slope of the current ramp decreased. In record C two traces are shown superimposed, both generated with the same current ramp slope. In one trace the cell responded to the current (level marked with short lines) and in the other trial it did not respond. In the current-latency graph, current at initiation of the first spike is denoted with an X. The open circles denote the maximum current reached by ramps of differing slope which did not elicit cell firing (as in C).

greater than the minimum level at shorter latencies, as in Figures 3 and 4, accommodation will be considered as 'present'. Given the arbitrary nature of this definition, it should be clear that a description of motoneurons as showing no accommodation cannot be taken to indicate that such cells had no definable minimal current gradient requirement which could have been demonstrated with ramps of much longer duration than those utilized in this study (Ushiyama et al., 1966).

The results of the present study are summarized in Table 1. This shows the number of type F and type S motoneurons which either exhibited or did not exhibit accommodation according to the definition above. All of the 40 motoneurons included had action potential amplitudes greater than 60 mV. One half of the 22 type F units studied showed accommodation to linear current ramps while the other half did not. Of those type F cells with accommodation, 7 had current-latency curves indicating definite minimal current gradient requirements with short time constants of accommodation (as in Figure 4). The remaining 4 type F cells with accommodation had indeterminate current-latency curves as illustrated in Figure 3. Of the 18 type S units in the total sample, 16 showed no accommodation and 2 had current-latency curves of the indeterminate type. One of the latter type S cells was found in the gastrocnemius pool and the other was a soleus unit. The paucity of type S motor units exhibiting accommodation is striking, compared with the larger proportion of type F cells showing such behavior. The statistical significance of this difference between the type F and type S groups of units was tested by the chi-square method. The probability that such a distribution of units arose by chance, without a qualitative difference between the two groups of motor units, was less than 0.025 ($\chi_Y^2 — 5.167$; P < 0.025), using the conservative

Yates correction for small sample size (see Croxton, 1953).

As mentioned above, the condition of the motoneurons tested for accommodation is critical to interpretation of the results of current-latency measurements such as obtained by the present methods. We felt that the simplest and most reliable measure of cell condition was the amplitude of the antidromic action potential and the stability of this amplitude through the testing period. In Table 1 the mean action potential amplitudes are included for each grouping of units in the 2 × 2 table. The mean action potential amplitudes of type F and type S groups of units were different and this difference was statistically significant (P < 0.005 by the t test). Such differences in action potential amplitudes of motoneurons innervating slow and fast muscle have been noted by Kuno (1959) and confirmed previously in this laboratory (Burke, unpublished observations). Comparing the mean spike amplitudes of motoneurons with and without accommodation, regardless of twitch type, shows less difference; this was not statistically significant (P > 0.1). Thus there is little indication that the group of motoneurons exhibiting accommodation did so because of a systematic difference in the amount of electrode injury and depolarization (see Bradley and Somjen, 1961).

As a further test of the latter possibility, we selected from the total sample of motor units a subset of cells having action potential amplitudes > 75 mV. These data are shown in Table II. Although the number of cells was smaller, most of those excluded being type F units, the same pattern of presence and absence of accommodation was evident in the subset as was seen in the total sample of units. Again, the statistical significance of this distribution of results was tested by the chi-square test and found to be significantly different from a

TABLE 1

2 × 2 table showing the presence or absence of accommodation in type F and type S motoneurons with action potential amplitudes greater than 60 mV

| | | Accommodation | | Totals | Mean AP (± S.D.) |
|---|---|---|---|---|---|
| | | + | − | | |
| Unit Type | F | 11 | 11 | 22 | 76.6 ± 8.1 mV |
| | S† | 2 | 16 | 18 | 85.0 ± 6.1 mV |
| | Totals | 13 | 27 | 40 | |

Mean AP (± S.D.) 77.2 ± 8.3 mV      81.8 ± 8.1 mV

$$\chi_Y^2 = 5.167; P < 0.025$$

† 10 soleus and 8 gastrocnemius type S units

TABLE 2

2 × 2 table showing the presence or absence of accommodation in type F and type S motoneurons with action potential amplitudes greater than 75 mV

| | | Accommodation | | | |
| --- | --- | --- | --- | --- | --- |
| | | + | − | Totals | Mean AP ($\pm$ S.D.) |
| Unit Type | F | 7 | 6 | 13 | 82.5 $\pm$ 4.0 mV |
| | S | 2 | 15 | 17 | 85.7 $\pm$ 5.5 mV |
| | Totals | 9 | 21 | 30 | |
| Mean AP ($\pm$ S.D.) | 81.9 $\pm$ 4.4 mV | | 85.3 $\pm$ 5.1 mV | | |

$$\chi^2_Y = 4.369; \ P < 0.05$$

chance distribution at the 5% level ($\chi^2_Y - 4.369$; $P < 0.05$).

It has been observed that the shapes of current-latency curves obtained in cortical pyramidal cells are apparently related to the time course of membrane non-linearities demonstrable during long square-wave current pulses (Koike, Okada, Oshima and Takahashi, 1969). These membrane non-linearities have been described as 'overshoot' and 'undershoot' by Ito and Oshima (1965). The presence of such non-linear membrane behavior during long current pulses has been examined in many of the motoneurons in the present series of cells. The results of this can be summarized by stating that over- and undershoot of the membrane potential during long current pulses was found in motoneurons of both type F and type S motor units, and has also been found in motoneurons without accommodation as well as in cells exhibiting accommodation with both types of current-latency curves (see Figures 3 and 4). There were no significant correlations apparent between the presence of over- and undershoot and either the twitch grouping or the accommodation responses of the tested units.

## DISCUSSION

The method used in the present experiments to assess accommodative responses in motoneurons was selected because it was technically relatively simple to apply to many motoneurons and because, given certain arbitrary criteria, it permitted the cells tested to be divided into two groups, those 'with' and those 'without' accommodation. Motoneurons with flat current-latency curves (as in Figures 1 and 2) can be considered as showing no accommodation to linearly rising currents of the durations tested (maximum usually about 400

msec) but this does not imply that the same cells would not have shown some accommodation to more prolonged depolarizing ramps. Cat motoneurons with very long time constants of accommodation have been described (Ushiyama et al., 1966) and some of these may well be represented in the groups of units here described as showing no accommodation. The current-latency curves obtained from these cells resemble the 'low-ceiling' type of Bradley and Somjen (1961) and the 'second group', or 'tonic' cells, found by Sasaki and Otani (1961, 1962). Whether the present group of units without accommodation also includes some examples of what Bradley and Somjen termed the 'high-ceiling' group is not clear, since in some cases we did not define the earliest portion of the current-latency curves in enough detail to rule out the presence of a short period with minimum current threshold values well below the later stable level.

Among the 13 motoneurons which did exhibit accommodation, 7 cells showed definite evidence of a minimal current gradient requirement with short time constant of accommodation. Such responses have been found in cat motoneurons in other studies (Ushiyama et al., 1966; Yamashita, 1966) but Bradley and Somjen have attributed this pattern of response to injury and depolarization of the tested cells (Bradley and Somjen, 1961). The other motoneurons in the group with accommodation showed some increase in current threshold with slowly rising current ramps but in the relatively short time periods investigated did not exhibit a clear minimal gradient requirement (Figure 3). Nevertheless, it seems likely that such cells may well have had time constants of accommodation which were moderately long and examples of such motoneurons have been given by Ushiyama and colleagues (1966).

It has been found that depolarization of motoneurons, either with injected current or through injury, can greatly change the response to current ramps, increasing the minimal current gradient requirement and shortening the time constant of accommodation (Bradley and Somjen, 1961; Ushiyama et al., 1966). The data given in Tables I and II, however, suggest that the factors of cell injury and depolarization do not account for the presence of accommodation in the motoneurons selected for the present report. In fact, it would be expected that the smaller cells, which innervate slow twitch muscle units (Burke, 1967), should on the average suffer more damage by intracellular penetration than the larger type F cells. The data clearly show the opposite pattern of both mean action potential amplitudes and presence of accommodation. No type S unit was encountered which showed a clear minimal current gradient with the ramp durations used. It can be concluded that accommodation to linearly rising currents is present in some cat alpha motoneurons at normal resting potential levels and these cells are primarily the ones which innervate rapidly contracting muscle units.

It has been reported that barbiturate anesthetics tend to increase the minimal current gradient requirement of cat motoneurons (Sasaki and Otani, 1962; Yamashita, 1966), although such an effect was not found in a small sample of cells in another study (Løyning, Oshima, and Yokota 1964). In the present study, 14 of the total sample of 40 units were obtained in unanesthetized spinal animals (see Method). The pattern of the presence or absence of accommodation in these cells was essentially the same as that in the units studied in pentobarbital-anesthetized preparations. It is therefore concluded that the presence of barbiturate anesthesia did not materially affect the results of the present study.

The present work has provided a direct test of the hypothesis that alpha motoneurons innervating slowly contracting muscle units tend to exhibit less accommodation to slowly increasing depolarizing currents than do the cells innervating rapidly contracting muscle units. Although there are exceptions and some overlap in this relation, the hypothesis is supported by the present evidence. It appears that accommodation is one of the several intrinsic motoneuron properties which are quantitatively different in type F and type S motoneurons (Burke, 1967). The pattern of accommodation found in the present study supports the idea that type S motoneurons are specialized for steady 'tonic' firing to steady synaptic input while at least a significant proportion of type F cells exhibit accommodation responses which would fit with

specialization for responding to steady or slowly increasing input only with bursts of firing (see Mishelevich, 1968).

It is, however, clear that many type F cells do not show accommodation according to the definition adopted here. This accords with the fact that a significant proportion of type F motor units can produce sustained repetitive firing in decerebrate preparations (Burke, 1968) as well as in response to steady depolarizing current (Mishelevich, 1968). Recent evidence from this laboratory (Burke, Tsairis, Levine, Zajac and Engel, in preparation) has shown that the type F group of motor units can be divided into two subgroups, one of which appears to contain muscle units specialized for sustained tension output with repetitive motoneuron firing. It is unfortunately impossible to assess with the data at hand whether the type F units in the present study without accommodation innervated such muscle units but it seems reasonable to postulate that this may be the case.

## SUMMARY

1. The response of triceps surae motoneurons to linearly rising ramps of depolarizing current injected through intracellular micropipette electrodes has been studied in cats. Most of the data were obtained in animals anesthetized with pentobarbital but some units were studied in spinal unanesthetized preparations with anemic destruction of the brain. There were no essential differences in results from the two types of preparations.

2. The speed of contraction of the muscle fibers innervated by the motoneurons studied was measured following stimulation of the cell through the intracellular micropipette. Motor units could then be classified into two groups, fast twitch or type F and slow twitch or type S, on the basis of the speed of contraction of the muscle fibers innervated.

3. The presence or absence of accommodation in each motoneuron was determined based on the shape of the current-latency curve obtained during passage of linearly rising depolarizing currents of widely varying slope. About one half of the total sample of 22 type F units studied exhibited accommodation to linearly rising current ramps according to the criteria adopted in this study, and the other one half did not. Only 2 of the 18 type S units similarly examined showed accommodation to current ramps. The difference in accommodative behavior between types F and S motoneurons was

statistically significant and did not appear to be due to any systematic difference in cell injury caused by microelectrode penetration.

## ACKNOWLEDGMENT

Some of the present experiments were done while one of the authors (REB) was a guest worker in the Department of Physiology, University of Göteborg, Sweden. Thanks are due to Prof. Anders Lundberg for his generous support during this work, to Dr. G. ten Bruggencate who participated in some of the experiments and to Mr. Erling Eide who provided valuable technical support. Dr. David Mishelevich participated in some of the experiments performed at NIH.

## REFERENCES

Bradley, K., and Somjen, G. G., 1961, Accommodation in motoneurones of the rat and cat. *J. Physiol.* **156,** 75–92.

Burke, R. E., 1967, Motor unit types of cat triceps surae muscle. *J. Physiol.* **193,** 141–160.

Burke, R. E., 1968, Firing patterns of gastrocnemius motor units in the decerebrate cat. *J. Physiol.* **196:** 631–654.

Croxton, F. E., 1953, *Elementary Statistics with Application in Medicine and Biological Sciences.* New York: Dover Publications.

Eide, E., 1968, Input amplifier for intracellular potential and conductance measurements. *Acta Physiol. scand.* **73:** 1A.

Frank, K., and Fuortes, M. G. F., 1956, Stimulation of spinal motoneurones with intracellular microelectrodes. *J. Physiol.* **134:** 451–470.

Frank, K., and Fuortes, M. G. F., 1960, Accommodation of spinal motoneurones of cats. *Arch. ital. Biol.* **98:** 165–170.

Hill, A. V., 1936, Excitation and accommodation in nerve. *Proc. Roy. Soc. B.* **119:** 305–355.

Ito, M. and Oshima, T., 1965, Electrical behaviour of the motoneurone membrane during intracellularly applied current steps. *J. Physiol.* **180:** 607–635.

Koike, H., Okada, Y., Oshima, T., and Takahashi, K., 1968, Accommodative behaviour of cat pyramidal tract cells investigated with intracellular injection of current. *Exp. Brain Res.* **5:** 173–188.

Kuno, M., 1959, Excitability following antidromic activation in spinal motoneurones supplying red muscles. *J. Physiol.* **149:** 374–393.

Løyning, Y., Oshima, T., and Yokota, T., 1964, Site of action of thiamylal sodium on the monosynaptic reflex pathway in cats. *J. Neurophysiol.* **27:** 408–428.

Mishelevich, D. J., 1968. Repetitive firing to current in cat motoneurons as a function of muscle unit twitch type. *Expt. Neurol.* **25:** 401–409.

Nelson, P. G., and Burke, R. E., 1968, Motoneuron accommodation and motor unit twitch properties. *Proc. Int. Union Physiol. Sci.* **7:** 316.

Sasaki, K., and Oka, H., 1963, Accommodation, local response and membrane potential in spinal motoneurons of the cat. *Japan. J. Physiol.* **13:** 508–522.

Sasaki, K., and Otani, T., 1961, Accommodation in spinal motoneurons of the cat. *Japan. J. Physiol.* **11:** 443–456.

Sasaki, K., and Otani, T., 1962, Accommodation in motoneurons as modified by circumstantial conditions. *Japan. J. Physiol.* **12:** 383–396.

Stoney, S. D. Jr., and Machne, X., 1969, Mechanisms of accommodation in different types of frog neurons. *J. Gen. Physiol.* **53:** 248–262.

Ushiyama, J., Koizumi, K., and Brooks, C. McC., 1966, Accommodative reactions of neuronal elements in the spinal cord. *J. Neurophysiol.* **29:** 1028–1045.

Yamashita, H., 1966, Factors which decide the excitability of a spinal motoneuron in the cat. *Japan. J. Physiol.* **16:** 684–701.

# INTRACELLULAR CONDUCTANCE OF *APLYSIA* NEURONS AND SQUID AXON AS DETERMINED BY A NEW TECHNIQUE

DAVID O. CARPENTER, MARTIN M. HOVEY and ANTHONY F. BAK

*Laboratory of Neurophysiology*
*National Institute of Mental Health*
*Bethesda, Maryland* 20014

A metal microelectrode can be used to record the conductivity of intracellular fluid in small volumes. The equivalent capacitance recorded from such an electrode, when subjected to alternating current at 100 kHz, varies linearly with the conductance of the solution in contact with the electrode tip.

The conductance of the interior of *Aplysia* neurons is less than 10% that of sea water, whereas the conductance of squid axoplasm is approximately equal to that of sea water. These results suggest that there is extensive binding of water and small ions in *Aplysia* neurons but not in squid axoplasm.

## INTRODUCTION

It is remarkable that there is still so much controversy surrounding such a basic problem as what is the state of water and small ions in living cells. The more generally accepted view is that small ions and complex organic molecules are in a free, aqueous solution which is restricted by the physical barrier of the cell membrane. This conclusion is supported by the finding that in red blood cells the $K^+$ is totally exchangeable and is characterized by a single rate constant (Raker, Taylor, Weller and Hastings, 1950). In squid axon the longitudinal mobility and diffusion coefficients of $K^+$ (Hodgkin and Keynes, 1953) and $Na^+$ (Hodgkin and Keynes, 1956) are about the same in axoplasm as in sea water, whereas in muscle the mobilities are 50–100% that of an aqueous solution, with the low mobility being attributed to a viscosity effect (Harris, 1954; Kushmerick and Podalsky, 1969). Moreover the axoplasmic resistance of squid axon is equal to or only slightly less than the resistance of sea water (Cole, 1968).

In extreme contrast to this view that ions in cells are free are the 'association-induction hypothesis' of Ling (1962) and the 'sorption theory' of Nasanov and Troshin (Troshin, 1960). In both of these theories the living cell is viewed as a proteinaceous fixed charge system which forms a three-dimensional network onto which ions and non-electrolytes are absorbed. They believe that the protein matrix polarizes and orients the successive layers of water molecules which fill in the spaces between the protein chains.

The most direct support for at least some binding of ions has come from studies which have shown that ionic concentration differences do not obey the laws of the Donnan equilibrium (Shaw, Simon and Johnstone, 1956; Ling and Ochsenfeld, 1966; Robertson, 1961; and Koketsu and Kimura, 1960). The experiments of Robertson are of particular interest because of their simplicity. He expressed fluid from lobster muscle and compared the concentrations of ions in the fluid with that of the intact muscle. On the assumption that the differences between the fluid and the intact muscle reflected binding of the ions to proteins, he concluded that the approximate percentage of binding was 82% for $Na^+$, 26% for $K^+$, 100% for $Ca^{++}$, 64% for $Mg^{++}$ and 22% for phosphate with all $Cl^-$ being free.

In the last several years a considerable body of information from a variety of tissues has accumulated which supports the view that whereas a portion of a given species of ion in a cell is 'free', there is also a 'bound' fraction. There is little disagreement that $Ca^{++}$ and $Mg^{++}$ are bound to a sizable extent (Hodgkin and Keynes, 1957;

Murdock and Henton, 1968; Kushmerick and Podalsky, 1969). The study of the state of $Na^+$ and $K^+$ ions has been enormously aided by the development of ion-specific glass electrodes which measure the activity of a given ionic species. When activity is related to the total ion content determined spectrophotometrically, the difference represents the portion of that ionic species which is bound or inaccessible by reason of compartmentilization. Experiments using this technique in crab and lobster muscle (Hinke, 1959), frog muscle (Lev, 1964) and barnacle muscle (McLaughlin and Hinke, 1966) gave results indicating that 50–100% of the $Na^+$ present was bound, whereas most $K^+$ was free. Hinke (1961) has used this technique in a study of ions in squid axon, where intracellular activities of $Na^+$ and $K^+$ were compared to total content of the ions in extruded axoplasm. His results support the view that all $K^+$ in squid axon is free, but indicate that 24% of $Na^+$ is bound.

Another technique recently applied to the study of $Na^+$ and water in cells is nuclear magnetic resonance (NMR). Free $Na^+$ appears as a narrow absorption band under NMR, but interactions between $Na^+$ and certain anions or molecules give rise to a broadening of the line to the point of non-detection for very strong interactions (Jardetzky and Wertz, 1956a). Using this technique and comparing the $Na^+$ signal in the intact tissue with the total $Na^+$ concentration chemically determined, evidence for a binding of 40–70% of $Na^+$ has been obtained from muscle, liver, kidney, brain and frog skin (Cope, 1967, 1970; Rotunno, Kowalewski and Cerijido, 1967; Martinez, Silvidi and Stokes, 1969). In contrast, all $Na^+$ in blood, plasma and red blood cells is detectable by NMR (Jardetzky and Wertz, 1956b).

Ling and Cope (1969) have attempted to study the state of $K^+$ in muscle cells using NMR by the indirect method of replacing $K^+$ in muscle by $Na^+$. They report that when the major portion of intracellular $K^+$ is reversibly replaced by $Na^+$, the extra $Na^+$ gained cannot be detected by NMR. Therefore they conclude that the bulk of intracellular $K^+$, like $Na^+$, is adsorbed onto cellular macromolecules.

In this paper we present details of a new technique which has the potential of being another important tool in the study of the state of ions in cells. This method is based on a procedure described by Bak (1967) for testing metal microelectrodes. When a high frequency alternating current is passed between an uninsulated microelectrode tip and a Ag–AgCl reference, the equivalent capacitance measured varies with the concentration of ions in the layer of fluid surrounding the microelectrode tip. For equivalent concentrations the equivalent capacitance varies with the mobility of the species of ion present. This method does not detect a single ionic species, but does give a measurement which is correlated with the local conductance. In experiments in squid axon and *Aplysia* neurons we find that while the internal conductance of squid axon approaches that of sea water the conductance in *Aplysia* neurons is less than 10% that of sea water.

## METHODS

Glass insulated electrodes were constructed by fusing glass tubing around 0.001, 0.002 or 0.003 in. gold, silver or platinum wire. Most work was done using the 0.001 in. gold wire. The fused end was ground flat with an emory wheel presenting a circular metal surface approximately $25\mu$ in diameter for the 0.001 in. wire. For the gold and platinum wires the surface was usually coated with platinum black by passing 10 $\mu$amps of current and alternating polarity twice for 15 sec. duration in a solution of 3% platinum chloride containing a trace of lead acetate. Silver wires were coated with silver chloride by passing 10 $\mu$amps for 15 seconds in a concentrated KCl solution.

Metal microelectrodes were prepared from 0.009 in. stainless steel wire which was etched to a fine point ($\leq 1\mu$ at the tip) in concentrated HCl (Green 1958). The microelectrodes were gold-plated by the method of Boudreau, Bierer and Kaufman (1968) and the tips were platinized in the same manner as the wire electrodes. The microelectrodes were then insulated by dipping in DuPont primer paint by the method of Kinnard and MacLean (1967). The finished microelectrodes had tips about $1\mu$ in diameter at the tip and were 10 to $100\mu$ in length to the beginning of the insulation. Electrodes were tested for the integrity of the insulation by the method of Bak (1967).

Equivalent capacitance was measured with both types of electrodes using an operational amplifier possessing an internal constant frequency (100 kHz) in connection with an AC–DC converter and a Hewlett-Packard 3430A digital voltmeter (Figure 1). The summing point at the input of the operational amplifier was calibrated with a series of capacitors (0–400 pF). Each electrode was calibrated by inserting the electrode in 9 dilutions of sea water and referenced to a Ag–AgCl electrode (6 mil.).

Specific conductivity of salt solutions was determined with a Radiometer Type CDM2e conductivity meter with cell constant 0.61 cm.

Salt solutions were prepared at concentrations (0.5 M for univalent salts and 0.34 M for divalent salts) osmotically equivalent to that of normal sea water (1000 mosmols). On occasion the concentrations of chloride salts were determined titrimetrically by the method of Schales and Schales (1941). All other salts were carefully weighed, and the osmolarity of the respective solutions determined on an osmometer (Advanced Instrument Co.).

Specimens of *Aplysia californica* were obtained from Pacific Bio-Marine Supply Company, Venice, California and were maintained in artificial sea water. The visceral and/or pedal ganglia were removed from the animal and pinned to a layer of paraffin in a lucite chamber. The connective tissue sheath covering the cell bodies was slit, and the preparation maintained under flowing sea water. Individual identified and unidentified neurons were penetrated with glass micropipettes filled with 3 M KCl and/or with the gold-plated metal microelectrodes. For intracellular recordings of equivalent capacitance the output of the digital voltmeter was fed into a Tektronix 3A3 preamplifier (DC coupled) which then connected to a Beckman Type R Dynograph pen recorder. Simultaneous recordings of electrical activity through the metal microelectrode were made by connection to a differential DC high impedance electrometer which fed into a Tektronix 3A3 preamplifier (AC coupled) and then to the pen recorder. Due to the shunting effects of the operational amplifier in this circuit, voltage calibration was not meaningful for these AC recordings. Recordings from the glass micropipettes were made simultaneously from the same cell to monitor the resting membrane potential and magnitude of the action potential. All other recordings were as previously described (Carpenter and Alving, 1968).

Squid axon experiments were done in the laboratory of Dr. I. Tasaki at the Woods Hole Marine Biological Laboratory. Axons were cleaned, ligated at both ends and removed to the recording chamber. Equivalent capacitance and electrical activity were simultaneously recorded through a metal microelectrode.

## RESULTS

*Evidence that this technique measures local conductivity*

Figure 1 is a block diagram of the instrument used

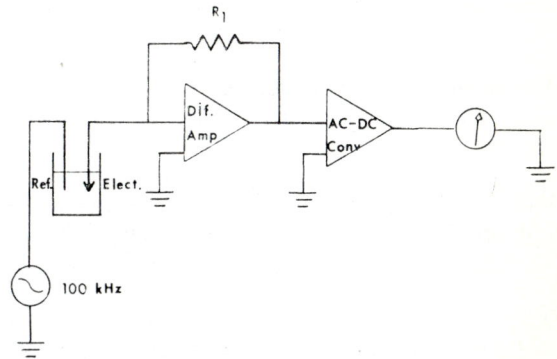

FIGURE 1  Block diagram of the method used to measure the equivalent capacitance of salt solutions and *Aplysia* neurons with metal electrodes. $R_1$ is the feedback resistance. The arrow represents a metal electrode in solution and the reference is a Ag-AgCl electrode connected to a 100 kHz oscillator.

in these studies. For the circuit of Fig. 1 we have

$$V_0 = \frac{R_1}{|Z_x|} V_s$$

where $V_0$ is the output voltage, $R_1$ is the feedback resistance, $V_s$ is the source voltage and $|Z_x|$ is the absolute value of the input impedance and independent of its phase angle. So for a pure resistance or conductance $G_x$

$$V_0 = G_x R_1 V_s$$

or a pure capacity $C_x$

$$V_0 = C_x \omega R_1 V_s$$

where $\omega$ is the angular frequency $2\pi f$. For the same values of $V_0$ the equivalent capacity

$$C_x \equiv 1/|Z_x|\omega \tag{1}$$

But for $Z_x = R_x + jX_x$, this equivalent capacity

$$C_x \equiv 1/\omega \sqrt{R_x^2 + X_x^2}$$

will approach

$$C_x \equiv G_x/\omega \tag{2}$$

as $X_x/R_x \to 0$.

The concept of a single electrode for the measurement of biological impedances was developed from potential theory (Kaufman and Johnston, 1943 and Cole, 1968). The technique was modified and further applied by Schwan and Kay (1956, 1957). Potential theory has treated a number of cases for the resistance, R in $\Omega$, of a perfect electrode in a medium of specific resistance $\rho$ in $\Omega$ cm. In the case of a small spherical electrode, we have for the resistance of a sphere with radius $a$ cm

$$R = \rho/4\pi a$$

For an electrode with a specific surface polarization impedance $\gamma \, \Omega \, cm^2$, the impedance for a sphere is,

$$Z = \gamma/4\pi a^2$$

and for a well platinized electrode in a given electrolyte and at constant temperature (Cole, 1968),

$$\gamma = \bar{\gamma}(j\omega)^{-\alpha}$$

where $\alpha$ is the slope and is sometimes about 0.5.

Then for a spherical polarizable electrode in an electrolyte we would have

$$Z_x = \frac{\rho}{4\pi a} + \frac{\bar{\gamma}}{4\pi a^2 \, (j\omega)^\alpha} \qquad (3)$$

For the equivalent capacity as defined by equation (1)

$$C_x \equiv 1/|Z_x|\omega$$

where $Z_x$ is given by equation (3). If the second term of equation (3) is negligible

$$C_x = \frac{1}{\rho} \cdot \frac{4\pi a}{\omega} \qquad (4)$$

where the first factor is the specific conductance $g_x$, and the second is the calibration factor. Then in this ideal condition the equivalent capacity is inversely proportional to frequency and indeed this is a test for an ideal arrangement.

Figure 2 is a log-log plot of equivalent capacity versus frequency. It is readily apparent that over the high frequency range of 50–250 kHz, the

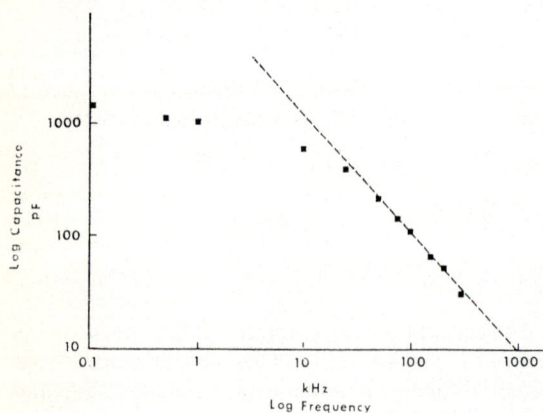

FIGURE 2 Log-log plot of equivalent capacity (pF) versus frequency (kHz). The equivalent capacity of 0.9% NaCl solution was recorded with a platinized, glass-insulated electrode at constant temperature over a wide range of frequencies. The dashed line is that portion of the plot in which the impedance is constant.

conditions of equation (4) are satisfied, i.e. a straight line with slope of –1 is obtained. At the lower frequencies, below 25 kHz, the polarization impedance becomes large and equation (4) is no longer valid.

So it is obvious that the second term of equation (3) must be made small as compared to the first, or the error (the second term of equation (3) over the first)

$$\Delta = \frac{\bar{\gamma}g_x}{a(j\omega)^\alpha} \ll 1$$

Therefore one or a combination of the following conditions must be met:

The polarization impedance, $\bar{\gamma}$, shall be small.
The test frequency, $\omega/2\pi$, shall be large.
The solution conductance, $g_x$, shall be small.
The electrode dimension, $a$, shall be large.

The first two conditions have been dealt with. Heavy platinization can reduce the polarization impedance of a metal electrode by $10^4$ (Schwan, 1963) and the frequency, 100 kHz, at which we have made our measurements is well within the range where the electrode polarization impedance is negligible (Figure 2 and equation 3).

As a demonstration that the equivalent capacitance recorded in this fashion is a relative measure of conductance we have compared independent measurements of equivalent capacitance and specific conductivity as a function of concentration of salts. These results are shown in Figure 3. It can be seen that both equivalent capacitance and specific conductivity increase linearly with concentration over the concentration ranges between distilled water and normal sea water (0–1000 mosmols).

Figure 4 compares the effect of temperature on the equivalent capacitance–concentration and specific conductivity–concentration plots. It is apparent that both types of measurements vary as a function of temperature in a similar fashion.

If the equivalent capacitance is a valid indicator of conductance it should vary with the mobilities of the salts present in the solution at a given concentration. This was carefully tested by using the $Cl^-$ salts of five univalent cations, as shown in the first part of Table I. Stock solutions were made at 0.5 M, then dilutions were made with 1, 5, 10, 15, 20, 25, 50 and 75% of that concentration. The concentration of each was carefully measured by a titrimetric determination of $Cl^-$ by the method of Schales and Schales (1941). The dilutions were made volumetrically and checked on an osmometer. Independent measurements were made in each solution of the equivalent capacitance and the

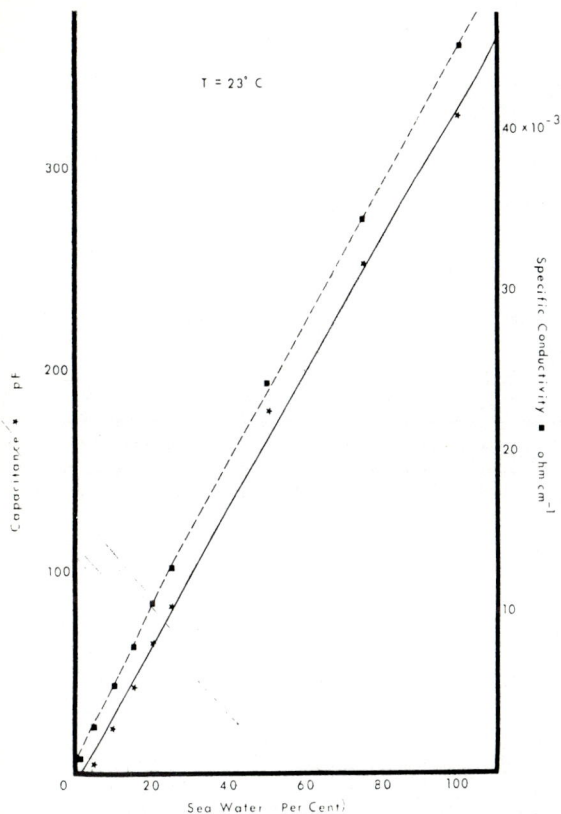

FIGURE 3 Equivalent capacitance and specific conductivity as a function of the concentration of sea water at 23°C. Apparent capacitance was recorded with a glass insulated gold wire electrode and at a frequency of 100 kHz. Specific conductivity was measured with a conductivity cell. Sea water dilutions were obtained by mixing appropriate volumes of distilled water and sea water.

specific conductivity at 23°C. For each salt plots were made of equivalent capacitance against concentration. The values in Table I were read from these plots for concentrations of exactly 0.5 M.

For both equivalent capacitance and specific conductivity the values for the Cl⁻ salts increased in the order Li < Na < K < Cs < Rb. This is the order of the mobilities of these ions (Robinson and Stokes, 1959).

The equivalent capacitance varied with specific conductivity for all of the remaining salts in Table I. The relative values in general vary with ionic mobilities. There are several exceptions, but they probably result from the fact that the concentrations and room temperature were not so well controlled in these experiments as with the univalent Cl⁻ salts. The most important conclusion from the experiments on this group of salts is that the equivalent capacitance always varied with the

independent specific conductivity measurement.

Table II shows the effect of added non-ionic solute (sucrose) on the specific conductivity of a salt solution. The conductivity decreased dramatically as the concentration of sucrose was increased. The fact that equivalent capacitance changed proportionally with conductivity further substantiates the use of our technique as a measure of conductance. Similar results were obtained with saturated mannitol solutions (about 15%). The depression of specific conductivity by mannitol and sucrose is known to result from changes in viscosity (Robinson and Stokes, 1959). Unlike sucrose, colloids such as egg albumin, which does not bind Na⁺ or K⁺ (Carr, 1956), polyglutamic acid or 5% agar-agar did not significantly depress the conductance as measured by either techniques, although these substances cause the solution to become a very viscous gel.

### Characteristics of the metal electrode

All of the measurements discussed up to this point were taken with 1 mil. gold wire electrodes insulated with glass with a layer of platinum black deposited on the exposed surface. Similar results were obtained with platinized platinum wire or chlorided silver wire electrodes. These electrodes all gave linear readings of equivalent capacitance with concentration in dilutions of 0.5 M univalent salts or normal sea water (1000 mosmols). The equivalent capacitance varied approximately with the surface area, although it was difficult to determine effective area with the platinum black deposit.

The bare metal surfaces had very different properties. A polished wire surface when inserted into salt solutions exhibited lower equivalent capacitance and instability relative to a platinized electrode. The plot of capacitance against concentration rose linearly up to dilute sea water solutions of about 200 mosmols and reached a plateau with little or no increase beyond that concentration.

The microelectrodes which we have used all exhibited properties intermediate between those of the bare metal surfaces and the platinized or chlorided metal surfaces. Most of the microelectrodes were made of etched stainless steel and were gold plated. Without platinizing, these electrodes behaved similarly to the bare wire electrodes in that they saturated almost completely at solutions more concentrated than 200 mosmols. With platinization the microelectrodes improved considerably, but in only a few instances did they ever exhibit a strict linear relation between equivalent capacitance and concentration over the whole

TABLE I

Comparison of Capacitance and Specific Conductivity of a Variety of Salt Solutions

| Salt (a) | Capacitance (pF) | Conductivity (ohm cm$^{-1}$ × 10$^{-3}$) | Salt (a) | Capacitance (pF) | Conductivity (ohm cm$^{-1}$ × 10$^{-3}$) |
|---|---|---|---|---|---|
| LiCl | 335 | 39.2 | NaF | 274 | 35.2 |
| NaCl | 360 | 46.0 | KF | 360 | 46.4 |
| KCl | 423 | 56.6 | NaBr | 336 | 43.9 |
| RbCl | 440 | 59.3 | KBr | 418 | 55.2 |
| CsCl | 432 | 58.0 | NaI | 346 | 44.3 |
| MgCl$_2$ | 365 | 53.3 | KI | 425 | 57.0 |
| CaCl$_2$ | 386 | 55.0 | Na acetate | 215 | 26.2 |
| SrCl$_2$ | 400 | 55.7 | Na propionate | 192 | 23.8 |
| BaCl$_2$ | 397 | 55.4 | Na benzoate | 189 | 23.0 |

(a) Concentration of univalent salts 0.5 M, divalent salts 0.34 M.

Recordings of equivalent capacitance and specific conductivity of individual salt solutions at 23°C. Values were obtained for each salt at a standard concentration (0.5 M for univalent salts, 0.34 M for divalent salts) from plots of apparent capacitance or specific conductivity as a function of concentration. Concentrations of the monovalent Cl$^{-}$ salts were determined titrimetrically. All other salts were weighed out and the osmolarity of the respective solutions checked on an osmometer.

TABLE II

Effect of Added Solute on Capacitance and Specific Conductivity of NaCl Solutions

| Solution | Capacitance (pF) | Specific Conductivity (ohm cm$^{-1}$ × 10$^{-3}$) |
|---|---|---|
| NaCl (0.43$M$) | 285 | 42.6 |
| NaCl (0.43$M$)—1% sucrose | 269 | 40.2 |
| NaCl (0.43$M$)—5% sucrose | 249 | 37.0 |
| NaCl (0.42$M$)—10% sucrose | 222 | 32.5 |
| NaCl (0.38$M$)—20% sucrose | 158 | 24.0 |
| NaCl (0.34$M$)—30% sucrose | 96 | 16.2 |
| NaCl (0.31$M$)—40% sucrose | 47 | 9.8 |
| NaCl (0.26$M$)—50% sucrose | 13 | 5.3 |
| NaCl (0.43$M$)—Saturated with Mannitol | 171 | 26.6 |
| NaCl (0.43$M$)—Saturated with egg albumen | 256 | — |
| NaCl (0.43$M$)—Saturated with poly-glutamic acid | 255 | 45.0 |
| NaCl (0.5$M$) | 316 | 46.0 |
| NaCl (0.5$M$)—5% agar gel | 315 | — |

Recordings of apparent capacitance and specific conductivity of NaCl solutions to demonstrate the effect of added solute. Both types of measurements were recorded as in Table I. The solutions were prepared in per cent by weight and the concentration of salt determined by titration for total chloride.

range of concentrations studied.

We do, however, feel justified in using these electrodes, particularly in the range in which the relationship between concentration and equivalent capacitance is linear. In so far as we could test, the slope is the same in this region whether or not the electrode is platinized. Because of this complication, however, we found it necessary to calibrate each electrode over the entire range of dilutions and to refer readings made in neurons to the dilution of sea water which gave a comparable reading.

Because the platinum-black and AgCl coats are somewhat fragile the readings often changed with repeated use. When the equivalent capacitance measured in normal sea water changed the electrode was re-calibrated.

*Intracellular conductance recorded in nerve cells*

Figure 5 shows records from an experiment on the giant cell ($R_2$ in the classification of Frazier, Kupfermann, Coggeshall, Kandel and Waziri,

A

B

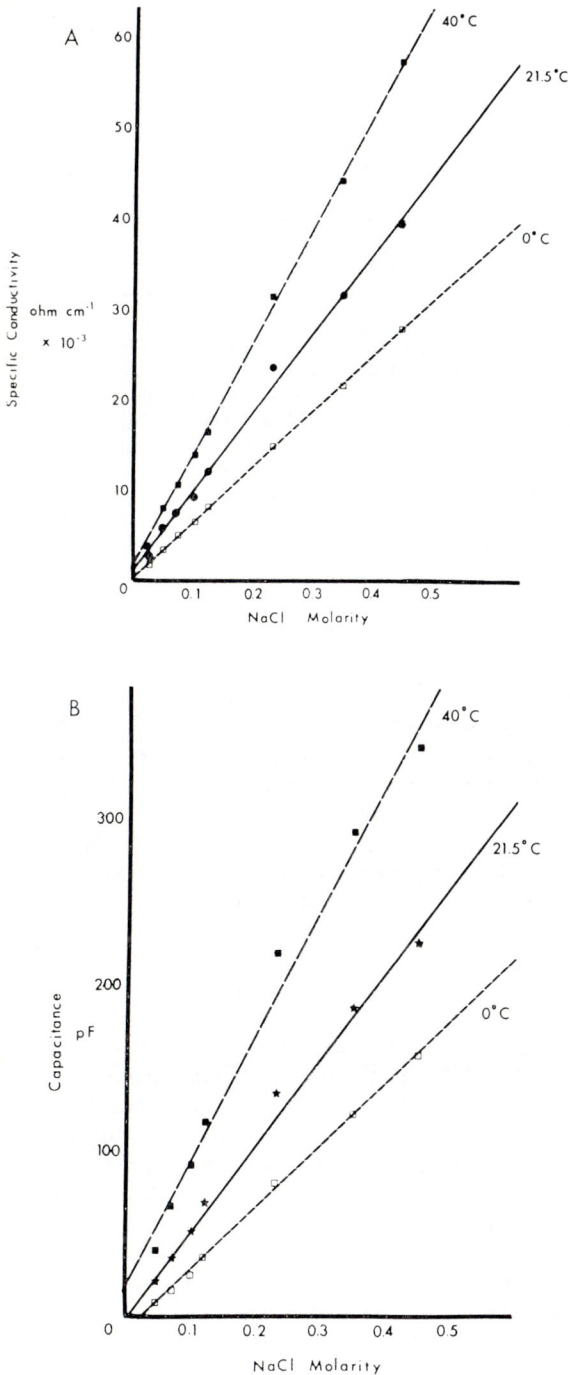

FIGURE 4   Recordings of specific conductivity (A) and apparent capacitance (B) of NaCl solutions at various temperatures. Measurements were taken as in Figure 1. For both measurements all solutions were maintained in a constant temperature bath with vigorous agitation for at least two hours before taking measurements.

1967) penetrated by both glass and metal microelectrodes. At the beginning of part A the glass electrode is in the cell (trace 1), which is silent and has a resting potential of −70 mv. The metal electrode at the beginning of the trace is in the sea water and reads 290 pF (trace 3). On lowering the metal electrode the equivalent capacitance falls suddenly as the cell is partially penetrated and small action potentials are recorded on the AC coupled voltage channel from the same metal microelectrode which is measuring equivalent capacitance. On further advancement the equivalent capacitance decreases further and the amplitude of the AC spikes increases to a stable value at the end of trace A. This indicates that all of the uninsulated tip of the metal microelectrode has been inserted into the cell. In B, 15 minutes later, the membrane potential in the DC trace has recovered from the injury of penetration with the metal electrode, while the AC trace and capacitance readings are stable as before. On withdrawal of the metal electrode the capacitance increased from the value of 90 pF in the cell to 290 pF in the sea water. The glass microelectrode records some loss of membrane potential on withdrawal but still shows an action potential of 100 mv on the oscilloscope. Part C shows the calibration curve for the microelectrode used. The value of equivalent capacitance recorded intracellularly (90 pF) corresponds to a solution of 10 % sea water and 90 % distilled water.

Table III shows averaged data from 48 neurons studied with 6 electrodes, each carefully calibrated in dilutions of sea water. For each electrode the averaged stable intracellular equivalent capacitance recorded was compared to the dilution of sea water which gave a similar reading. The average equivalent concentration is about 5 % sea water—95 % distilled water.

Figure 6 shows another experiment on an unidentified neuron from the lower left quadrant of the *Aplysia* visceral ganglion. At the beginning of the trace both electrodes are in the cell. The action potential in the DC trace was 80 mv with a 40 mv resting potential. The equivalent capacitance at 20°C in the cell was 27 pF, equivalent to a dilution of 4 % sea water. The temperature of the sea water perfusing the ganglion was changed by flow through an ice bath and was recorded by a small thermistor placed near the ganglion. On cooling to 11°C the equivalent capacitance fell from 27 to 25 pF (equivalent to about 6 % sea water). At the end of A the metal electrode was withdrawn from the cell with a return to a reading in sea water of 218 pF at 12°C. Part B shows the calibration of

TABLE III

| Electrode No. | No. of Cells | Average Reading in Sea Water (pF) | Average Reading inside Cell (pF) | Equivalent Sea Water Concentration | Action Potential (mV) |
|---|---|---|---|---|---|
| 1 | 27 | 52 | 10 | 3% | — |
| 2 | 3 | 217 | 55 | 6% | 90 |
| 3 | 5 | 177 | 33 | 5% | 85 |
| 4 | 9 | 278 | 75 | 6.5% | 100 |
| 5 | 3 | 250 | 49 | 6% | 70 |
| 6 | 1 | 224 | 28 | 4% | 70 |

Equivalent capacitance recorded in *Aplysia* neurons with gold-plated stainless steel microelectrodes. All electrodes except no. 1 were coated with a thin layer of platinum black. The uninsulated tips ranged in length from 10 to 100μ. Each electrode was calibrated by taking readings at nine dilutions of sea water. The average reading in the neurons was compared to an equivalent sea water concentration by reading this value from the plot. In all experiments, except with electrode no. 1, simultaneous recordings were made with a glass microelectrode inserted in the same cell. The average action potential amplitude refers to the action potential recorded with the glass microelectrode.

the microelectrode at two temperatures.

In all experiments where temperature was changed the value of intracellular apparent capacitance changed little or not at all even though the calibration curve of the electrode changed significantly over the temperature range used.

Because the very low readings in *Aplysia* neurons were surprising and as a further test of the validity of this technique for measuring intracellular conductance, experiments were performed on squid axons, where conductivity of the axoplasm has been measured. Hodgkin and Cole (1939) report that the resistivity of axoplasm (the inverse of conductivity) is $1.4 \times$ that of sea water. However, Cole (1968) reports unpublished experiments with R. E. Taylor on extruded axoplasm where the resistivity was found to be equal to that of sea water. Therefore on the basis of these results the values for equivalent capacitance in squid axon should be equivalent to 70–100% sea water if our technique is a valid measure of conductance.

Thirteen penetrations were made with metal microelectrodes in 6 squid giant axons. The average internal equivalent capacitance reading was 104% that of the sea water control, with a range of 87–117%. One such penetration is shown in Figure 7. In sea water this electrode read 80 pF. After the transients resulting from the electrode being forced against the membrane, the equivalent capacitance fell to 75 pF when the electrode entered the axon. On withdrawal, at the second arrow, the equivalent capacitance recovered to its control value.

## DISCUSSION

*The validity of using this technique to study the state of ions in cells*

The method which we have described is a new application of the properties of metal microelectrodes to biology. Since the equivalent capacitance measured from a very small layer of fluid at the electrode tip varies with the conductance of that fluid, this technique has considerable potential in the study of the conductance of the interiors of cells or even in different parts of the same cell. We feel that we have presented ample evidence as to the validity of the technique as a measure of relative intracellular conductance.

Support for the conclusion that this is a valid measure of conductance is obtained from the fact that the increase in equivalent capacitance with concentration is linear, the change in equivalent capacitance with temperature parallels the change in specific conductivity, and the relative reading for different ions varies with their mobilities. Furthermore, addition of a non-ionic substance such as sucrose depresses the equivalent capacitance and specific conductivity equally, probably as a result, primarily, of the change in viscosity (Robinson and Stokes, 1959).

We believe that the results obtained in *Aplysia* neurons reflect a real and important difference between these cells and some other tissues, such as red blood cells and squid axon. The readings obtained in *Aplysia* neurons fall in the region of the equivalent capacitance-concentration curve

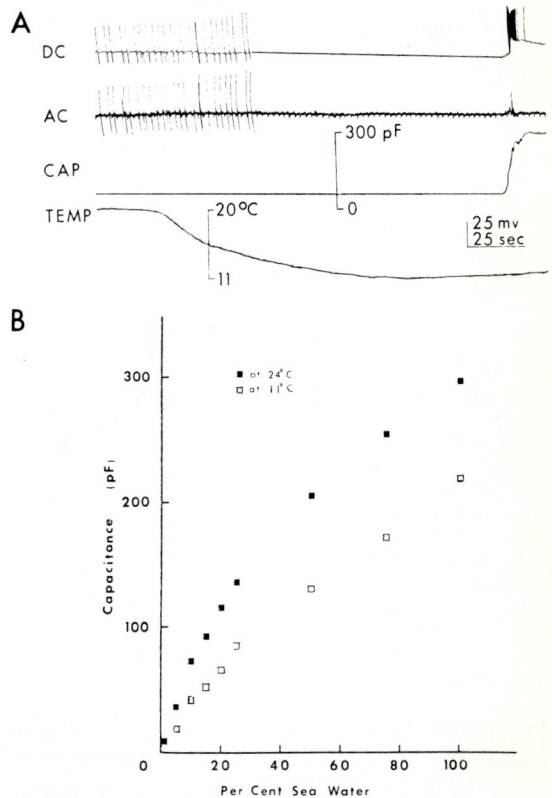

FIGURE 6 The effect of temperature upon equivalent capacitance in an unidentified *Aplysia* neuron. In A the DC recording was through a glass micropipette, while the AC voltage and apparent capacitance recordings were through a metal microelectrode in the same cell, Temperature was varied by flowing the sea water through an ice bath before perfusing the recording chamber. At the beginning of the trace both microelectrodes were in the cell. At the end of the trace the metal microelectrode was withdrawn from the cell, leaving the glass microelectrode in place.
The calibration of the metal microelectrode as a function of temperature and concentration is shown in B. The equivalent capacitance recorded in the cell corresponds to a 6% sea water solution. The voltage calibration applied only to the DC recording.

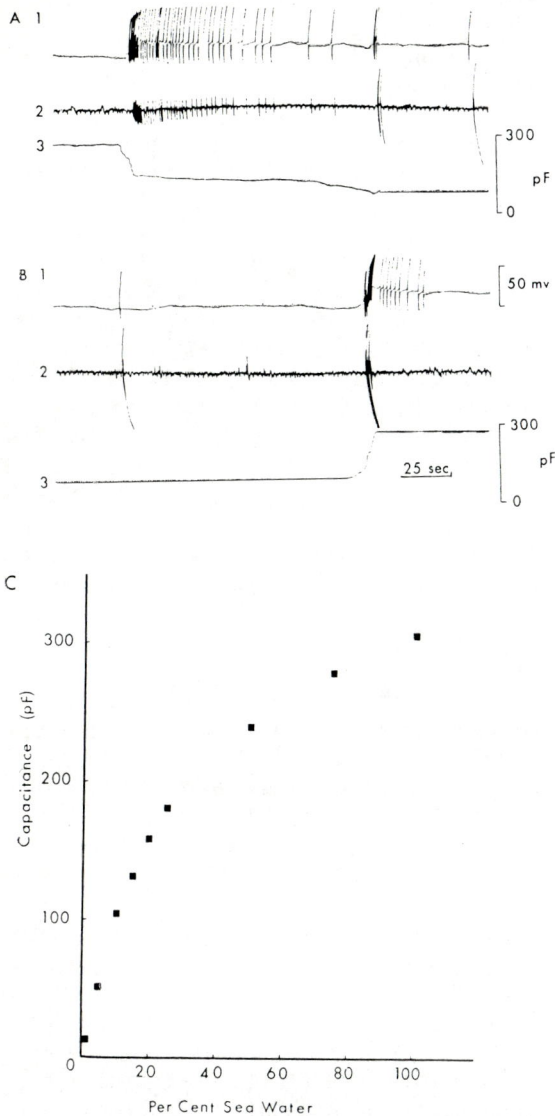

FIGURE 5 Equivalent capacitance recorded in the giant cell ($R_2$) of *Aplysia* with a gold-plated stainless steel microelectrode and a glass micropipette. Trace 1 shows the DC recording from the glass microelectrode. Initial resting membrane potential was –70mv. Trace 2 is an AC coupled electrical recording from the metal microelectrode, while trace 3 is the equivalent capacitance recorded simultaneously through the metal microelectrode. The metal microelectrode was withdrawn from the cell near the end of trace 3. In C the calibration curve of this electrode is shown as a function of sea water concentration. The reading inside this cell was equivalent to about 10% sea water. Voltage calibration applied only to the DC recording.

where the response is linear. Moreover, we have calibrated each electrode over the entire range of sea water dilutions, and a comparison of the readings in and out of the cell therefore has meaning even when the curve is not linear, as in the case of the experiments on squid axons. In this portion of the curve, however, it is not possible to be as confident about the numerical significance of the reading.

The low readings obtained in *Aplysia* neurons do not result from the physical properties of the membrane. At the frequency used (100 kHz) the membrane resistance is not detected. The membrane capacitance of *Aplysia* neurons is about 23 nF (Fessard and Tauc, 1956). The range of equivalent capacitance recorded from our microelectrodes (5–400 pF) is small relative to this value, and there-

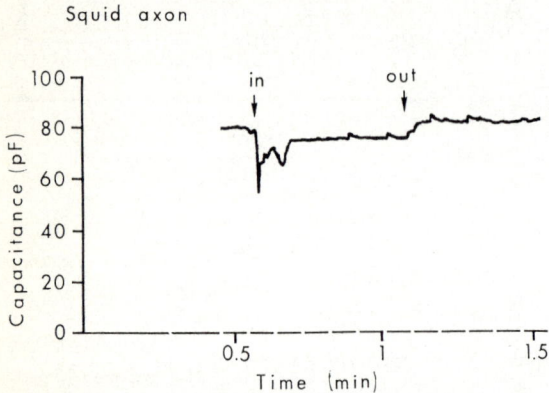

Squid axon

FIGURE 7 Equivalent capacitance recorded in squid axon with a gold-plated stainless steel microelectrode. The initial reading in sea water with this electrode was 80 pF. The first arrow marked the beginning of insertion. The transient decrease in reading reflects the period of time when the electrode was indenting the membrane but had not yet penetrated. The electrode has penetrated at the point where the recording becomes stable. Simultaneous recording of the electrical activity was made through the metal electrode, and the action potential had an amplitude of 100mv. The electrode was withdrawn from the cell at the second arrow.

fore membrane capacitance does not affect our measurement. The only membrane property which would cause a low equivalent capacitance would be a resistance in series with the membrane capacitance. There is no evidence for a series resistance in *Aplysia* neurons (Alving, 1969) and we have ruled out the possibility that a series resistance is the explanation for the low values recorded in cells by experiments in which the equivalent capacitance was recorded differentially between two metal microelectrodes inserted into the same cell. On insertion of the second microelectrode the equivalent capacitance fell further, whereas if the low reading were due to a series resistance it should increase on insertion of the second microelectrode.

Another possible explanation for the low values recorded in *Aplysia* neurons is that the tip of the microelectrode becomes coated with a layer of protein or other macromolecular material inside the cell, with the result that the effective surface area is greatly reduced. This possibility is hard to rule out, but appears very unlikely. The change in equivalent capacitance occurred very rapidly on penetration of the cell and was always immediately reversible on withdrawal. If some substance were adhering to the tip, it is unlikely that it would attach and wipe off so quickly. We also found repeatedly that the same values of equivalent capacitance were recorded in sea water and in sea

water containing egg albumin, 5% agar-agar, or polyglutamic acid. These results indicate, at least, that just any protein does not coat the tip and change its recording properties.

The results of our experiments on squid axon provide additional support for the use of this technique to measure intracellular conductance. In squid axon the conductivity of axoplasm is known to be 70–100% that of sea water (Cole and Hodgkin, 1939; Cole, 1968), the total concentration of ions are known (Keynes and Lewis, 1951) and K+, at least, is probably all in a free state (Hinke, 1961). Our experiments recorded an average conductance of axoplasm of 104% that of sea water. However, confidence in this value is less than that for recordings in *Aplysia* neurons since these readings all fall in that portion of the concentration—equivalent capacitance plot for the electrode that is least sensitive because of saturation. It must be pointed out that a conductance equal to or even greater than sea water is not incompatible with some binding of Na+ (Hinke, 1961) and anions. In sea water Na+ is the predominate cation, whereas in axoplasm there is about eight times the amount of K+ as Na+ (Keynes and Lewis, 1951). Because of its greater mobility, K+ will contribute more to the conductance than an equal concentration of Na+.

*Possible reasons for the low conductance of Aplysia neuronal cytoplasm*

Our results indicate that the conductance recorded in *Aplysia* neurons is less than 10% that of sea water. We have as yet been unable to confirm these results by a conventional measure of resistivity because of the technical difficulties inherent in the insertion of a fixed array of four electrodes into the same cell (Li, Bak and Parker, 1968). However, the squid axon experiments add greatly to our confidence in the results from *Aplysia* neurons.

There are several factors which may contribute to the low conductance in *Aplysia* neurons. Although the local conductance might vary in different parts of the same cell, we have no evidence that this is the case. Most of the electrodes we used were very large—usually with about 100$\mu$ exposed tip. Such an electrode could not be exclusively in the nucleus or other localized area which might have very different properties from the cytoplasm. We feel that our readings reflect an average conductance of the cell interior, since the results were very similar in all the cells studied and in multiple penetrations of the same cell. Moreover, in a few experiments we used small electrodes with 2–10μ exposed tips, and did not measure significant differences on advancing the electrode through the cell.

The low conductance could result from a simple lack of small ions, perhaps as a result of the osmotic pressure contributed by a large concentration of uncharged molecules. Although this seems unlikely, there are no good measurements of total ion content in these cells.

The low conductance could also result from a very high content of a substance such as sucrose, which does drastically decrease both the apparent capacitance and specific conductivity in solution in concentrations of 10% or more. This effect would appear to be due to changes of viscosity, although colloids such as 5% agar-agar, do not similarly depress the specific conductivity. It seems likely that the sucrose molecules impede water and ion movements as a result of extensive hydrogen binding to a much greater extent than occurs with the more coarse organization of the agar gel.

A final possibility is that the low conductance results from extensive binding of ions. This would apply particularly to $K^+$, which is usually the predominant intracellular small ion, but might equally apply to other cations and possibly even anions. It seems to us that this is the most likely explanation of our results simply because of the evidence from other preparations that binding of ions to varying degrees does occur, whereas no tissues are known to have very low free ion concentrations as a result of osmotic forces alone, and no low conductivities have been measured resulting from the viscosity of the intracellular fluid.

The effect of temperature is the only experimental manipulation which we have performed which might allow us to choose among these alternatives. We found very little change in the equivalent capacitance on changing temperature. If the low readings resulted from a lack of ions or from a viscosity effect the readings should decrease with cooling in the cell at least as much as in solution. Temperature might affect the amount of binding, but to explain the observation that the readings did not change significantly with temperature we must suggest that the extent of binding is somewhat less in the cold than in the warm to a degree which balances the effect of temperature on conductance at a constant ion concentration. Wesson, Cohn and Brues (1949) suggested that $K^+$ binding is decreased on cooling, although the results of Harris (1952) and Whittam and Davies (1954) on $K^+$ binding and of Elford (1970) on water binding suggest that binding increases on cooling.

*The nature of the differences between squid axon and Aplysia neurons*

It would appear most likely that the dramatic difference in intracellular conductance between squid axon and *Aplysia* neurons results from differences in the organic content and the presence of internal membraneous structures. The somata of molluscan neurons contain large nuclei (frequently polypodial), Nissel bodies, Golgi apparatus, mitochondria, granular endoplasmic reticulum, free ribosomes and inclusion granules (Rosenbluth, 1963; Coggeshall, 1967; Bullock and Horridge, 1965). At the origin of the axon there is an abrupt and dramatic change in the character of the cytoplasm, with a drastic reduction in the content of identifiable cell structures and changes in the staining reaction and density of ground substance (Bullock and Horridge, 1965). The axoplasm of squid axon is by no means free of internal structuring, but contains a filamentous network, microtubules, mitochondria and agranular endoplasmic reticulum (Metuzals, 1969). The density of internal structures is, however, very much less than is characteristic of cell bodies. Although, as yet, we have not tested conductance of other nerve cell bodies or other axons it is very possible that the differences which we have recorded represent basic differences between nerve somata and axons.

There is precedent for supposing that the extent of binding of ions varies with the concentration of organic constituents of the cell as well as the nature of the specific macromolecules present. In frog eggs Naora, Naora, Izawa, Allfray and Mirsky (1962) found that there was 5 times the amount of $Na^+$ and 4 times the amount of $K^+$ in the nucleus as in the cytoplasm. Furthermore they found the $Na^+/K^+$ ratio to be different in nucleus and cytoplasm.

The nature of the binding of small ions has been studied in several laboratories. Jardetzky and Wertz (1956, 1959) have presented NMR evidence that $Na^+$ forms specific complexes with lactate, pyruvate, citrate, phosphates, keto-acids and alcohols. $Na^+$ is known to bind strongly to deoxyribonucleic acid (Ascoli, Botre and Liquori, 1961). Both $Na^+$ and $K^+$ bind to phospholipids (Kirschner, 1958) and a variety of proteins (Carr, 1956; Lewis and Saroff, 1957; Ho and Waugh, 1965). Ling and Ochsenfeld (1965) have suggested that in muscle the $\beta$ or $\gamma$ carboxyl groups of proteins bind ions in the sequence $RB^+ > K^+ > Na^+$, and that this selective binding is the primary factor in determining the development of the intracellular concentration profiles.

It is difficult to explain our results solely on the basis of ion binding without also suggesting that the water is bound or organized in such a fashion that there is little or no free solution in the part of the cell where our electrode measures conductance. Although we have no other evidence that this is the case than the very low conductance measurement, our results in *Aplysia* are in agreement with Ling's suggestion that 'all or nearly all water molecules in the living cell can be considered to exist in polarized multilayers oriented on the surfaces of cell proteins' (Ling, 1965). Such a situation does not appear to be true for squid axon, probably because the content of organic macromolecules is too small to result in the same degree of organization of water and ions.

Our results raise many more questions than they solve, and we must ask how they can be reconciled with what has become known as 'membrane theory'. One possibility is that there are at least two compartments to the cell: a central portion which includes most of the somatoplasm in large neurons in which water and ions are tightly organized as a result of covalent and hydrogen binding, and a peripheral ring in which ions are free. It is possible that these two compartments might correspond to the exo- and endoplasm which can be distinguished in giant neurons by histologic and electron microscopic techniques (Bullock and Horridge, 1965). The endoplasm is the more darkly staining cytoplasm filling most of the cell, while the ectoplasm is a thinner and clearer layer near the cell membrane which has many of the same structural characteristics of the axon (Rosenbluth, 1963). Our metal microelectrodes, by virtue of their size, would almost certainly be primarily in endoplasm, although we did not detect any small peripheral region of higher conductance in the few experiments using much smaller tips. It is interesting to note that in squid axon ecto- and endoplasm can also be distinguished; however, here the ectoplasm contains the more dense packing of intracellular elements (Metuzals and Izzard, 1969).

Another possibility is that all of the cytoplasm of the cell is structured and that free solutions exist only as a thin layer on the inside of the cell membrane or perhaps as a thin layer around all intracellular membranes. The conductance we measure might reflect these criss-crossing paths of low resistance which run through regions of very high resistance.

The existence of two compartments containing bound and free ions respectively was proposed by Kurella (1960) in a discussion of the relative merits of the sorption theory of Troshin and the more conventional membrane theory. Although the existence of two such intracellular compartments is not the only possible explanation of our results, these experiments do suggest that, if they exist, the relative size of these compartments may vary in different cells, presumably depending on the amount and specific binding properties of intracellular macromolecules.

## ACKNOWLEDGMENTS

We are grateful to Dr. Ichiji Tasaki for making his facilities at Woods Hole, Massachusetts, available to us for the squid axon experiments, and to Dr. K. S. Cole for many helpful discussions and who was kind enough to provide the theoretical background on metal electrodes.

## REFERENCES

Alving, B. O., 1969, Differences between pacemaker and non-pacemaker neurons of *Aplysia* on voltage clamping, *J. Gen. Physiol.* **54**: 512–531.

Ascoli, F., Botré, C., and Liquori, A. M., 1961, Irreversible changes of ionic activities following thermal denaturation of sodium deoxyribonucleate, *J. Mol. Biol.* **3**: 202–207.

Bak, A. F., 1967, Testing metal micro-electrodes, *Electroenceph. Clin. Neurophysiol.* **22**: 186–187.

Boudreau, J. C., Bierer, P. and Kaufman, J., 1968, A gold plated, platinum tipped, stainless steel micro-electrode, *Electroenceph. Clin. Neurophysiol.* **25**: 286–287.

Bullock, T. H., and Horridge, G. A., 1965, *Structure and Function in the Nervous Systems of Invertebrates*. San Francisco: Freeman and Co.

Carpenter, D. O., and Alving, B. O., 1968, A contribution of an electrogenic Na$^+$ pump to membrane potential in *Aplysia* neurons, *J. Gen. Physiol.* **52**: 1–21.

Carr, C. W., 1956, Studies of the binding of small ions in protein solutions with the use of membrane electrodes. VI. The binding of sodium and potassium ions in solutions of various proteins, *Arch. Biochem.* **62**: 476–484.

Coggeshall, R. E., 1967, A light and electron microscope study of the abdominal ganglion of *Aplysia californica*, *J. Neurophysiol.* **30**: 1263–1287.

Cole, K. S., 1968, *Membranes, Ions and Impulses*. Berkeley and Los Angeles: Univ. of California Press.

Cole, K. S., and Hodgkin, A. L., 1939, Membrane and protoplasm resistance in the squid giant axon, *J. Gen. Physiol.* **22**: 671–687.

Cope, F. W., 1967, NMR evidence for complexing of Na$^+$ in muscle, kidney and brain, and by actomyosin. The relation of cellular complexing of Na$^+$ to water structure and to transport kinetics, *J. Gen. Physiol.* **50**: 1353–75.

Cope, F. W., 1970, Spin-echo nuclear magnetic resonance evidence for complexing of sodium ions in muscle, brain and kidney, *Biophys. J.* **10**: 843–858.

Elford, B. C., 1970, Non-solvent water in muscle, *Nature* **227**: 282–283.

Fessard, A., and Tauc, L., 1956, Capacité, résistance et variations activés d'impédance d'un soma neuronique, *J. Physiol. (Paris.)* **45**: 541–544.

Frazier, W. T., Kupfermann, I., Coggeshall, R. E., Kandel, E. R., and Waziri, R., 1967, Morphological and functional properties of identified neurons in the abdominal ganglion of *Aplysia californica*, *J. Neurophysiol.* **30**: 1288–1351.

Green, J. D., 1958, A simple microelectrode for recording from the central nervous system, *Nature* **182**: 962.

Harris, E. J., 1952, The exchangeability of the potassium of frog muscle, studied in phosphate media, *J. Physiol.* **117**: 278–288.

Harris, E. J., 1954, Ionophoresis along frog muscle, *J. Physiol.* **124**: 248–253.

Hinke, J. A. M., 1959, Glass micro electrodes for measuring intracellular activities of sodium and potassium, *Nature* **184**: 1257–8.

Hinke, J. A. M., 1961, The measurement of sodium and potassium activities in the squid axon by means of cation selective glass micro-electrodes, *J. Physiol.* **156**: 314–335.

Ho, C. and Waugh, D. F., 1965, Interactions of bovine $\alpha$-casein with small ions, *J. Am. Chem. Soc.* **87**: 110–117.

Hodgkin, A. L., and Keynes, R. D., 1953, The mobility and diffusion coefficient of potassium in giant axon from *Sepia*, *J. Physiol.* **119**: 513–528.

Hodgkin, A. L., and Keynes, R. D., 1956, Experiments on the injection of substances into squid giant axons by means of a microsyringe, *J. Physiol.* **131**: 592–616.

Hodgkin, A. L., and Keynes, R. D., 1957, Movements of labelled calcium in squid giant axons, *J. Physiol.* **138**: 253–281.

Jardetzky, O., and Wertz, J. E., 1956a, The complexing of sodium ion with some common metabolites, *Arch. Biochem.* **65**: 569–572.

Jardetzky, O., and Wertz, J. E., 1956b, Detection of sodium complexes by nuclear spin resonance, *Am. J. Physiol.* **187**: 608,

Jardetzky, O., and Wertz, J. E., 1960, Weak complexes of the sodium ion in aqueous solution studied by nuclear spin resonance, *J. Am. Chem. Soc.* **82**: 318–323.

Kaufman, W. and Johnston, F. D., 1943, The electrical conductivity of the tissues near the heart and its bearing on the distribution of the cardiac action currents, *Am. Heart J.* **26**: 42–54.

Keynes, R. D., and Lewis, P. R., 1951, The sodium and potassium content of cephalopod nerve fibres, *J. Physiol.* **114**: 151–182.

Kinnard, M. A., and MacLean, P. D., 1967, A platinum microelectrode for intracerebral exploration with a chronically fixed stereotaxic device, *Electroenceph. Clin. Neurophysiol.* **22**: 183–186.

Kirschner, L. B., 1958, The cation content of phospholipids from some erythrocytes, *J. Gen. Physiol.* **42**: 231–241.

Koketsu, K., and Kimura, Y., 1960, The resting potential and intracellular potassium of skeletal muscle in frogs, *J. Cell. Comp. Physiol.* **55**: 239–244.

Kurella, G. A., 1960, Polyelectrolytic properties of protoplasm and the character of resting potentials. In: *Membrane Transport and Metabolism.* Kleinzeller, A., and Kotyk, A. eds. Prague: Academic Press, pp. 54–68.

Kushmerick, M. J., and Podolsky, R. J., 1970, Ionic mobility in muscle cells, *Science* **166**: 1297–1298.

Lev, A. A., 1964, Determination of activity and activity coefficients of potassium and sodium ions in frog muscle fibres, *Nature* **201**: 1132–1134.

Lewis, M. S., and Saroff, H. A., 1957, The binding of ions to muscle proteins. Measurements on the binding of potassium and sodium ions to Myosin A, Myosin B and Actin, *J. Am. Chem. Soc.* **79**: 2112–2117.

Li, C. L., Bak, A. F., and Parker, L. O., 1968, Specific resistivity of the cerebral cortex and white matter, *Exp. Neurol.* **20**: 544–557.

Ling, G. N., 1962, *A Physical Theory of the Living State*. New York: Blaisdell.

Ling, G. N., 1965, The physical state of water in living cell and model systems, *Ann. N. Y. Acad. Sci.* **125**: 401–417.

Ling, G. N., and Cope, F. W., 1969, Potassium ion: Is the bulk of intracellular $K^+$ adsorbed?, *Science* **163**: 1335–1336.

Ling, G. W., and Ochsenfeld, M. M., 1965, Studies on the ionic permeabilities of muscle cells and their models, *Biophys. J.* **5**: 777–807.

Ling, G. N., and Ochsenfeld, M. M., 1966, Studies on the ion accumulation in muscle cells, *J. Gen. Physiol.* **49**: 819–843.

McLaughlin, S. G. A., and Hinke, J. A. M., 1966, Sodium and water binding in single striated muscle fibres of the giant barnacle, *Canad. J. Physiol. Pharmacol.* **44**: 837–848.

Martinez, D., Silvidi, A. A., and Stokes, R. M., 1969, Nuclear magnetic resonance studies of sodium ions in isolated frog muscle and liver, *Biophys. J.* **9**: 1256–1260.

Metuzals, J., 1969, Configuration of a filamentous network in the axoplasm of the squid (*Loligo pealii L.*) giant nerve fiber, *J. Cell. Biol.* **43**: 480–505.

Metuzals, J., and Izzard, C. S., 1969, Spatial patterns of threadlike elements in the axoplasm of the giant nerve fiber of the squid (*Loligo pealii L.*) as disclosed by differential interference microscopy and by electron microscopy, *J. Cell. Biol.* **43**: 456–479.

Murdock, J. A., and Henton, F. W., 1968, Intracellular distribution of metals (sodium, potassium, calcium and magnesium) in rat brain, kidney and intestinal neurons, *Comp. Biochem. Physiol.* **26**: 121–128.

Naora, H., Naora, H., Izawa, M., Allfrey, V. G., and Mirsky, A. E., 1962, Some observations on differences in composition between the nucleus and cytoplasm of the frog oocyte, *Proc. Nat. Acad. Sci.* **48**: 853–859.

Raker, J. W., Taylor, I. M., Weller, J. M., and Hastings, A. B., 1950, Rate of potassium exchange of the human erythrocyte, *J. Gen. Physiol.* **33**: 691–702.

Robertson, J. D., 1961, Studies on the chemical composition of muscle tissue. II. The abdominal flexor muscle of the lobster, *Nephrops norbeguus, J. Exp. Biol.* **38**: 707–728.

Robinson, R. A., and Stokes, R. H., 1959, *Electrolyte Solutions*. London: Butterworths.

Rosenbluth, J., 1963, The visceral ganglion of *Aplysia californica, Z. Zellforsch.* **60**: 213–236.

Rotunno, C. A., Kowalewski, V., and Cereijido, M., 1967, Nuclear spin resonance evidence for the complexing of sodium in frog skin, *Biochim. Biophys. Acta* **135**: 170–173.

Schales, O., and Schales, S. S., 1941, A simple and accurate method for the determination of chloride in biological fluids, *J. Biol. Chem.* **140**: 879–884.

Schwan, H. P., 1963, Determination of biological impedances. In: *Physical Techniques in Biological Research, Vol. VI*, Nastuk, W. L. ed. New York: Academic Press, pp. 323–406.

Schwan, H. P. and Kay, C. F., 1956, Specific resistance of body tissues, *Circulation Research* 4: 664–670.

Schwan, H. P. and Kay, C. F., 1957, Capacitative properties of body tissues, *Circulation Research* 5: 439–443.

Shaw, F. H., Simon, S. E., and Johnstone, B. M., 1956, The non-correlation of bioelectric potentials with ionic gradients, *J. Gen. Physiol.* **40:** 1–17.

Troshin, A. S., 1960, Sorption properties of protoplasm and their role in cell permeability. In: *Membrane Transport and Metabolism*. Kleinzeller, A., and Kotyk, A. eds. Prague: Academic Press, pp. 45–53.

Wesson, L. G., Jr., Cohn, W. E., and Brues, A. M., 1949, The effect of temperature on potassium equilibria in chick embryo muscle, *J. Gen. Physiol.* **32:** 511–523.

Whittam, R., and Davies, R. E., 1954, Relations between metabolism and the rate of turnover of sodium and potassium in guinea pig kidney-cortex slices, *Biochem. J.* **56:** 445–453.

# THE EFFECT OF EXPERIMENTAL HEAD INJURY ON ONE-TRIAL LEARNING IN RATS

AYUB K. OMMAYA, ANNE GELLER,† and L. CLAIRE PARSONS‡

*From the Branch of Surgical Neurology, National Institute of Neurological Diseases and Stroke, National Institutes of Health, Public Health Service, U.S. Department of Health, Education and Welfare, Bethesda, Maryland*

The amnesic and convulsive effects of experimental cerebral concussion were observed in rats. Marked convulsions were associated with only partial retrograde amnesia in one trial passive avoidance when the head injury was given less than 7 seconds after learning the task. These results are compared with the complete retrograde amnesia associated with a lesser severity of convulsions produced by electroconvulsive shock in rats under identical learning situations.

The most frequently used method of disrupting learning in experimental studies of memory has been electroconvulsive shock [ECS] (Weiskrantz, 1966). The mechanism of these amnesic effects of ECS is not known and conflicting opinions exist on the role of convulsive phenomena in the causation of the memory deficits (Cronholm and Ottosson, 1963a, 1963b; Holmberg, 1955; Weissman, 1963; and Ottosson, 1960). Analogies between the post-ECS state and reversible cerebral concussion after head injury have often been drawn, particularly with regard to the production of retrograde amnesia (Williams, 1966).

One of the leading hypotheses for the mechanism of cerebral concussion is that advanced by A. E. Walker who on the basis of experiments with cats, dogs and monkeys suggested that 'the physiological basis of concussion consists of the traumatic discharge of the polarized cell membranes of the neurons of the central nervous system by the shaking up or commotion of the brain and that the subsequent course of events is that which would be expected when large masses of nerve cells discharge such as is seen in the spontaneous or electrically

induced convulsive seizures. . . . Thus we conceive of cerebral concussion as the traumatic excitation of nervous tissue, producing in its wake the train of electroencephalographic, chemical and clinical findings which are known to characterize intense stimulation of the nervous system by electrical, chemical or other agents' (Walker *et al.*, 1944).

This hypothesis was developed on the basis of the experiments cited above as well as two sets of earlier observations. The first was the description by Dusser de Barenne and McCulloch (1939) of the phenomenon of 'extinction' or inhibition of neural activity following a primary intense stimulation. Secondly it had frequently been observed that an immediate generalized muscular spasm, described by Duret (1920) as the tetanic stage of cerebral concussion, often occurred when the head was struck by blows of sufficient severity to produce unconsciousness. As early as 1892, Miles had described convulsive phenomena after mechanical brain trauma in the rabbit, pig and cock (Miles, 1892).

If extinction after intense neural stimulation is indeed the prime physiologic mechanism of cerebral concussion then it would be logical to predict that the intensity or duration of the tetanic phenomena (i.e. convulsions) would be associated in some manner with the amnesic sequelae of head injury. It is important to test this prediction from the hypothesis because if such an association were demonstrated, a potentially useful analogy between

† Department of Pharmacology, Albert Einstein College of Medicine of Yeshiva University, Bronx, New York.

† On leave from the Dept. of Pharmacology, Albert Einstein Medical College, New York.

‡ Supported by Special Postdoctoral Fellowship, Division of Nursing, USPHS and by Developmental Grant No. –69–620, American Nurses' Foundation.

the mechanisms of cerebral concussion and retrograde amnesias induced by ECS could be suggested. Our understanding of traumatic unconsciousness as well as the study of amnesic and other sequelae of head injury could consequently be greatly facilitated by availability of the extensive data and research strategies from the experience with ECS (Weiskrantz, 1966).

In order to test the stimulation-extinction hypothesis for traumatic unconsciousness we have begun a series of experiments comparing the amnesic effects of ECS and cerebral concussion. Tonic or tonic-clonic seizure-like activity has been observed more frequently during experimental cerebral concussion (ECC) in mice, rats, rabbits, and cats than in primates (Walker *et al.*, 1944; Ommaya, 1966). This report consists of our observations on the effect of cerebral concussion produced by head impact on one-trial conditioned passive avoidance in the rat. The effects of ECS on this type of learning in mice and rats have been well documented (Schneider *et al.*, 1969; Robustelli *et al.*, 1969) and the procedural conditions could be duplicated precisely, substituting ECC by head impact in place of ECS. The experimental data indicate a dissociation between the convulsive and amnesic effects of concussive head injury.

## MATERIALS AND METHODS

### Subjects

Subjects were 54 male Sprague Dawley rats, 250–350 gms housed individually in wire cages.

### Apparatus for One-Trial Avoidance Learning

Animals were trained in a step-through apparatus patterned after the one designed for mice (Jarvik and Kopp, 1967). It consisted of a trough shaped two compartment box made of 6.4 mm plexiglass plates. The smaller, transparent compartment (22 cm × 20 cm) was connected with the larger darkened compartment (42 cm × 20 cm) by an opening 6.25 cm wide, 10 cm high and flush with the floor. The opening could be closed by a guillotine door. The floor of both compartments consisted of pairs of parallel metal plates 1.25 cm apart, one pair in the smaller and two in the larger compartment. The plates were bent up to form the side walls. A photocell was placed in the side wall of the larger compartment 23 cm from the opening.

### Learning Procedure

The subjects were placed in the smaller compartment facing away from the opening. A timer was started manually when the subject was placed in the apparatus and stopped automatically when he stepped spontaneously into the dark compartment and crossed the photocell. This time interval measured the training (TL) and restest latencies (RL) for the passive avoidance response. Foot shock of 1.5 mA was triggered by the break in the photobeam and was delivered across the plates in the dark compartment. The subjects escaped from the foot shock back into the smaller transparent compartment from which they were removed. Retest trials were run using the same procedure and the subjects were observed for 600 sec.

### Preparation of Animals for ECC

Anesthesia was produced and maintained by intraperitoneal injection of sodium pentobarbital, 35 mg/kg. The animal's fur was clipped from the bridge of the nose to the shoulder. The anesthetized rat was then placed in the head holder of a stereotaxic frame. The operative site was cleaned and a transverse incision made in the scalp midway between the eyes and the ears. The skin and underlying connective tissue was freed from the top of the calvarium by blunt dissection. Two drill holes were placed diagonally across the calvarium from each other. Stainless steel screws size 000/20 and approximately 1/32″ in length were threaded into the skull to anchor subsequently placed layers of dental cement. The dental cement was used to build up an artificial calvarium and hold a plexiglass disc (16 mm × 6 mm high) placed horizontally on top of the cement in a uniform location. This disc ensured uniformity of impact and acted as a load distributor to prevent skull fracture. Sulfadiazine powder U.S.P. was placed in the incision as a prophylactic measure. The skin edges were pulled over the cement and sutured closed against the disc with 3–0 black silk. Each animal was then returned to its individual cage and allowed to recover for at least 7 days before training and concussion.

### Apparatus for ECC

A coil spring gun designed specifically for experimental head injury in the rat was used to produce the *ECC* (Parsons, 1968). The concussion gun was rigidly fixed to a semicircular arc previously mounted on a solid platform; this ensured reproducible impact conditions and minimal energy dissipation by recoil. As shown in Figure 1, *ECC* was produced by fitting the barrel of the concussion gun over the disc implanted on the rat skull, the animal being held around the forelimbs. When the trigger was pulled the impact of the piston on the disc and underlying unsupported head produced the desired cerebral concussion with the head free to move. This gun was calibrated by repeated impacts against a force transducer and the results for three settings of the thumb screw (which determines the degree of spring compression) is shown in Figure 2. It will be noted that the range of force available from this gun was quite small being 26 lbs at zero coil compression (thumb screw out all the way) and 42 lbs at maximum coil compression. The duration of the impacts ranged between 15 and 30 milliseconds.

*Criteria for ECC in the Rat*

The following changes occurring in the unanesthetized drug-free animal were arbitrarily chosen to define the *onset* of *ECC* after head impact.

(a) Tetanic phenomena; in most animals these consisted of prolonged tonic-clonic seizures. In a few the tonic components were predominant. In all cases these phenomena were noted immediately on impact.

(b) Loss of corneal and palpebral reflexes; these were readily produced prior to the head injury and immediately absent after onset of *ECC*.

(c) Loss of normal maintenance of posture and voluntary movement. The *duration* of concussion was defined as the time from impact to the time when the animal spontaneously righted itself.

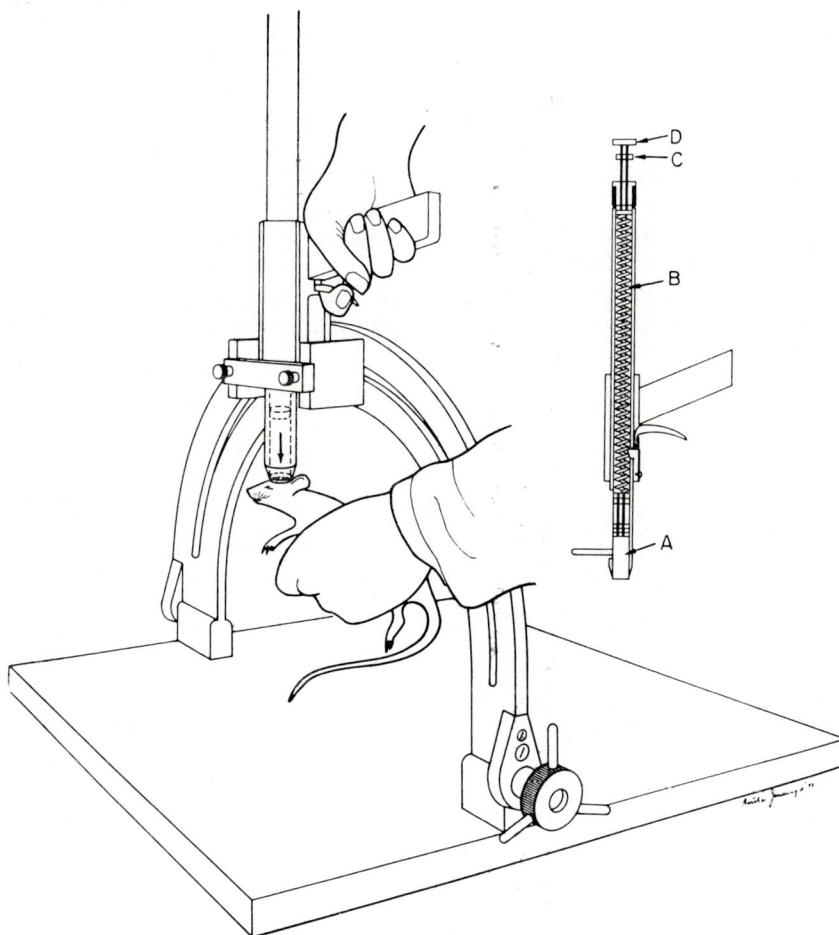

FIGURE 1   Diagram of apparatus for experimental head injury in the rat.
A. Piston and Cocking Bolt.   B. Spring.   C. Thumb screw stop.   D. Thumb screw.

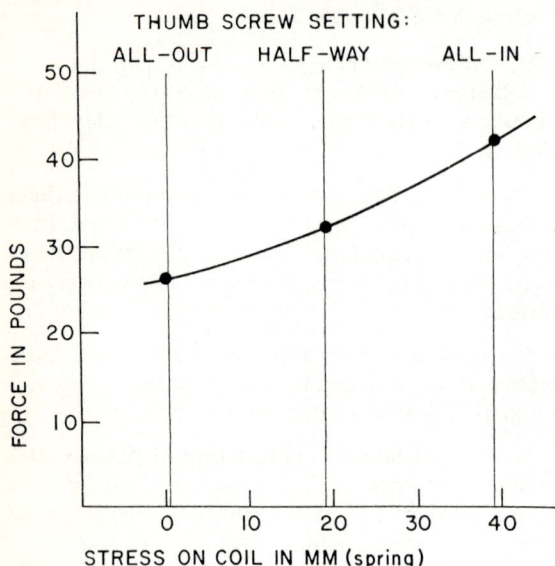

FIGURE 2    Calibration curve for concussion gun.

*Behavioral Recording System*

The behavioral changes during learning and memory testing as well as those accompanying concussion, e.g. length of clonic convulsions, return of righting and voluntary movements were monitored by a television system allowing video-tape storage of this data. Behavior after concussion was observed in a plexiglass chamber with a digital clock providing the time base. Neurological examinations were also conducted within this chamber.

*Experimental Protocol*

The subjects were randomly divided into four groups. Group I (control) subjects were removed from the transparent compartment of the conditioning apparatus immediately after training and given a sham concussion. For this they were grasped firmly around the forequarters and held beside the concussion apparatus which was then fired. They were then placed in the clear plexiglass box and observed before being returned to their home cages for the training-retest interval of 24 hours. Subjects of Groups II and III were trained as Group I but were concussed immediately after training and observed in the clear plexiglass box until they had righted and moved spontaneously. Group II subjects were returned to their home cages to be retested in 24 hr as for Group I, but Group III subjects were retested as soon as they showed spontaneous coordinated running movements. Subjects of Group IV were concussed as naive

animals, i.e. *prior* to training. Following the concussion they were observed in the plexiglass box until they had righted and were showing coordinated running movements. They were then placed in the conditioning apparatus, trained in the same manner as the other groups and returned to their home cages for retesting 24 hrs after *ECC*.

RESULTS

In pilot experiments aimed at producing head injuries associated with *ECC* reproducibly there appeared to be a very narrow range of force levels for reversible *ECC* after head injury. *ECC* was reliably obtained with the thumb screw set between 17–19 mm, i.e. at forces = 31–32 lbs applied for 15–30 milliseconds. Impacts given below these levels would usually be sub-concussive while heavier impacts would be fatal. Even under optimal conditions concussive levels were very close to lethal levels of impact. Thus in the experimental groups of 42 rats, twelve animals died following concussion (28.5%). Of the remaining 30 concussed animals, 29 had clonic convulsions following the impact. The median duration of the convulsions was 23 sec (mean 22.1 ± 24), the median time to righting was 55 sec (mean 242 ± 410) and the first spontaneous running movement occurred at a median of 190 sec (mean 325 ± 486). These data are tabulated in Tables I and II. There appeared to

TABLE I

| Experimental Group | Total Number in Group | Number Dying | Procedure | Duration of Concussion (Time to righting) – [sec] |
|---|---|---|---|---|
| I | 12 | 0 | Controls with sham concussion after training & 24 hrs. retest | Nil (Sham concussion) |
| II | 14 | 2 | Concussion after training & 24 hrs. retest | 10-1350 |
| III | 14 | 8 | Concussion after training & immediate retest | 20-560 |
| IV | 14 | 2 | Concussion before training & 24 hrs. retest | 6-932 |

be no correlation between the length of convulsions, time to righting, the return of voluntary movement, and the 24 hr retest latencies. However, the duration of convulsions tended to increase 'pari-passu' with the duration for return of corneal reflexes (Figure 3).

The time intervals for training (TL) and retest latencies (RL) in seconds were subjected to the 'median test' (Spiegel, 1956). This statistical

TABLE II

| No. of Experimental Animals | Duration of Convulsion in seconds | | Return of Righting in seconds | | Spontaneous Voluntary Movement in seconds | |
|---|---|---|---|---|---|---|
| | median | mean | median | mean | median | mean |
| 30 | 23 | 22.1 ± 24 | 55 | 242 ± 410 | 190 | 325 ± 486 |

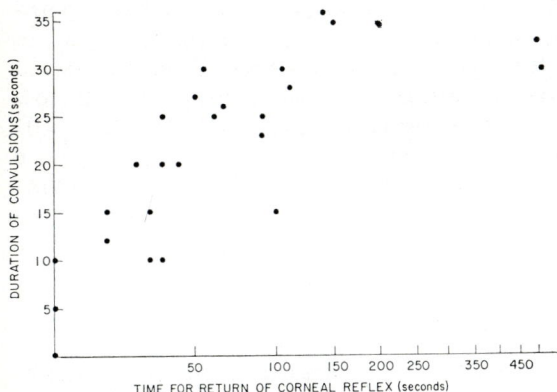

FIGURE 3   Relation of convulsion durations to time for return of corneal reflexes after *ECC*.

procedure tests whether two or more independent groups differ in central tendencies, i.e., whether two or more samples have been drawn from populations having the same median and, therefore, yielding equal splits (dichotomies). The results are shown in Tables III, IV, and V.

TABLE III

| Group Latencies | Median $Q_2$ | Interquartile Range $Q_1 - Q_3$ | Common Median $Q_x$ | $\chi^2$* | d.f.[+] | P[‡] |
|---|---|---|---|---|---|---|
| Group I | | | | | | |
| Training Latency (TL) | 6.2 | 3.8–13.6 | 81.1 | 22.407 | 1 | <.0005 |
| Retest Latency (RL) | 600.0 | 600.0–600.0 | | | | |
| Group II | | | | | | |
| Training Latency (TL) | 12.3 | 5.6–22.5 | 71.4 | 13.500 | 1 | <.0005 |
| Retest Latency (RL) | 188.4 | 104.0–454.0 | | | | |
| Group III | | | | | | |
| Training Latency (TL) | 12.3 | 4.0–23.0 | 27.2 | 3.000 | 1 | <.05 |
| Retest Latency (RL) | 504.8 | 54.6–600.0 | | | | |
| Group IV | | | | | | |
| Training Latency (TL) | 8.4 | 4.8–20.6 | 18.2 | 4.167 | 1 | <.025 |
| Retest Latency (RL) | 34.6 | 14.2–163.9 | | | | |

*Chi-square median test
+Degrees of Freedom
‡Probability level of significance

Table III shows the results of this test for training latency versus retest latency for all four groups. For all four groups, the median retest latency was higher than the median training latency.

Table IV shows the results of this test extended to four groups, Group I vs II vs III vs IV. For training

TABLE IV

| Group Latencies | Common Median $Q_x$ | $\chi^2$ | d.f. | P |
|---|---|---|---|---|
| Latency | | | | |
| Training Group I Group II Group III Group IV | 8.4 | 2.667 | 3 | .25>P>.15 |
| Retest Group I Group II Group III Group IV | 264.3 | 15.667 | 3 | <.005 |
| Retest Group II Group II | 242.1 | .250 | 1 | .35>P>.25 |

## TABLE V

| Items | $\chi^2$ | d.f. | P |
|---|---|---|---|
| Rows (TL and RL) | 51.887 | 1 | <.0005 |
| Columns (I, II, II, IV) | 2.116 | 3 | .37P>.25 |
| Interaction (RxC) | 2.317 | 3 | .37P>.25 |
| Total | 56.320 | 7 | |

## Analysis of Variance of Column (C), Row (R) and Interaction (RxC) Effects of Groups I, II, III & IV

latency, for all four groups, and retest latency, Group II vs Goup III, there are no differences. Therefore, the population/populations, from which these groups are drawn, have equal medians. For retest latency, for all groups, it is indicated that the median retest latency is higher for Group I and III than the median for Group II and IV. Therefore, the population/populations, from which these groups are drawn, do not have equal medians.

Table V shows the results of a distribution-free analysis of variance, a further extension of the median test. For the column (Groups I, II, III, and IV) and the interaction (R × C) effects, there is no significance. However, the row (Training Latency and Retest Latency) effect is significant at P < .005) indicating, again, that the median retest latency is higher than the median training latency. Therefore, the population/populations from which these groups are drawn, do not have equal medians.

Table VI summarizes these data. It should be noted that there were no significant differences in training latencies over the four groups (Column 3 in Table VI) despite the fact that Group IV had received its training trials *after* the concussion, whereas the other three groups had received no treatment prior to training. Thus, although a trend is suggested, there was no statistically significant evidence of a proactive effect of concussion on the latencies of untrained animals.

As can be seen in Table VI and Fig. 4, retest

TABLE VI

| GROUP | N | Training Latency (Sec) | | Retest Latency (Sec) | | *Significance of Difference P< | Train — Concussion Interval (Sec) | Concussion Train or Retest Interval (Sec) |
|---|---|---|---|---|---|---|---|---|
| | | Median | IQR+ | Median | IQR+ | | | |
| I | 12 | 6.2 | 3.8–13.6 | >600 | 600–600 | 0.005 | --- | --- |
| II | 12 | 12.3 | 5.6–22.5 | 188.4 | 104.0–454.0 | 0.005 | 6.5 | --- |
| III | 6 | 12.3 | 4.0–23.0 | 504.8 | 54.6–600.0 | 0.05 | 5.8 | 420 |
| IV | 12 | 8.4 | 4.8–20.6 | 34.6 | 14.2–163.9 | 0.025 | --- | 405 |

*Siegel, S. (1956) Non-Parametric Statistics
New York: McGraw-Hill, pp. 111-116, 174-184

+IQR = Interquartile range

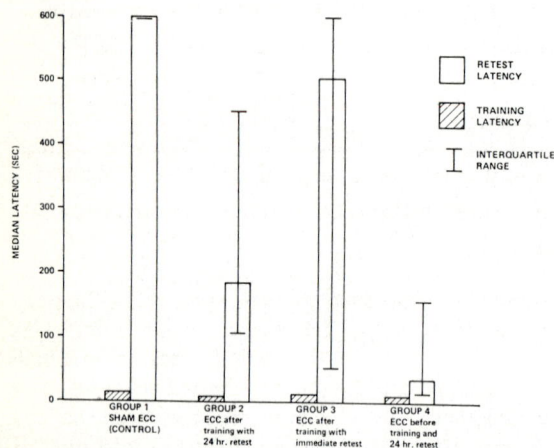

FIGURE 4 Durations of median latencies (± interquartile ranges) for training and 24 hour retest in the four experimental groups. Note the partial retrograde amnesia produced in Groups II and III and the much more drastic decrease in 24 hour retest latency in Group IV (*ECC before* training).

latencies were significantly higher than training latencies for all four groups indicating that the conditioned response was formed and retained in all treatment conditions. However, the three experimental groups, II, III, and IV had retest latencies significantly lower than the non-concussed controls (p < 0.005, 0.05 and 0.025 respectively) indicating the development of partial retrograde amnesia (RGA.) Groups II and III concussed immediately after training but retested respectively 24 hr and 420 sec after the concussion did not differ significantly, indicating within the limits of the experimental situation neither a development of, nor a recovery from, concussion induced

amnesia with time although a trend towards development was suspected from Group III to Group II. Even though retest latencies for Groups II and III were lower than the non-concussed Group I, the conditioned response was clearly retained to a great extent in both groups, i.e. the amnesia was only partial. At the 24 hr retest (Group II) only two animals had retest latencies below 100 sec (median training latency was 12.3 sec) and both of these animals had residual neurological deficits at the time of their retest (hemiparesis and ataxia) which may have contributed to their relatively poorer performance.

The retest latencies of Group IV subjects trained after concussion, were significantly lower than those of all the other groups (p < 0.005, 0.01, and 0.01 respectively) and in fact, 25% of the subjects in this group showed latencies lower on the retest than on the training trial, a phenomenon not observed in any subject in the other three groups.

*Pathologic Results*

Of the 42 animals concussed 12 died as a result of the blow to the head. No animal in this group or the surviving group received a fracture of the skull. Autopsy of animals receiving a lethal blow revealed extensive subarachnoid hemorrhages over the dorsal and ventral surfaces of the brain stem in all 12; of these 3 also had subarachnoid hemorrhages over both cerebral hemispheres. Efforts to resuscitate these animals proved fruitless with death resulting from respiratory arrest. Heart rate continued in each animal for at least 1 to 2 minutes following apnea.

Twenty-eight of the 30 surviving animals demonstrated no residual neurologic deficits when sacrificed 2 to 3 weeks post-trauma. No gross pathological lesions were visible. Histologic examination of all brains awaits processing of the paraffin embedded tissue and will be reported later.

DISCUSSION

These data show that only *partial* retrograde amnesia was obtained *when ECC was produced less than 7 seconds after training in a passive avoidance learning situation.* Under identical procedural conditions, Schneider *et al* (1969) obtained complete retrograde amnesia with ECS given at 10 seconds after training and partial amnesia extending to a six hour training—ECS interval. Kopp *et al* (1966) and Robustelli *et al* (1969) have observed similar gradients after ECS with mice in this training situation. Bohdanechky *et al* (1968) have also noted such memory impairment after treatment with Flurothyl. This relatively lesser disruptive effect of *ECC* on memory compared to that of ECS and flurothyl stands in sharp contrast to the severity of tetanic phenomena produced by such treatments. Thus, the convulsions produced by *ECC* are of longer durations ($23 \pm 24$ seconds) than with either of the other methods and in particular after ECS, where convulsions barely persist 4–5 seconds after the passage of the ECS (15 mA for 200 msec). It would appear paradoxical therefore that the apparently much more long-lasting convulsions and neurological deficits of the early post-*ECC* state should be less productive of retrograde amnesia than the shorter convulsive state after ECS. The suggested trend for a pro-active effect of *ECC* given before training noted in Group IV of our experiments has also been observed for ECS when given 1–4 hours before learning (Kopp *et al.*, 1968). The initial training latency of Group IV animals was not significantly different from that of the other groups, although the retest latencies were the lowest of all the experimental groups and in 25% of the animals were less than the training latencies. This indicates that animals trained in the period following *ECC* can perform the step through response on the training trial within the same time as the control animals but do show impaired performance when retested 24 hours later. The design of our experiment did not permit us to determine whether this partial retrograde amnesia was due primarily to a defect in the acquisition, retention, or retrieval of the passive avoidance response.

Is there any way in which one could explain the apparent dissociation between the tetanic and amnesic effects of *ECC*? The simplest explanation that convulsive activity bears no relation to the neural mechanism disrupting the learned response in the passive avoidance situation is not a very productive avenue for further work.

We would propose therefore the following hypothesis to explain why ECS is more disruptive for memory with lesser convulsive activity than that produced by *ECC*. It may be that the electrophysiologic 'noise' generated by 15 mA passing through the brain (Weiskrantz, 1966) is distributed diffusely and in effect much more uniformly within the entire brain as compared to the 'noise' generated by *ECC* and its associated tetanic phenomena. It would seem unlikely that a lesser intensity or duration of 'noise' is generated by *ECC* when its tetanic phenomena are so prominent. In order to test our hypothesis it would have to be shown that some specific neural structures are more involved than others in the neurophysiologic effects of *ECC*. This asymmetric, non-uniform distribution of the effects of blunt *mechanical* trauma is supported by numerous studies in various animal species on the distribution of gross and microscopic lesions in the brain after head injury (Ommaya and Corrao, 1970; Ommaya, 1970; Oppenheimer, 1968; Chason *et al.*, 1966; Grcevic and Jacob, 1965; Windle *et al.*, 1944). We have previously discussed the difficulties in making morphologic-functional correlations in *ECC* (Ommaya and Corrao, 1970) particularly if one selects chromatolytic changes alone as an index of *ECC* severity (Windle, 1944; Ommaya and Corrao, 1970).

Our data do not support an association between the tetanic and amnesic phenomena of *ECC*. It certainly does not follow however that the stimulation-extinction hypothesis for cerebral concussion is invalidated. Indeed if the asymmetric 'noise' effects postulated above can be shown to correlate with the distribution of mechanical stresses and strains (lesions) in the brain after *ECC* than it would be logical to suggest that varying degrees of selective stimulation/inhibition within certain structures may counteract the amnesic effect of the 'noise' generated by the tetanic phenomena. This interpretation is especially worthy of consideration because it has been suggested that hippocampal lesions exert a weakening effect on ECS induced RGA in rats (Hostetter, 1968). Thus a relatively greater or selective stimulation of the hippocampal complex by *ECC* could reasonably explain the demonstrated severity of the tetanic phenomena combined with only partial retrograde amnesia after *ECC*. A variety of recent observations on the inhibitory and facilitatory functions of telencephalic structures, and in particular of the septal and cingulate areas, suggest that many response modulating systems exist in the brain, each of which could exert different effects on the type of behavioral response chosen for measurement (McCleary, 1966). It may well be therefore that the deficit in an active avoidance test situation would reveal very different degrees of memory impairment after *ECC* than that shown in the passive avoidance type of situation chosen for these experiments. A combination of such experiments with the effect of small lesion in specific limbic structures on both passive and active avoidance after *ECC* is now planned as the next step in our investigations.

74  A. K. OMMAYA, A. GELLER AND L. C. PARSONS

ACKNOWLEDGMENT

Valuable assistance was received from Mrs. Doris Sadowsky for statistical testing of the data.

REFERENCES

Bohdanechky, Z., Kopp, R., and Jarvik, M. E., 1968, Comparison of ECS and flurothyl-induced retrograde amnesia in mice, *Psychopharmacologie* (Berlin) **12**: 91–95.

Chason, J. L., Fernando, O. U., Hodgson, V. R., Thomas, L. M., and Gurdjian, E. S., 1966, Experimental brain concussion: Morphologic findings and a new cytologic hypothesis, *J. Trauma* **6**: 767–779.

Cronholm, B., and Ottosson, J. O., 1963a, The experience of memory functions after electroconvulsive therapy, *Brit. J. Psychiatr.* **109**: 251–258.

Cronholm, B., and Ottosson, J. O., 1963b, Ultrabrief stimulus technique in electroconvulsive therapy. I. Influence on retrograde amnesia of treatments with the elther ES electroshock apparatus, Siemens Konvolsator III and of lidocaine-modified treatment, *J. Nerv. Ment. Dis.* **137**: 117–123.

Duret, H., 1920, Commotions graves, mortelles, sans lesions (commotions pures) et lesions cerebrale etendues sans commotion dans les traumatisme cranio-cerebraux, *Rev. neurol.* **35**: 888–900.

Dusser de Barenne, J. G., and McCulloch, W. S., 1939, Factors for facilitation and extinction in the central nervous system, *J. Neurophysiol.* **2**: 319–355.

Grcevic, N., and Jacob, N., 1965, Some observations on pathology and correlative neuroanatomy of sequels of cerebral trauma. Late sequelae of head injuries. *Proceedings of the 8th International Congress of Neurology*, Vienna, **1**: 369–373.

Holmberg, G., 1955, The effect of certain factors on the convulsions in electroconvulsive therapy, *Acta Psychiat. Neurol. Scand.* **30**: 98....

Hostetter, G., 1968, Hippocampal lesions in rats weaken by the retrograde amnesic effect of ECS, *J. Comp. Physiol. Psychol.* **66**: 349–353.

Jarvik, M. E., and Kopp, R., 1967, An improved one-trial passive avoidance learning situation, *Psychological Reports* **21**: 221–224.

Kopp, R., Bohdanechky, Z., and Jarvik, M. E., 1966, A long temporal gradient of retrograde amnesia for a well discriminated stimulus. *Science* **153**: 1547–1549.

Kopp, R., Bohdanechky, Z., and Jarvik, M. E., 1968, Proactive effect of a single electroconvulsive shock (ECS) on one-trial learning in rats, *J. Comp. Physiol.* **65**: 514–517.

McCleary, R. A., 1966, Response-modulating functions of the limbic system: Initiation and suppression. In: *Progress in Physiological* Psychology, Vol. 1, edited by E. Stellar and J. M. Sprague, pp. 209–272.

Miles, A., 1892, On the mechanism of brain injuries, *Brain* **15**: 153–189.

Ommaya, A. K., 1970, Physiopathology of head injuries. Report to Working Group 58. CHABA. Committee on Hearing, Bioacoustics, and Biomechanics, National Research Council.

Ommaya, A. K., 1966, Trauma to the nervous system. *Annals of the Royal College of Surgeons of England* **39**: 317–347.

Ommaya, A. K., and Corrao, P., 1970, Pathologic biomechanics of central nervous system injury in head impact and whiplash trauma. *Proceedings of the International Conference on Accident Pathology*, Washington, D.C., Government Printing Press.

Oppenheimer, D. R., 1968, Microscopic lesions in the brain following head injury, *J. Neurol. Neurosurg. Psychiat.* **31**: 299–306.

Ottosson, J. O., 1960, Experimental studies of memory impairment after electroconvulsive therapy. The role of the electrical stimulation and of the seizure studied by variation of stimulus intensity and modification of lidocaine of seizure discharge. *Acta Psychiat. Neurol. Scand.* **35** (Suppl. 145): 103–131.

Parsons, L. Claire, 1968, Electrophysiological changes in sleep and wakefulness following experimental head trauma in rats. *Ph. D. Thesis*, Dept. of Physiology, University of Texas Medical Branch, Galveston, Texas.

Robustelli, F., Geller, A., Aron, C., and Jarvik, M. E., 1969, The relationship between the amnesic effect of electroconvulsive shock and strength of conditioning in a passive avoidance task. *Communications in Behavioral Biology*, Part A, 3, Nos. 5/6.

Schneider, A. M., Kapp, B., Aron, C., and Jarvik, M. E., 1969. Retroactive effects of transcorneal and transpinnate ECS on step-through latencies of mice and rats, *J. Comp. Physiol. Psychol.* **69**: 506–509.

Siegel, S., 1956, *Non-parametric Statistics*. New York: McGraw-Hill pp. 111–116, 174–184.

Walker, A. E., Kollros, J. K., and Case, T. J., 1944, The physiological basis of concussion. *J. Neurosurg.* **1**: 103–116.

Weiskrantz, L., 1966, Experimental studies of amnesia. Chap. 1, In: *Amnesia* edited by C. W. M. Whitty and O. L. Zangwill, London: Butterworths.

Weissman, A., 1963, Effect of electroconvulsive shock intensity and seizure pattern on retrograde amnesia in rats. *J. Comp. Physiol. Psychol.* **56**: 806–810.

Williams, M., 1966, Memory disorders associated with electroconvulsive therapy. In: *Amnesia* edited by C. W. M. Whitty and O. L. Zangwill, London: Butterworths.

Wilson, K. V., 1956, A distribution-free test of analysis of variance hypotheses. *Psych. Bull.* **53**: 96–101.

Windle, W. F., Groat, R. A., and Fox, C. A., 1944, Experimental structural alterations in the brain during and after concussion. *Surg. Gynec. & Obstet.* **79**: 561–572.

# BEHAVIORAL AND NEUROPHYSIOLOGICAL STUDIES ON CAT COLOR VISION

ALAN L. PEARLMAN and NIGEL W. DAW

*Departments of Neurology, Physiology and Ophthalmology*
*Washington University School of Medicine, St. Louis, Missouri 63110*

The mechanisms of cat color vision were investigated with behavioral and neurophysiological techniques. Initial spectral sensitivity measurements of responses from single optic tract and lateral geniculate cells uncovered input from only one type of cone, with peak sensitivity at 556 nm. Behavioral studies then demonstrated that cats could distinguish colors in the photopic range, above the level of rod saturation, thus clearly indicating the presence of more than one cone type. A subsequent microelectrode survey of 434 optic tract and lateral geniculate units disclosed rare units in layer B of the geniculate that receive input from two types of cone in an opponent-color fashion.

Color vision in cats has been the subject of controversy for many years. A number of behavioral studies presented evidence on both sides of the question as to whether cats could discriminate colors at all. This question was finally settled by two different sets of investigators (Mello and Peterson, 1964; Sechzer and Brown, 1964). Both groups were able to show that cats could discriminate colors after prolonged periods of training, if great care was taken to eliminate other cues. On the other hand, the neurophysiological mechanisms underlying this ability were less well defined.

The two best examples of animals with behaviorally demonstrable color vision that have been studied neurophysiologically are the goldfish and the rhesus monkey. Both have been shown to have three types of cones, each type with a different spectral sensitivity (Marks, 1965; Marks *et al.*, 1964; Brown and Wald, 1964) and both have cells at a later stage of the system which respond differentially to wavelength (Daw, 1968; DeValois *et al.*, 1958; Wiesel and Hubel, 1966). The situation in the cat was much less clear: Granit, (1945) suggested the presence of more than one cone type many years ago, but the techniques employed to obtain these spectral sensitivity curves have met with a number of objections (Remberg, 1953; Daw and Pearlman, 1969, 1970). Receptive field analyses of cat ganglion cells and lateral geniculate cells by several investigators (Barlow, Fitzhugh and Kuffler, 1957; Hubel and Wiesel, 1961) failed to demonstrate the familiar opponent color cells of monkey optic nerve and lateral geniculate, or the double opponent cells of goldfish retina (Daw, 1968) and monkey cortex (Hubel and Wiesel, 1968). We therefore undertook a series of neurophysiological and behavioral studies in an attempt to clarify this situation; the present paper is a summary of these experiments.

The first task was to determine the spectral sensitivity of cat cones under conditions that eliminated any contribution from the rods (Daw and Pearlman, 1969). Ideally, one would like to record from single cones and determine their spectral response curves directly. Since this is not technically feasible in the cat retina at present, we elected to record from single optic nerve fibers and lateral geniculate cells. We eliminated the rod contribution to their spectral sensitivity curves by applying a steady background illumination (10–30 cd/m²) that we found to be more than adequate to produce rod saturation. Spectral sensitivities were determined for the receptive field centers of 64 units under these conditions. In all cases the spectral sensitivity curve had approximately the same shape, with a peak at about 556 nm. The means of the measurements of the 64 units is represented by circles in Figure 1; the bars indicate the standard deviations. The spectral sensitivity curve is slightly narrower than the Dartnall nomogram (dashed curve in Figure 1) for a visual pigment with peak sensitivity at 556 nm; in no case could the spectral sensitivity be shifted by chromatic adaptation with blue, green, yellow or red backgrounds. Furthermore, the spectral sensitivities for the receptive field centers were identical to those for the receptive field surrounds. Thus the evidence pointed to only one cone type providing input to optic tract and lateral geniculate units.

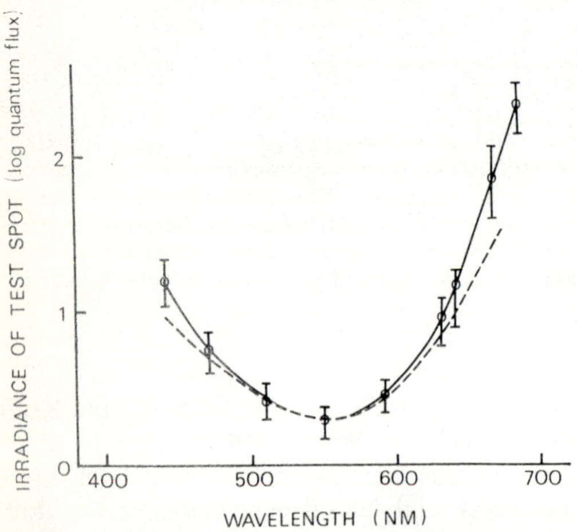

FIGURE 1  Solid curve is the average spectral sensitivity of 64 lateral geniculate units, measured against a white background sufficient to saturate the rods. Bars indicate the standard deviations. Dashed curve is the Dartnall nomogram for a pigment with peak sensitivity of 556 mn.

At this stage we were faced with an obvious dilemma: color vision requires at least two types of receptor with different spectral sensitivities; we had evidence for only one type of cone. This finding led to the suggestion that the two types of receptor underlying cat color vision might be the rods, with a peak sensitivity of 500 nm, and a single type of cone, with a peak sensitivity at 556 nm. If that were the case, the cat should be able to make wavelength descriminations in the mesopic range, where both rods and cones are functioning, but lose this ability at light intensities sufficient to saturate the rods. If the cat retains the ability to distinguish colors above the level of rod saturation, then a further neurophysiological search for another cone type would be indicated. Behavioral studies were therefore undertaken to test this hypothesis (Daw and Pearlman, 1970).

Two cats were trained in a behavioral test box designed by Dews and Wiesel (1970). At one end of the rectangular box there were two translucent plexiglas doors separated by a plywood divider. Each door was illuminated by a steady white back-

ground, and by a stimulus slide projected from a Kodak Carousel projector. If the cat pushed open the door lit by the positive stimulus, it was automatically given a food reward; if it chose the door with the negative stimulus, it got no reward, and returned to the other end of the box to begin again. Slide presentations were programmed to avoid any consistent position strategy (Dews and Wiesel, 1970; Daw and Pearlman, 1970).

Before training the cats to distinguish colors, it was first necessary to eliminate brightness as a cue. We did this by determining the relative brightness, for the cat, of the stimulus colors to be used. The cats were trained to make a brightness discrimination with white circles of different brightness, the brighter of the pair being the positive stimulus. Orange stimuli (Wratten 22) were then compared with cyan (Wratten 64), and red stimuli (Wratten 24) with blue (Wratten 47) at several levels through the mesopic and photopic ranges. Figure 2 shows the result of one such comparison. When the orange stimulus was much brighter than the cyan, the cat chose it more than 90% of the time; when cyan was much brighter than orange, it was chosen 90% of the time. The pair of stimuli in which orange and cyan were chosen with equal frequency (the 50% point in the curve of Figure 2) was taken to be of equal brightness for the cat. We were then able to prepare a series of stimuli for training the cats to make color discriminations. In $\frac{1}{3}$ of the stimulus pairs the orange member of the pair was slightly brighter (for the cat) than the cyan; in $\frac{1}{3}$ orange and cyan were equal in brightness; and in $\frac{1}{3}$ cyan was slightly brighter than orange. Brightness was therefore eliminated as a cue.

Both cats were trained to make color discriminations in the mesopic range. One was trained to make an orange/cyan discrimination; the other was initially trained to distinguish red from blue and later switched to a red/cyan discrimination.

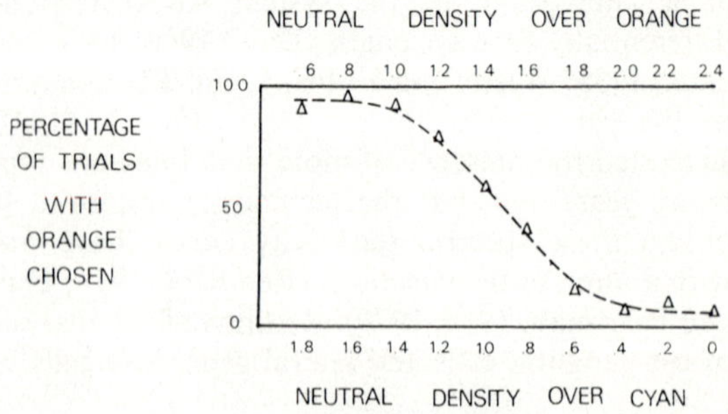

FIGURE 2  Each point represents the percentage of times orange was chosen over cyan for each of ten pairs of stimuli. Each pair was presented 30 times in a period of 4 days.

Training cats to distinguish wavelength proved to be a rather formidable task, as others have reported (Sechzer and Brown, 1964; Mello and Peterson, 1964). Each cat required more than 1500 trials to reach criterion (90% choice of the orange stimulus in the orange/cyan discrimination, for example).

When the cats were able to reliably distinguish the colored stimuli against a white background in the mesopic range, the brightness of the backgrounds was increased by 2 log units to raise it above the level of rod saturation. The brightness of the stimuli was also increased by approximately 2 log units in order to maintain equivalent contrast between the background and the stimuli. After a few hundred more trials, both cats were able to reliably distinguish colors against the bright backgrounds. Figure 3 compares series of trials made alternately against bright and dim backgrounds. Each triangle represents a series of 30 trials against a dim background (2 cd/m²) and each cross is a series against a bright background (200 cd/m²). Thus the cat trained to make color discriminations in the mesopic range retains this ability in the photopic range, above the level of rod saturation. A number of controls designed to eliminate optical cues such as chromatic aberration, or cues generated by the test apparatus convinced us that this was so. We therefore had firm behavioral evidence that the cat has more than one cone type, and were encouraged to look again for neurophysiological evidence.

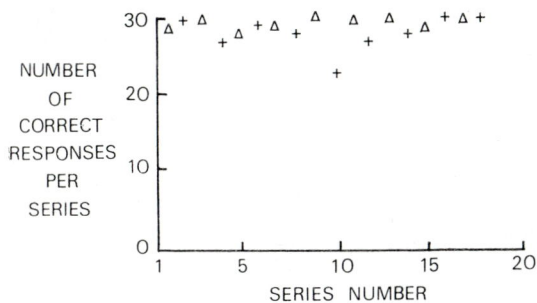

FIGURE 3  Each point represents the number of correct color discriminations (orange rewarded) in a series of 30 trials. Triangles indicate a series at a mesopic level, and crosses a series in the photopic range, above the level of rod saturation.

We undertook a further microelectrode study of the lateral geniculate and optic tract (Pearlman and Daw, 1970; Daw and Pearlman, 1970), and again found that the vast majority of units have input from a single cone type, with peak spectral sensitivity at 556 nm. There is an occasional unit, however, which clearly has input from two cone types. From a total of 434 geniculate and optic tract units, we found four with opponent-color features: each responded with an 'on' response to blue light in the center of the receptive field, and with an 'off' response to green and red light. The spectral sensitivities of these two processes, as measured against colored adapting lights, are shown in Figure 4. Thresholds for 'on' responses (plusses) were measured against an orange background, and for 'off' responses (minuses) against a blue background. The curve for the 'off' process is the same as the curve for the non-color coded geniculate cells, with a peak spectral sensitivity at 556 nm. The curve for the 'on' process is the Dartnall nomogram for a pigment peaking at 445 nm.

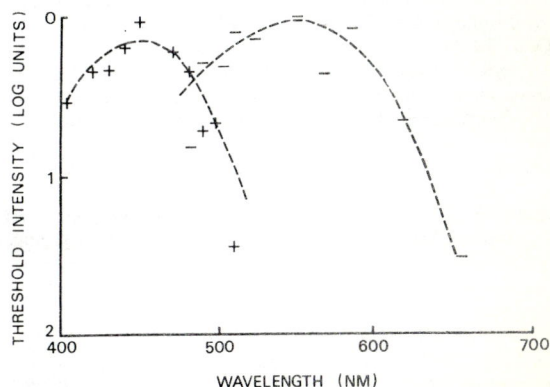

FIGURE 4  Spectral sensitivities of an opponent-color lateral geniculate unit. Plusses indicate thresholds for "on" responses, and are measured against an orange background (Wratten 22) covering the receptive field. Minuses are the thresholds for "off" responses as measured against a blue background (Wratten 47).

Two of the four opponent color cells had double-opponent features similar to those described in the goldfish retina (Daw, 1968) and the monkey cortex (Hubel and Wiesel, 1968). They were excited by blue and inhibited by green in the center of the receptive field, and were also excited by green and inhibited by blue in the receptive field periphery. One of the more interesting features of the opponent color cells is an anatomical one: histological reconstructions of the electrode tracts demonstrated that all four were in layer B (Thuma, 1928) of the lateral geniculate. They all responded to stimulation of the contralateral eye, and thus were in layer C in the new classification of Guillery (1970). None of the 254 cells recorded in layers A and $A_1$ were color coded (Daw and Pearlman, 1970).

## ACKNOWLEDGMENT

We wish to thank Drs. Torsten N. Wiesel, David H. Hubel, and Peter D. Dews for advice and encouragement, and for providing space and equipment in their laboratories at Harvard Medical School. The portion of the work carried out at Harvard was supported by NIH grants NB 2260 and NB 05554, and by NIH special fellowships NB 20241 USN (Dr. Daw) and NB 1864 NSRA (Dr. Pearlman). The portion of the work carried out at Washington University was supported by NIH Program Project Grant NBO 4513 and General Research Funds RR 05389. We thank the editors of the *Journal of Physiology* for permission to reproduce the figures in this report.

## REFERENCES

Barlow, H. B., Fitzhugh, R., and Kuffler, S. W., 1957, Change of organization in the receptive fields of the cat's retina during dark adaptation, *J. Physiol.* **137**: 338–354.

Brown, P. K., and Wald, G., 1964, Visual pigments in single rods and cones of the human retina, *Science* **144**: 45–51.

Daw, N. W., 1968, Color-coded ganglion cells in the goldfish retina: extension of their receptive fields by means of new stimuli, *J. Physiol.* **197**: 567–592.

Daw, N. W., and Pearlman, A. L., 1969, Cat color vision: one cone process or several?, *J. Physiol.* **201**: 745–764.

Daw, N. W., and Pearlman, A. L., 1970, Cat colour vision: Evidence for more than one cone process, *J. Physiol.* **211**: 125–137.

DeValois, R. L., Smith, C. J., Karoly, A. J., and Kitai, S. T., 1958, *J. Comp. Physiol. Psychol.* **51**: 662–668.

Dews, P. D., and Wiesel, T. N., 1970, Consequences of monocular deprivation on visual behaviour in kittens, *J. Physiol.* **206**: 437–455.

Granit, R., 1945, The color receptors of the mammalian retina, *J. Neurophysiol.* **8**: 195–310.

Guillery, R., 1970, The laminar distribution of retinal fibres in the dorsal lateral geniculate nucleus of the cat: a new interpretation, *J. comp. Neurol.* **138**: 339–368.

Hubel, D. H., and Wiesel, T. N., 1961, Integrative action in the cat's lateral geniculate body, *J. Physiol.* **155**: 385–398.

Hubel, D. H., and Wiesel, T. N., 1968, Receptive fields and functional architecture of monkey striate cortex, *J. Physiol.* **195**: 215–243.

Marks, W. B., Dobelle, W. H., and MacNichol, Jr., E. F., 1964, Visual pigments of single primate cones, *Science* **143**: 1181–1183.

Marks, W. B., 1965, Visual pigments of single goldfish cones, *J. Physiol.* **178**: 14–32.

Mello, N. K., and Peterson, N. J., 1964, Behavioral evidence for color discrimination in cat, *J. Neurophysiol.* **17**: 289–294.

Pearlman, A. L., and Daw, N. W., 1970, Opponent color cells in the cat lateral geniculate nucleus, *Science, N. Y.* **167**: 84–86.

Remberg, H., 1953, Die spektrale empfindlichkeit des tauben penzipierunden mechanisms in zwielichtsehen und ihre bedeutung für den primat der helligheit serregung, *Pflügers Arch. ges. Physiol.* **258**: 324–342.

Sechzer, J. A., and Brown, J. L., 1964, Color discrimination in the cat, *Science N. Y.* **144**: 427–429.

Thuma, B. D., 1928, Studies on the diencephalon of the cat, I. The cyto-architecture of the corpus geniculatum laterale, *J. comp. Neurol.* **46**: 173–200.

Wiesel, T. N., and Hubel, D. H., 1966, Spatial and chromatic interactions in the lateral geniculate body of the rhesus monkey, *J. Neurophysiol.* **29**: 1115–1156.

# RESPONSES OF BULBORETICULAR UNITS TO SOMATIC STIMULI ELICITING ESCAPE BEHAVIOR IN THE CAT.†

KENNETH L. CASEY

*Department of Physiology, University of Michigan, Ann Arbor, Michigan* 48104

Units were recorded from the brainstem of awake, unrestrained cats trained to escape electrical stimulation of a cutaneous nerve; stimuli sufficient to elicit escape were associated with natural pain behavior. Of the 104 units histologically localized, 59, recorded from the bulboreticular area of nucleus gigantocellularis (NGC), show maximal or exclusive response to noxious stimuli (36 units) and/or to nerve stimuli eliciting escape (39 units). Increased inter-stimulus activity commonly precedes escape behavior and is the only stimulus-related response observed from some units not driven by nerve stimulation. However, the pooled data from 39 of 53 units tested with nerve stimulation shows that responsiveness (chi-square, Kolmogorov-Smirnov statistic, and average spikes/ stimulus) increases with stimulus strength, reaching maximum levels at escape-producing intensities. Responses of some units to non-somatic inputs can be altered by repetitive noxious stimulation.

## INTRODUCTION

One of the principal aims of sensory neurophysiology is to identify the neural representation of various aspects of sensory experience. Our understanding of the neural basis of pain has, until recently, relied largely upon clinical observations which necessarily limit insight into the underlying neurophysiology. The use of modern electrophysiological and behavioral techniques, especially in combination, holds promise for the future, but there is continuing debate regarding pain mechanisms (Casey, 1970).

Current evidence indicates that activity in somatic sensory afferents with conduction velocities in the range of A-delta and, possibly, C fibers is necessary for pain sensation. Stimulation of cutaneous nerve at or above A-delta threshold is required to elicit reports of pain in man (Collins, Nulsen and Randt, 1960), and the loss of small-diameter fibers is associated with elevated pain threshold (Swanson, Buchan and Alvord, 1965). Furthermore, in cat and monkey, both A-delta and C fiber populations include afferents responding only to mechanical or thermal stimuli which are presumably noxious to the awake animal (Burgess and Perl, 1967; Perl,

1968; Bessou and Perl, 1969). The fine fibers responding to light tactile stimuli may also play a role in pain sensation, for Noordenbos (1959) reports that a shift in cutaneous fiber composition, toward a predominance of small-diameter fibers, is found in patients with post-herpetic neuralgia.

Identifying the essential elements and pathways for pain has proved more difficult so far as the central nervous system is concerned. The thalamic projections of the antero-lateral spinal cord are known (Mehler, Feferman and Nauta, 1960), but electrophysiological studies of these areas reveal few, if any, cells responding exclusively to noxious stimuli (Poggio and Mountcastle, 1960; Pearl and Whitlock, 1961; Poggio and Mountcastle, 1963) although some neurons in the medial thalamus of the awake squirrel monkey respond most vigorously to stimuli eliciting withdrawal (Casey, 1966). Identification of potential nociceptive elements has proved equally difficult in the cortex (Carreras and Anderson, 1963) and the trigeminal sensory nucleus (Kerr, *et al.*, 1968). Recently, however, Christensen and Perl (1970) have identified dorsal horn cells receiving input from A-delta and C fiber afferents and responding exclusively or primarily to presumably noxious stimuli. A similar population, activated primarily by A-delta fibers, is found in the medullary reticular formation in the region of the nucleus gigantocellularis (NGC) (Casey, 1969).

Thus far, studies of bulboreticular unit activity

†Supported by NIH Grant NS 06588, U.S. Public Health Service. The technical assistance of John Matthews and Richard Macklin is gratefully acknowledged.

have been limited to anesthetized or decerebrate animals in which responses to somatic stimuli may be affected by drugs or surgery, and where there is no opportunity for correlating unit activity with behavioral responses operationally identifiable with pain sensation. The experiments described in this report reveal a relation between escape-producing cutaneous nerve stimuli and the responses of bulboreticular neurons recorded in the NGC region of awake cats. A brief report of these results appears elsewhere (Casey, 1971a).

## METHODS

Seventeen cats were used in these experiments. Each was anesthetized with sodium pentobarbital and placed in a stereotaxic instrument for placement of stimulating and recording leads (Figure 1). A

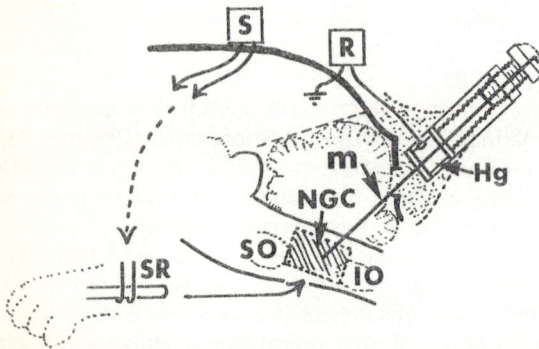

FIGURE 1  *Diagram of stimulating and recording arrangement*. Stimulating leads are passed subcutaneously from a stimulating receptacle (S) mounted on the skull to a rubber cuff electrode on the superficial radial nerve (SR). A lead from the recording receptacle (R) pierces the wall of a nylon nut filled with mercury (Hg) sandwiched between silastic membranes and embedded in silastic rubber (stipple) which covers the craniotomy. When the uninsulated part of the microelectrode (m) contracts the mercury, its tip is in the nucleus gigantocellularis (NGC). SO: superior olivary nucleus. IO: inferior olivary nucleus.

silastic rubber cuff with stainless steel contacts was placed around the superficial radial nerve and the two leads passed subcutaneously to a skull-mounted stimulating receptacle. Three nylon nuts, cemented together with dental acrylic, were positioned above an occipital burr hole (7 mm diameter) at an angle of 50° from the vertical. The bottom nut contained a pool of mercury sandwiched between silastic rubber membranes and connected to a separate recording plug by a lead piercing the walls of the nut. The top two nuts guided a hollow nylon screw holding a stainless steel microelectrode insulated

from near the tip to about 25 mm up the shaft. To insure electrical and physical isolation of the mercury, liquid silastic rubber was poured over the pool and guide assembly and between the skull surface and mercury pool nut. The pool assembly and stimulating and recording receptacles were fastened to the skull with screws and dental acrylic cement. Following surgery, the cats were maintained on parenteral antibiotics (50 mg sulfadimethoxine and 25 mg oxytetracycline HCl/day, I.M.) for 5 days; careful wound care, using topical applications of alcohol and nitrofurazone, prevented infection throughout the post-operative period.

### Training Procedures

Escape training was conducted with the cat inside a $40 \times 41 \times 50$ cm lighted chamber provided with a one-way observation window and divided by a 12 cm high barrier. Nerve stimuli of 0.5 to 5.0 mA (100/sec, 0.2 msec pulses) were required for the two-way escape training; stimulation was terminated when the cat crossed the barrier. Only stimuli which elicited natural pain behavior such as vocalization, biting or shaking the paw, and, in the earlier trials, scratching at the chamber walls, were effective in eliciting and maintaining escape performance. An inter-trial interval of at least 2 min was maintained throughout training. After the escape task had been learned (100% criterion, 10 to 15 trials per day for at least 3 days), escape latency ranged from 3 to 15 seconds. Additional trials were then given using 1/sec nerve stimulation to allow for analysis of post-stimulus neural activity. During and after each training session, the cats responded calmly to petting and accepted food, indicating that the stimuli used, while noxious, were not unnecessarily distressing to the animal.

### Data Acquisition

On the first day of recording, the microelectrode was passed through the mercury pool and into the brainstem until the electrical record showed an abrupt drop in background noise level and the appearance of unit activity, indicating contact between the uninsulated portion of the micro-electrode and the mercury pool. At this point, the microelectrode tip was calculated to be in the dorso-medial reticular formation near the A–P plane of the facial nucleus. The microelectrode was then cemented to the drive screw and advanced manually for the remaining 5 to 6 mm.

When isolated unit activity was encountered, it

was observed for several minutes to be sure that minor changes in spike amplitude, sometimes occurring when the cat moved, were not associated with changes in discharge rate and were not confused with the activity of nearby cells. Natural stimuli were applied by opening the chamber door and examining the effect of clicks, light flashes, moving objects in cat's visual field, and various somatic stimuli (hair, tap, pinch, pin-prick, and passive movements of the limbs and head). Continuous auditory and visual monitoring was used to relate unit activity to active movement.

The amplified unit activity was led into a window amplitude discriminator system combined with a stimulus-triggered series of counters for generation of post-stimulus histograms (PSH) (Spears, Smith and Casey, 1970). Before nerve stimulation, at least 30 100 msec samples of unit activity were accumulated at 1 sample/sec in ten 10 msec time bins; this was usually repeated during the study of each unit as a check on possible changes in inter-stimulus discharge rate. Similarly, a PSH was constructed for each trial of 30 nerve stimuli at several intensities of stimulation, beginning well below escape threshold and increasing with each trial of 30 stimuli until execution of escape. At escape current levels, PSH sample size was reduced since escape was usually completed before the 30th stimulus.

Inter-trial intervals were maintained at 2 minutes or more throughout all experiments. This may account, in part, for the lack of avoidance behavior at lower stimulus intensities, for the effective escape current did not vary by more than 20% in any one cat and was always associated with flexion or shaking of the paw and, frequently, with vocalization.

### Data Analysis

In addition to inspection of filmed records, pre- and post-stimulus histograms were analyzed by computing (IBM System 360) the chi-square and Kolmogorov-Smirnov statistics (Sokal and Rohlf, 1969) and the average number of spikes/stimulus within a 100 msec post-stimulus period. These response parameters were related to the amount of current used at each intensity level by computing the correlation coefficient and associated F-statistic.

### Histological Controls

Electrode position was estimated during the experiments by measuring the height of the drive screw above a fixed reference point on the skull assembly.

These readings were compared to the position of 2 to 3 electrolytic marks along each track, stained with the Prussian Blue technique (Green, 1958). The brain was perfused with 10% formalin and saline-agar and $50\mu$ frozen sections were stained with cresyl violet. Enlarged, projected images of the sections were inspected and traced.

## RESULTS

### Evoked Potential and Multiple-Unit Recording

As the microelectrode enters the bulboreticular formation in the region of the NGC, stimulation of the superficial radial nerve evokes slow potentials and multiple unit discharge at latencies of 15 to 20 msec; peak responses appear at approximately 25 to 30 msec (Figures 2 and 3). Evoked responses to the initial stimuli of a series at fixed intensity increase in amplitude as stimulus intensity increases toward escape levels (Figure 2). At higher intensities, however, some of the responses to succeeding stimuli are reduced in amplitude as the cat arises and moves in the process of preparing to escape. Passively moving the cat does not, by itself, produce this effect, indicating that movement of the electrode is not responsible and that neural activity associated with the behavior in some way attenuates the synchronized evoked response. Additional information was obtained by using an integrating device (Weber and Buchwald, 1965) which shows (Figure 3) that increased inter-stimulus multiple-unit activity depresses the amplitude of some synchronized population responses during active movement at the higher stimulus intensities. Even so, peak integrated response levels, as measured against a fixed pre-stimulus background, are highest near escape stimulus intensities and many of these integrated responses also attain maximal levels relative to the increased background activity. These observations suggested that single unit recording would permit more reliable quantification of changes in response to escape-producing stimuli.

### Unit Responses to Natural Stimuli

The anatomical distribution of 104 of the 115 units studied is shown in Figure 4. Eleven units could not be localized histologically and therefore will not be discussed. The activity of 45 units failed to show any relation to behavioral escape or noxious stimuli. Although the ongoing activity of these cells was not systematically studied, 10 reticular

82 K. L. CASEY

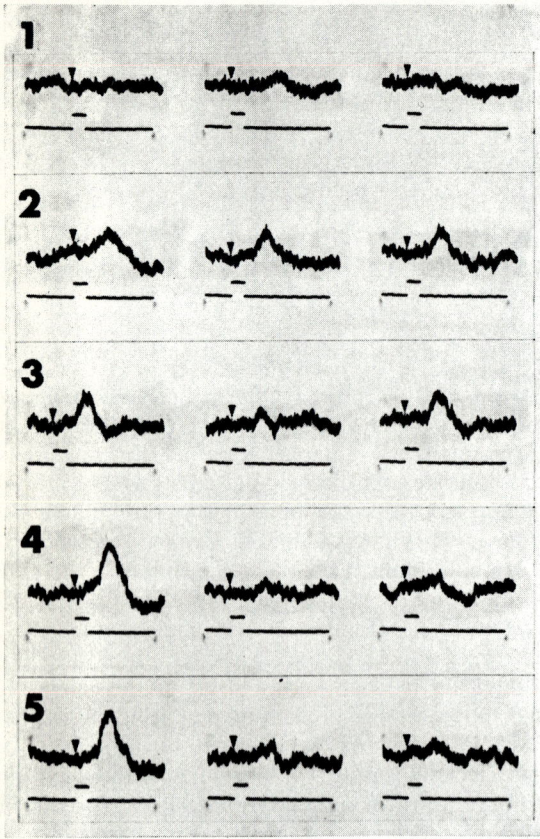

FIGURE 2  *Evoked potentials recorded from NGC in awake cat.* Superficial radial stimulation at 1/sec. Each row shows the first, middle, and last response to one of a series of stimuli delivered at intensities of: 1, 200$\mu$A; 2, 300$\mu$A; 3, 400$\mu$A; 4, 600$\mu$A; 5, 800$\mu$A. Note increasing amplitude of the first responses at higher intensities; diminished responses to subsequent stimuli as cat moves and orients toward barrier (4, 5). Cat moved only during middle response in 3, and was sitting quietly throughout 1 and 2. Pulse calibration below each trace: 50$\mu$V, 10 msec.

formation units showed very regular, sustained discharge in the absence of applied stimuli, as has been observed previously (Scheibel, *et al.*, 1955). Eight units failed to respond to any of the stimuli tested and 9 responded only to auditory (5) or visual (4) stimuli. Four cells, recorded near the medial longitudinal fasciculus, discharged in relation to eye movements and 2, recorded near the vestibular nuclei, discharged during active or passive head movement. Another group of 22 units responded to somatic stimuli, but their responses were not further increased by either strong natural stimuli (pinch, pin-prick) eliciting withdrawal and vocalization or by electrical stimuli eliciting escape. Hair movement within relatively restricted regions of

the face excited 6 trigeminal nucleus units. As with 15 reticular formation units responding to hair or light touch, their increased discharge did not continue throughout the application of continuous deep pressure or pinching. The reticular formation units appeared to respond equally to tactile stimulation of each limb, but this could not be studied systematically.

In contrast with the above units the discharge of 59 cells recorded from the reticular formation could be related to noxious somatic stimuli or to escape behavior. Thirty-six units tested for response to natural stimuli were excited by a brisk and clearly innocuous tap to any limb, but gave sustained responses only during the application of deep pressure or pinching (Figure 5). The effective stimulus nearly always elicited withdrawal of the stimulated limb; unit activity could not be related to active or passive limb movement in the absence of stimulation. Nine of these cells responded exclusively to clearly noxious stimuli, and 5 had stocking-like receptive fields over one or both forelimbs; none responded to auditory or visual stimulation.

*Unit Responses Related to Escape-Producing Nerve Stimuli*

The effect of nerve stimulation at several intensities was tested for 53 units. Six of these were not driven even at the higher intensities ($X^2$ and KS with $p \geq 0.05$), and 8 were driven well below escape levels but were either inhibited or discharged less frequently as stimulus intensity increased. The remaining 39 cells maintained or even increased their post-stimulus discharge as current was raised toward escape level. In most instances, the PSH reveals a peak response between 20 to 40 msec after stimulation (Figure 6). An example of a shorter latency response is shown in Figure 7 where the bi-modal character of the response is evident in the sample records and in the PSH (Figure 8).

In order to relate unit responses to stimulus-related behavior, stimulus intensity is expressed as a fraction of the current required to elicit escape during the study of each unit. This variable is shown in relation to three measures of unit response computed from the PSH at each stimulus intensity: the chi-square and Kolmogorov-Smirnov (KS) statistics and the average number of spikes/stimulus within the 100 msec post-stimulus period. The sample analyses in Figure 9 show all response parameters increasing with stimulus intensity, indicating that the higher post-stimulus spike

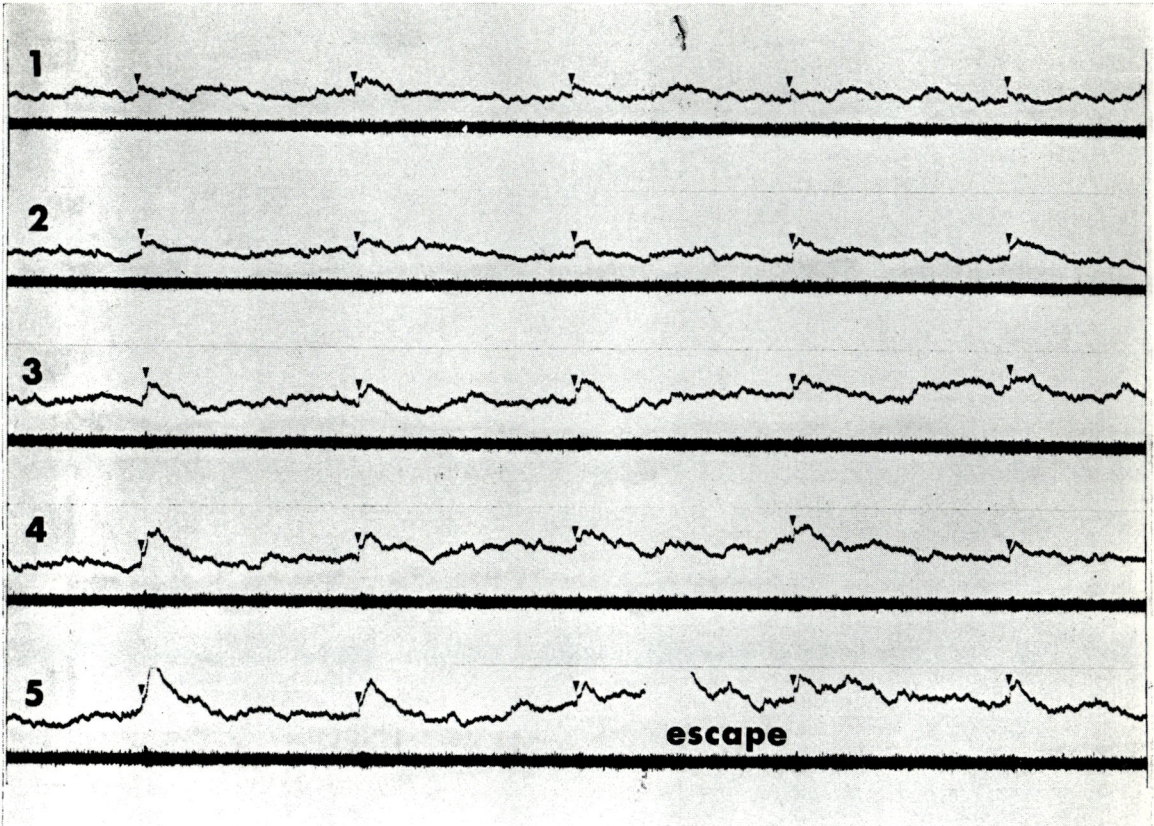

FIGURE 3 *Integrated multiple unit recording.* Recording from same position as in Figure 2. Each row shows multiple unit (bottom) and integrator (top) responses to 5 successive stimuli at intensities of: 1, 100μA; 2, 300μA; 3, 500μA; 4, 800μA; 5, 1.0mA, delivered at 1/sec. Note, in the integrator record, increased activity at the higher stimulus intensities. Many integrated population responses appear above the increased background activity, especially at escape intensity. Escape was executed at point indicated.

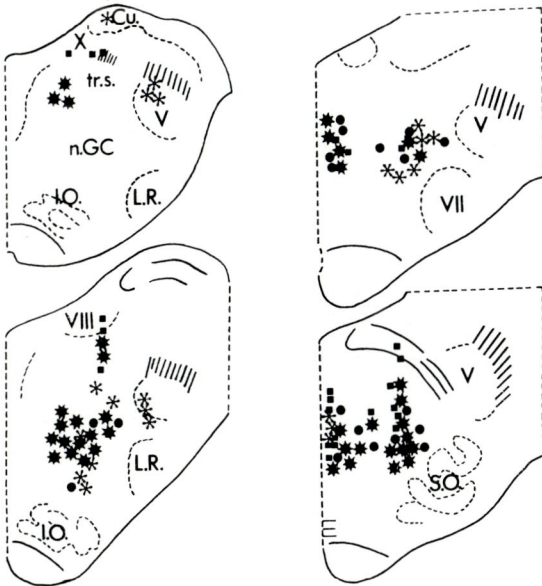

FIGURE 4 *Anatomical distribution of* 104 *units.* Filled squares: 23 units which were either unresponsive or responded only to non-somatic stimuli. Asterisks: 22 units responding to innocuous somatic stimuli and/or nerve stimulation, but without activity or response increase related to stimulus intensity. Filled circles: 20 units with escape-related activity or responding maximally or exclusively to natural somatic stimuli eliciting withdrawal and/or vocalization. Stars: 39 units with increasing responses to nerve stimuli eliciting escape (see text). Cu: external cuneate nucleus. 1.0: Inferior olivary nucleus. L.R.: lateral reticular nucleus. n.GC: nucleus gigantocellularis. S.O.: superior olivary nucleus. tr.s.: tractus solitarius. V: trigeminal nucleus. VII: facial nucleus. VIII: vestibular nucleus. X: vagal nucleus.

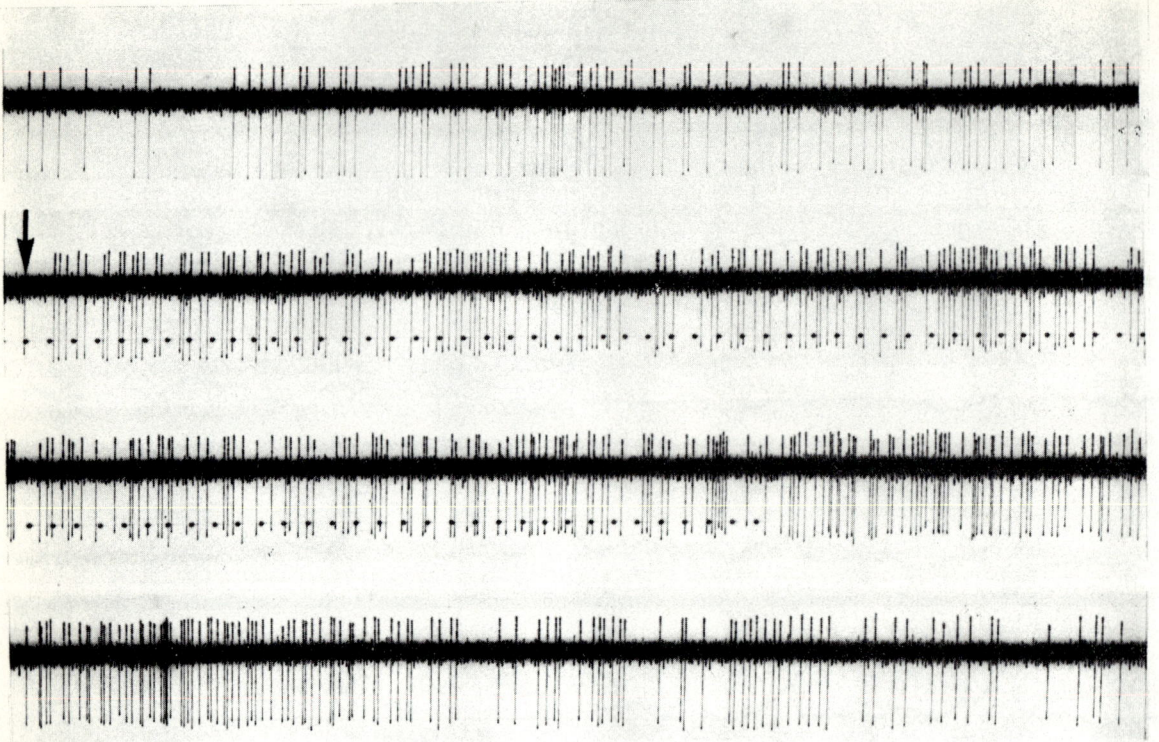

FIGURE 5. *NGC unit showing sustained response to noxious stimulus.* Continuous record. Strong pinch applied (arrow) to forepaw and continued during the 10/sec pulse signal below the record. Note continued increased discharge following stimulus termination. Positivity upward.

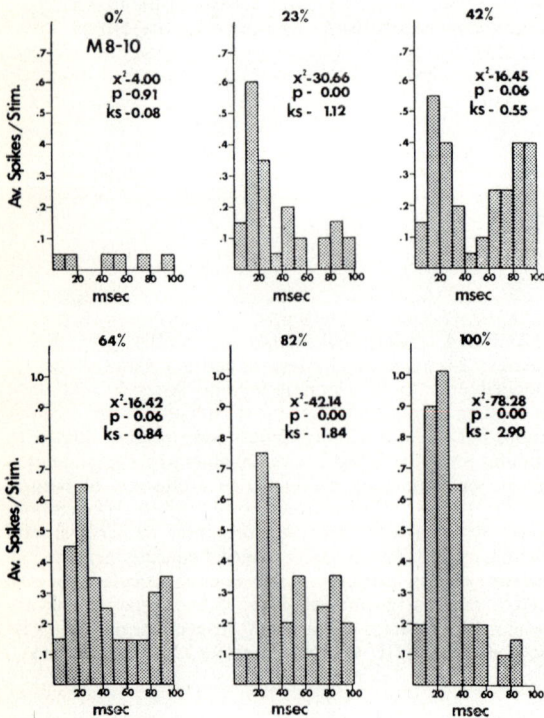

FIGURE 6 *Post-stimulus histograms of unit M*8-10. Stimulus intensity shown as a percentage of escape current above each histogram. Chi-square and associated significance level (p) is given above each Kolmogorov-Smirnov (ks) statistic.

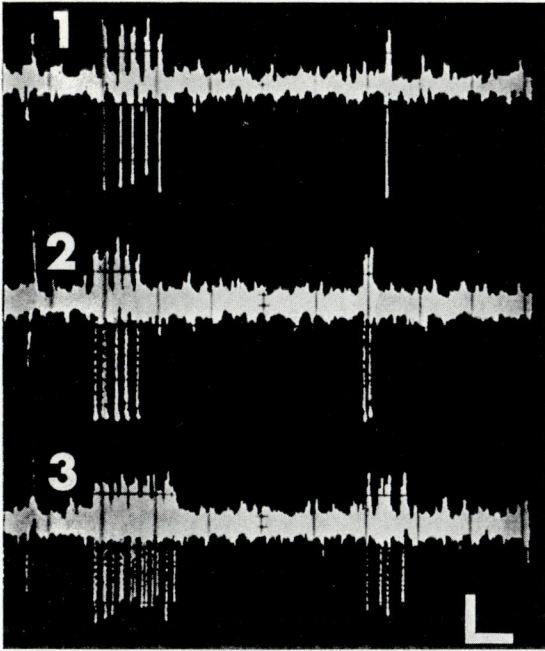

FIGURE 7 *NGC unit M32-11.* Note increased bi-modal response and decreased latency as stimulus intensity increases: 1, 500μA; 2, 800μA; 3, 1.0mA. Calibration: 5 msec, 100μV. Positivity upward.

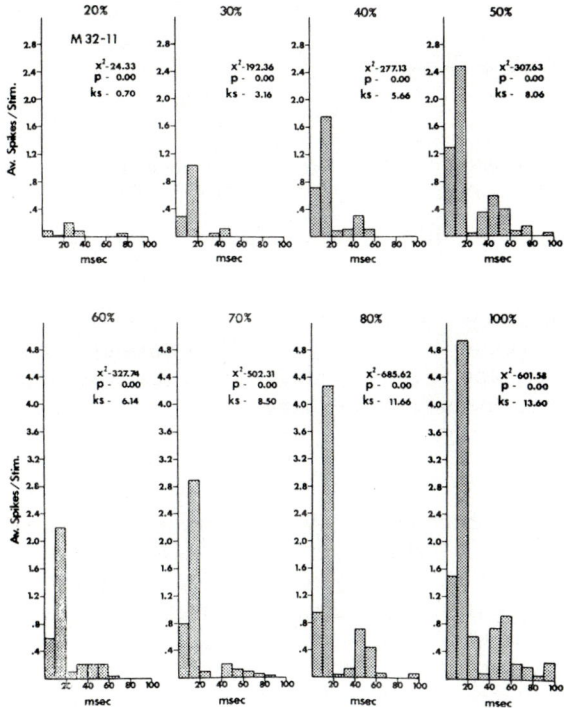

FIGURE 9 *Response parameters as a function of percentage escape current.* Analysis of units shown in Figure 6, 7, and 8. Solid line: chi-square. Dashed line: average number of spike/stimulus. Broken line: Kolmogorov-Smirnov (ks) statistic. Left ordinate combines chi-square and ks values.

FIGURE 8 *Post-stimulus histograms of unit M32-11.* Same conventions as Figure 6.

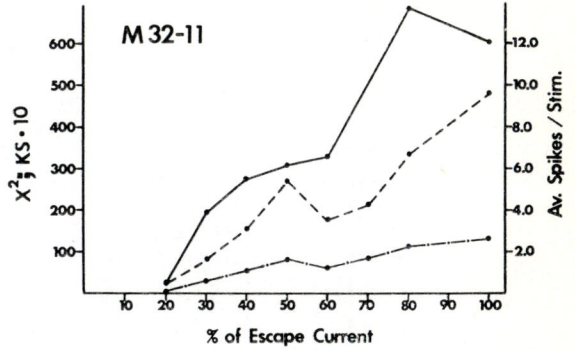

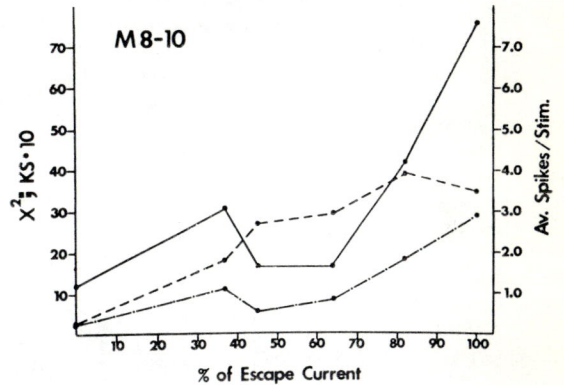

counts are in part due to augmented time-locked responses and are not solely attributable to increases in inter-stimulus discharge throughout each trial. This is the prevailing pattern within the sample of 39 cells as is shown by the correlated scatter plots of the pooled data in Figure 10. It is necessary, in comparing unit responses, to express each response parameter as a percentage of the maximum value observed while a unit was under study because actual parameter values varied widely among cells. In all cases, however, the associated significance levels indicated response to nerve stimulation ($p \leq 0.01$) during the series of 30-stimulus trials. Some units showed a drop in chi-square and KS values at escape-producing intensities even though the average post-stimulus spike count was maximal; this is presumably due to the smaller sample sizes obtained when stimulation was terminated by escape.

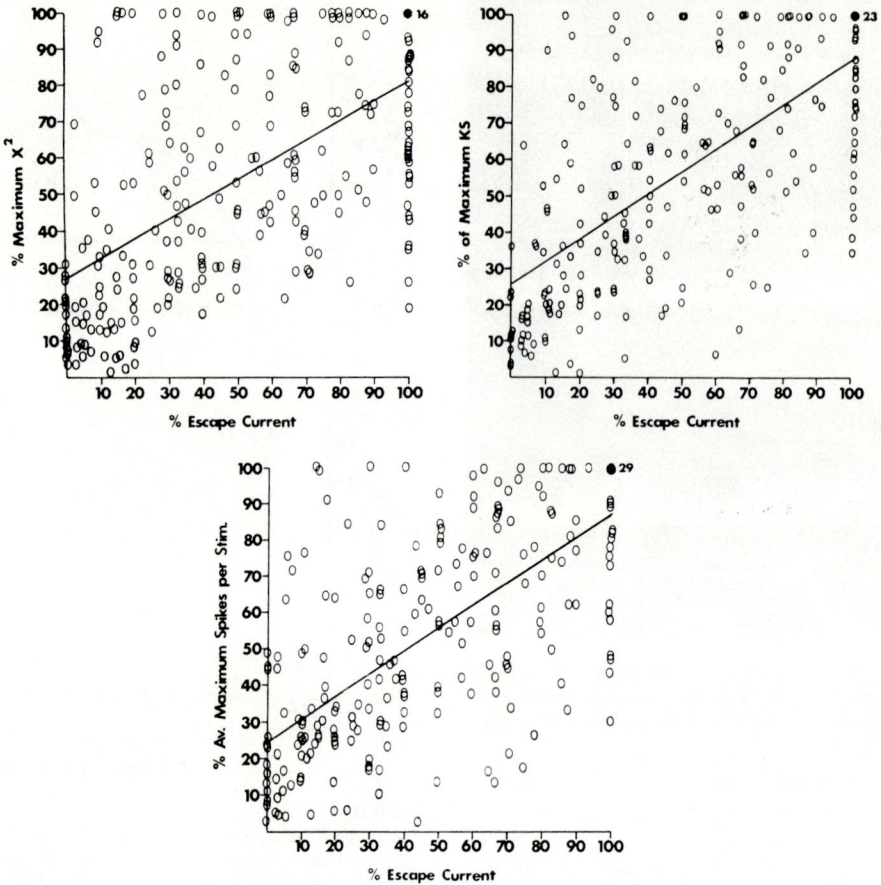

FIGURE 10 *Scatter plots of stimulus-response relationship for 39 units.* Separate plots for each parameter (chi-square, ks, and average number of spikes/stimulus) expressed as a percentage of the maximum value obtained for each unit. Number at upper right of each graph indicates overlapping values at the 100% intersection. Correlation coefficients for the 246 observations are all highly significant ($p < 10^{-10}$): % max. chi-square, r = 0.59; % max. ks, r = 0.67; % max. av. spikes/stim., r = 0.69.

The above analysis does not indicate the absence of increased inter-stimulus activity; indeed, as stimulus intensity is increased, there is commonly a marked rise in inter-stimulus discharge which accompanies the augmented responses to nerve stimulation (Figure 11). Rarely did it appear that the increased post-stimulus discharge was entirely limited to the 100 msec PSH analysis period. This increased background discharge is usually most pronounced several seconds before execution of escape and is not clearly correlated with a specific pattern of movement.

The effect of natural somatic stimulation was examined for 20 cells of this sample. Sixteen units were among those responding to a tap or innocuous pressure and showing more intense and prolonged discharges to stimuli eliciting withdrawal. The remaining 4 units fired only when noxious stimuli were applied; 3 had stocking-like receptive fields over one or both forelimbs.

*Other Escape-Related Unit Activity*

Included among the 59 units with activity related to escape behavior are 6 cells which failed to respond consistently to a specific movement or to any natural or electrical stimuli, but showed increased discharge only as the cat oriented toward the barrier and executed escape. An example of this phenomenon is shown in Figure 12. The PSH of this cell does not reveal driven activity within the 100 msec analysis period; however, the back-

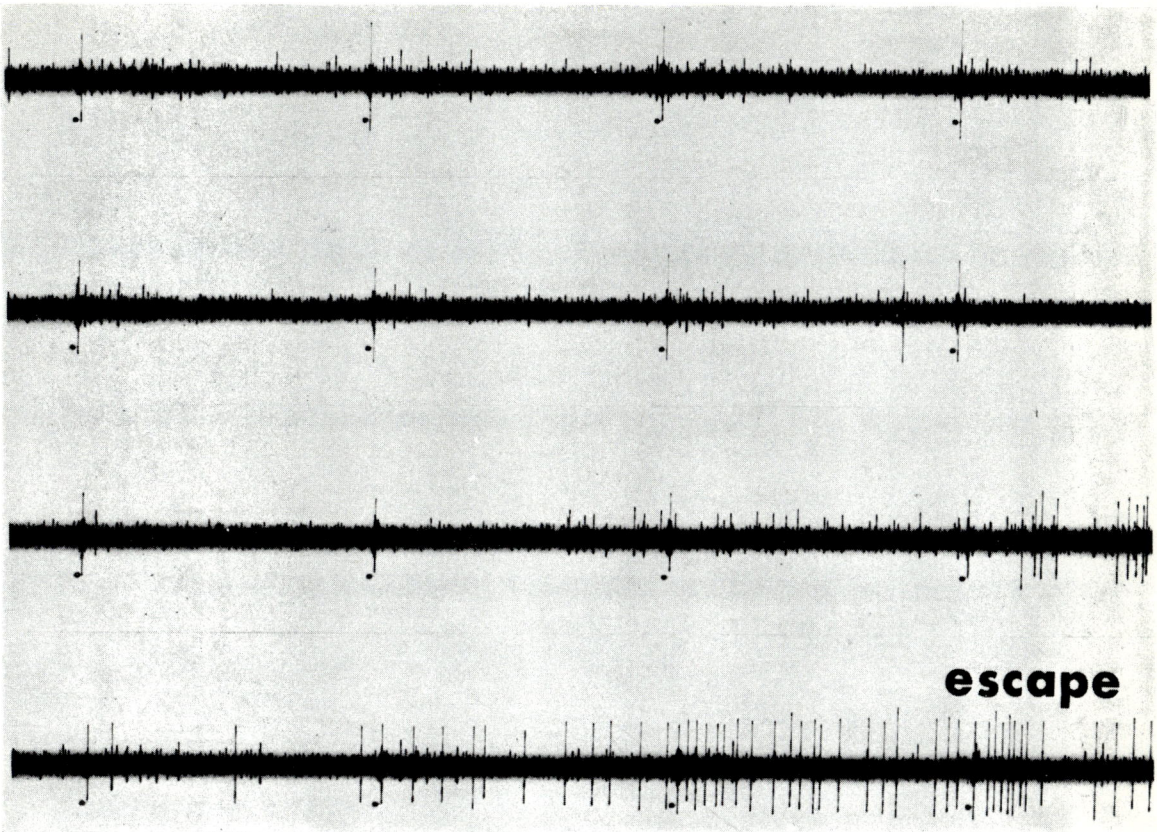

FIGURE 11 *Increased response and inter-stimulus activity preceding escape.* Continuous record of response to 1/sec stimulation (dots below record) at escape intensity. Cell responds witu one spike/stimulus and then shows increased response and inter-stimulus activity for 5 seconds preceding escape. Negativity upward.

ground discharge clearly increases with stimulus intensity and reaches a maximum as escape is executed. Each of these cells returned to a low level of pre-escape activity (usually less than 1/5 sec) even though the cat continued moving actively after crossing the barrier.

There was no systematic attempt to study changes in unit response due to habituation or stimulus pairing. However, 2 cells recorded near the medial aspect of NGC changed their discharge pattern as a result of repeated noxious stimulation. Figure 13 shows the essential features of the observation in both cases. As the chamber door is opened, a few spikes are seen, followed by an inactive period which is interrupted by a brief response to a pinprick. The door is closed and the procedure repeated 6 to 7 times at approximately 20 sec intervals. On each successive trial, the pin response remains approximately the same or slightly increased, but

the discharge associated with opening the door is progressively augmented until it exceeds the pin response in frequency and duration. During the development of this unit response pattern, the cat remained in a relaxed, sphinx-like posture; there were no gross changes in motor response to either the door opening or pin-prick. Neither active nor passive head movement affected spike discharge, indicating that the motor act of orientation was not the responsible factor. Eye movement, changes in pupil diameter, clicks, light flashes and moving objects in the visual field did not affect unit activity.

## DISCUSSION

The results support the hypothesis that a population of neurons in the bulboreticular formation form an important part of CNS pain mechanisms. Previous

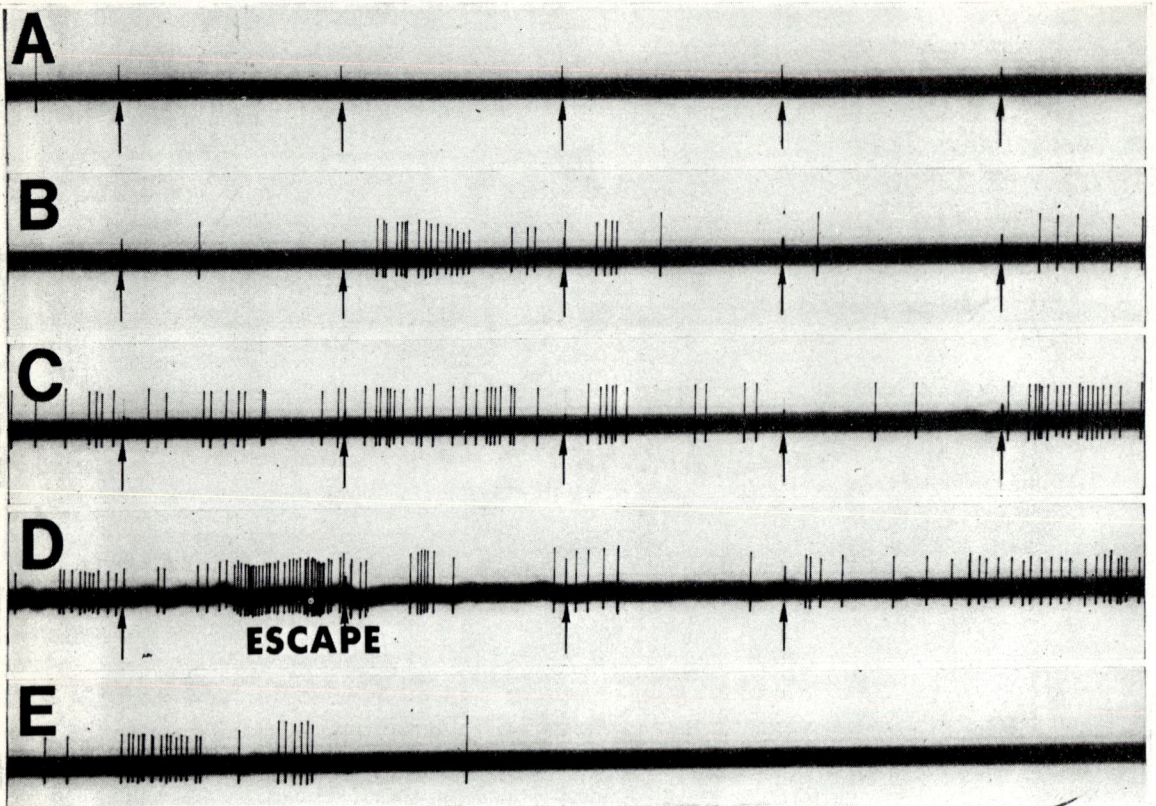

FIGURE 12 *Increased inter-stimulus activity associated with escape.* A: 1/sec stimulus (arrows) at 50% escape intensity. B: at 75% escape intensity. C, D: at escape intensity. Note progressive increase in inter-stimulus activity without response to each stimulus; maximum discharge during escape followed by return to pre-stimulus activity level when stimulation ends (D, E). C, D, and E are continuous records. Negativity upward.

studies on anesthetized (Bowsher, *et al.*, 1968) or decerebrate (Casey, 1969) cats had confirmed the anatomical findings (Bowsher, 1957; Mehler, Feferman and Nauta, 1960) that the NGC region received somatic input from the antero-lateral spinal cord and had revealed physiological properties suggesting a role in pain sensory mechanisms. However, there remained the possibilities that drugs or surgery significantly altered neural responses and that, even were this not so, the observed neural activity might be unrelated to pain behavior.

*Evoked Potential and Multiple-Unit Recording*

Had evoked slow potentials been used as the basis of this study, the conclusions might have been much different. Although the responses to the initial stimuli of a series increase in amplitude with stimulus intensity, subsequent responses decrease with orientation and movement toward the barrier.

Since the mechanism underlying the slow potential is not understood, changes in response amplitude cannot be explained on the basis of these data. However, the integrated multiple unit records reveal that an increase in background activity is associated with an apparent relative decrease in the synchronized population response to many o the stimuli near escape intensities. In contrast with the evoked potential data, multiple unit recordings also suggested that some units might show increased responses to the higher levels of nerve stimulation. These findings, together with the data obtained by unit recording, reveal some of the interpretive difficulties associated with recording population responses in the awake animal.

*Unit Responses to Natural Stimuli*

The results confirm previous reports (Bowsher, *et al.*, 1968) that a number of bulboreticular cells

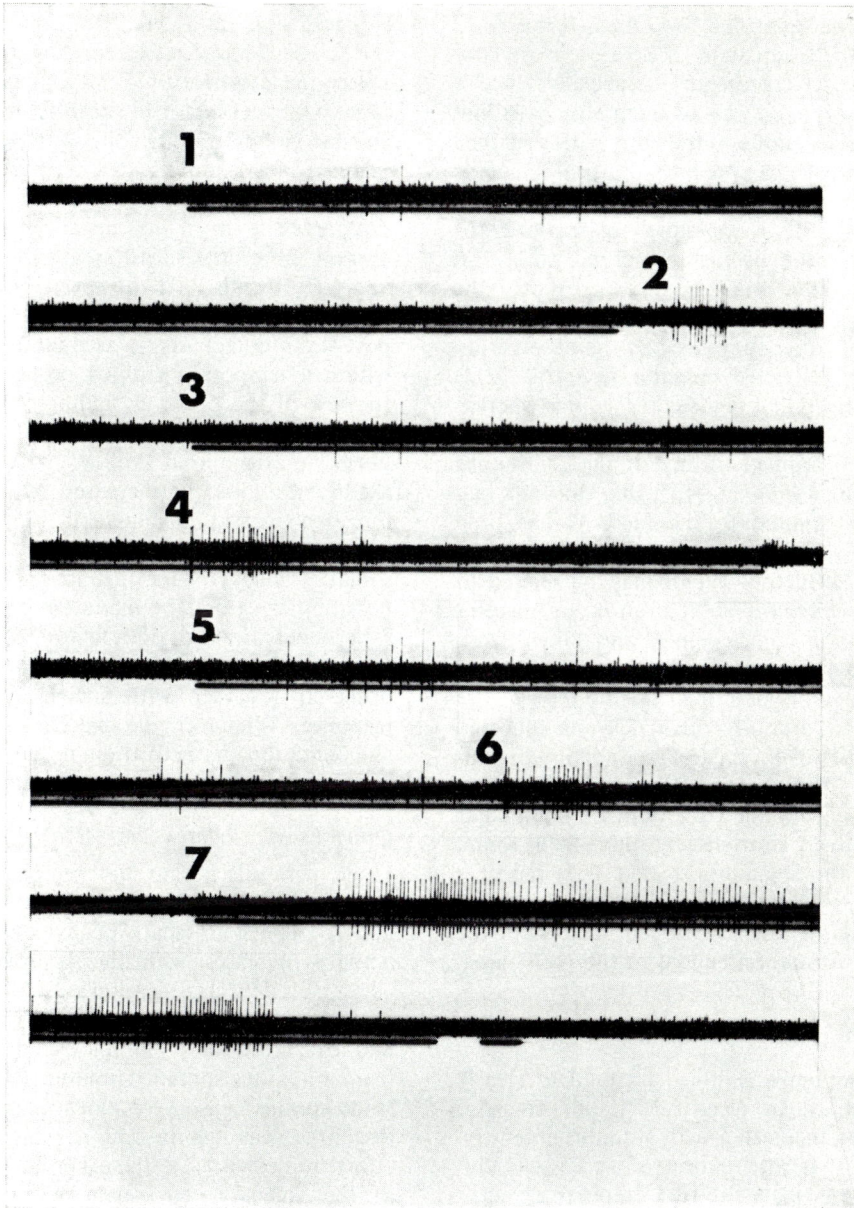

FIGURE 13 *Change in unit response pattern.* Numbers 1-6 indicate 3 sequential openings of the chamber door (1, 3, and 5) followed by a noxious stimulus (pin-prick) (2, 4 and 6). Note increase in discharge at 5 as compared with 1 and 3. The procedure was repeated 3 more times between 6 and 7 until, at the seventh door opening (number 7) unit discharge to door opening alone exceeded the previous responses to pin-prick.

respond to only one modality. Since this study focused on somatosensory responses, however, it is likely that the true extent of heterosensory convergence is under-represented, especially since a complete sensory analysis of each unit was not possible. However, none of the units with sustained responses limited to noxious natural somatic stimuli were excited by auditory or visual input.

The fact that it is possible to identify the sensory and motor events associated with unit discharge provides some assurance as to the reliability of unit response analysis in these experiments. This is further supported by the observation that relatively restricted somatic receptive fields are found in trigeminal nucleus recordings whereas bulboreticular units typically respond to a brisk tap or pressure applied to any of the four limbs (Bowsher, et al., 1968). The fact that no units had receptive fields limited to the face, as has been reported elsewhere (Segundo, Takenaka, and Encarbo, 1967; Burton, 1968), may be related to differences in the region of exploration. In Burton's (1968) study, the unit sample was taken from bulboreticular areas more caudal and lateral to the location of units described in this report. The high proportion of nociceptive neurons reported in his study, however, suggests a functional similarity between these areas. Regional differences may also account for the paucity of bulboreticular units with relatively restricted receptive fields found in this study, for Segundo, et al. (1967) report a much higher proportion of units with 'highly restricted' fields (a few square centimeters) in the bulboreticular formation caudal to the NGC area.

*Unit Responses to Nerve Stimulation*

The majority of units sampled respond to clearly innocuous natural or electrical stimuli but also show responses increasing with stimulus intensity. It is possible that attenuation of successive unit responses takes place during repetitive nerve stimulation and that this might account, in part, for the low response levels observed especially at the lower stimulus intensities (Segundo, Takenaka and Encabo, 1967). No effort was made to identify this phenomenon since, if present, it did not interfere with the principal objective of this study. The 2-minute inter-series intervals and progressively increasing stimulus strength may have prevented a significant amount of response habituation. Other aspects of unit responsiveness are also likely to be affected by the lack of anesthesia, immobilization and surgery.

The relation between unit response and intensity of nerve stimulation recalls the results obtained in the unanesthetized, decerebrate cat (Casey, 1969), where the presence of A-delta activity was shown to be a critical factor in determining cell response. In the present study, the additional factor of increased inter-stimulus activity required particular attention, for it is evident that 'background' discharge is, in many instances, augmented at current intensities near escape levels. Statistical analysis of each PSH, however, indicates that for many units, a significant proportion of the increased post-stimulus activity is attributable to discharge within a temporally limited post-stimulus period. In view of the previous findings (Casey, 1969), it is possible that this reflects the addition of A-delta fibers to the input volley. Alternatively, unit excitability could be increased by other inputs to this area (Rossi and Zanchetti, 1957; Scheibel and Scheibel, 1958), thus augmenting the response to somatic volleys. The distinction between these possibilities cannot be made on the basis of these experiments; indeed, both factors may be operative since increased inter-stimulus discharge is commonly observed together with increased post-stimulus response. Whatever the mechanism of response augmentation, its activation is temporally related to the subsequent execution of escape.

*Other Escape-Related Unit Activity*

Units with discharges not temporally locked to somatic stimuli are also included among those with activity related to escape behavior if their ongoing activity increases with the approach of escape. These cells defy classification as 'sensory' or 'motor', for their discharge is instead related to the imminence and execution of a complex behavior: escaping from a noxious somatic stimulus. It is possible that some specific sensory or motor event causing unit discharge was overlooked in spite of concerted efforts to detect it. On the other hand, the behavior of these cells is consonant with the increasing background discharge observed in multiple unit recordings and accompanying the responses of units driven by nerve stimulation. Furthermore, it is not surprising to find that unit activity recorded from a complex structure in an awake, behaving animal is best related to abstract elements of behavior rather than to the physical characteristics of a movement or of a stimulus in one modality. This argument is further supported by the dramatic changes in the response properties of 2 units after repeatedly associating noxious stimulation with a complex preceding event (opening of the chamber door and visual presentation of the pin). In this case, the units are excited by noxious stimulation but the subsequent development of response to the preceding event suggests a relation to an internal state associated with anticipation of pain. The infrequency with which such units were recorded is probably due, in large part, to their very low rate of 'spontaneous' activity and to a lack of familiarity with their properties.

*Significance for Pain Mechanisms*

It is possible that the unit activity described in this report has no significance so far as pain sensation or pain-related behavior is concerned. This region of the bulboreticular formation is a well-known source of reticulo-spinal (Nyberg-Hansen, 1965) as well as ascending (Brodal and Rossi, 1955) projections and receives input from a number of sources (Rossi and Zanchetti, 1957; Scheibel and Scheibel, 1958). Clearly, one cannot argue for functional homogeneity of the NGC area on the basis of current information. The output of NGC and its anatomically contiguous reticular nuclei may be related to autonomic or spinal reflex functions or to generalized arousal states coincident with pain sensation or response. Nonetheless, these results indicate that the activity of many bulboreticular neurons is not associated with a well-defined movement or posture as might be expected of a 'motor' area; nor do they respond equally well to inputs from several sensory modalities, suggesting a more generalized function. Rather, both their ongoing activity and response to electrical and natural somatic stimuli are systematically related to a behavior operationally identified with pain sensation. Moreover, it has been shown in cat that lesions in the NGC (Halpern and Halverson, 1967) or its thalamic projection (Mitchell and Kaelber, 1966) significantly increase escape latency without affecting motor performance and that stimulation of NGC in rat elicits escape (Keene and Casey, 1970). The following report will also document the aversive character of direct stimulation of this bulboreticular area in the cat (Casey, 1971b). Whatever additional functions it may serve, then, the evidence strongly suggests that the NGC area is an integral part of central pain mechanisms.

REFERENCES

Bessou, P., and Perl, E. R., 1969, Response of cutaneous sensory units with unmyelinated fibers to noxious stimuli. *J. Neurophysiol.* **32**: 1025–1043.

Bowsher, D., Mallart, A., Petit, D., and Albe-Fessard, D., 1968, A bulbar relay to the centre median. *J. Neurophysiol.* **31**: 288–300.

Bowsher, D., 1957, Termination of the central pain pathway in man: The conscious appreciation of pain. *Brain* **80**: 606–621.

Burgess, P. R., and Perl, E. R., 1967, Myelinated afferent fibres responding specifically to noxious stimulation of the skin. *J. Physiol.*, London. **190**: 541–562.

Burton, H., 1968, Somatic sensory properties of caudal bulbar reticular neurons in the cat (Felix domestica). *Brain Res.* **11**: 357–372.

Carreras, M., and Anderson, S. A., 1963, Functional properties of neurons of the anterior ectosylvian gyrus of the cat. *J. Neurophysiol.* **26**: 100–126.

Casey, K. L., 1971a, Bulboreticular formation and pain: Unit recording and brain stimulation in the awake, trained cat. *Fed. Proc.* **30**: 663.

Casey, K. L., 1971b, Escape elicited by bulboreticular stimulation in the cat. *Intern. J. Neurosci.* **2**: 29–34

Casey, K. L., 1969, Somatic stimuli, spinal pathways, and size of cutaneous fibers influencing unit activity in the medial medullary reticular formation. *Exp. Neurol.* **25**: 35–56.

Casey, K. L., 1970, Some current views on the neurophysiology of pain. In: *Pain and Suffering*, Crue, B. L., Jr., ed. C. C. Thomas, Springfield.

Casey, K. L., 1966, Unit analysis of nociceptive mechanisms in the thalamus of the awake squirrel monkey. *J. Neurophysiol.* **29**: 727–750.

Christensen, B. R., and Perl, E. R., 1970, Spinal neurons specifically excited by noxious or thermal stimuli: Marginal zone of the dorsal horn. *J. Neurophysiol.* **33**: 293–307.

Collins, W. F., Nulsen, F. E., and Randt, C. T., 1960, Relation of peripheral nerve fiber size and sensation in man. *Arch. Neurol. Psychiat.* **3**: 381–385.

Green, J. D., 1958, A simpler microelectrode for recording from the central nervous system. *Nature.* **181**: 962

Halpern, B. P., and Halverson, J. D., 1967, Elevated escape latencies after hindbrain lesions. *Physiologist.* **10**: 193.

Keene, J. J., and Casey, K. L., 1970, Excitatory connection from lateral hypothalamic self-stimulation sites to escape sites in medullary reticular formation. *Exp. Neurol.* **28**: 155–166.

Kerr, F. W. L., Kruger, L., Schwassmann, H. O., and Stern, R., 1968, Somatotopic organization of mechanoreceptor units in the trigeminal nuclear complex of the Macaque. *J. Comp. Neurol.* **134**: 127–144.

Mitchell, C. L., and Kaelber, W. W., 1966, Effect of medial thalamic lesions on responses elicited by tooth pulp stimulation. *Am. J. Physiol.* **210**: 263–269.

Noordenbos, W., 1959, *Pain*. Elsevier, Amsterdam.

Nyberg-Hansen, R., 1965, Sites and mode of termination of reticulo-spinal fibers in the cat. *J. Comp. Neurol.* **124**: 71–100.

Perl, E. R., and Whitlock, D. G., 1961, Somatic stimuli exciting spinothalamic projections to thalamic neurons in cat and monkey. *Exp. Neurol.* **3**: 256–296.

Perl, E. R., 1968, Myelinated afferent fibres innervating the primate skin and their response to noxious stimuli. *J. Physiol.*, London. **197**: 593–615.

Poggio, G. F., and Mountcastle, U. B., 1960, A study of the functional contributions of the lemniscal and spinothalamic systems to somatic sensibility. Central nervous mechanisms in pain. *Johns Hopkins Hosp. Bull.* **106**: 266–316.

Poggio, G. F., and Mountcastle, V. B., 1963, The functional properties of ventrobasal thalamic neurons studied in unanesthetized animals. *J. Neurophysiol.* **26:** 775–806.

Rossi, G. F., and Zanchetti, A., 1957, The brain stem reticular formation. Anatomy and physiology. *Arch. ital. Biol.* **95:** 199–435.

Scheibel, M., Scheibel, A., Mollica, A., and Moruzzi, G., 1955, Convergence and interaction of afferent impulses on single units of reticular formation. *J. Neurophysiol.* **18:** 309–331.

Scheibel, M. E., and Scheibel, A. B., 1958, Structural substrates for integrative patterns in the brain stem reticular core. In: *Reticular Formation of the Brain*, Jasper, H. H., et al., eds., Little, Brown, Boston, pp. 31–55.

Segundo, J. P., Takenaka, T., and Encabo, H., 1967, Somatic sensory properties of bulbar reticular neurons. *J. Neurophysiol.* **30:** 1221–1238.

Sokal, R. R., and Rohlf, F. J., 1969, *Biometry*. W. F. Freeman, San Francisco, pp. 571–575.

Spears, R., Smith, G., and Casey, K. L., 1970, A pulse height discriminator and post-stimulus histogram system using integrated circuits, *Physiol. Behav.* **5:** 1327–1329.

Swanson, A. G., Buchan, G. C., and Alvord, E. C., 1965, Anatomic changes in congenital insensitivity to pain. *Arch. Neurol. (Chicago).* **12:** 12–18.

Weber, D. S., and Buchwald, J. S., 1965, A technique for recording and integrating multiple unit activity simultaneously with the EEG in chronic animals. *Electroenceph. clin. Neurophysiol.* **19:** 190–192.

# ESCAPE ELICITED BY BULBORETICULAR STIMULATION IN THE CAT†

KENNETH L. CASEY

*Department of Physiology, The University of Michigan, Ann Arbor, Michigan* 48104

Electrical stimulation of the bulboreticular formation in the region of nucleus gigantocellularis (NGC) elicits escape and natural pain behavior in cats trained to escape electrical stimulation of the superficial radial nerve. The effect of bulboreticular stimulation is generalized so that escape is elicited on the initial stimulus trials at an effective NGC locus. The stimulus sites with lowest threshold (25μA, 0.2 msec. pulses, 100/sec.) are found in locations from which escape-related unit activity was recorded with the stimulating microelectrode. The effective bulboreticular stimuli can function as the unconditioned stimulus in auditory avoidance conditioning and can be used to shape operant behavior.

## INTRODUCTION

Since unit analysis in the awake cat has shown a correlation between bulboreticular neural activity and escape behavior (Casey, 1971), a study of the behavioral effects of electrical stimulation of the same area is especially important. If direct stimulation of the NGC region fails to elicit escape or natural pain behavior, it strengthens the possibility that the unit activity, although systematically related to escape-producing stimuli, has little or no bearing on pain sensory mechanisms. It is possible, of course, that a negative result could be attributed to the highly unnatural mode of stimulation, but the force of this argument is tempered by the fact that definite escape is elicited, in the rat, by electrical stimulation of the dorsal midbrain tegmentum, periventricular area, and non-specific thalamic nuclei (Olds and Olds, 1963). A positive result, however, would confirm previous observations (Halpern and Halverson, 1967; Keene and Casey, 1970) suggesting that NGC activity has behavioral significance beyond that of a purely reflex nature.

The results discussed in this report show that stimulation in the area where escape-related unit activity is recorded elicits escape and that this behavior reflects the aversive character of the stimulation.

†Supported by NIH Grant NS-06588, U.S. Public Health Service. The technical assistance of John Matthews and Richard Macklin is gratefully acknowledged.

## METHODS

The effect of bulboreticular stimulation was tested in the same cats used for unit recording in the previous study (Casey, 1971). An additional 5 animals, identically prepared, were tested for stimulation effects only. All observations were made with the cat in the observation chamber used for unit recording. A two-way escape paradigm was employed. Histological controls were as previously described (Casey, 1971).

### Data Acquisition

When unit recording was completed in each cat, the microelectrode was left in place at the end of the last electrode track. The effect of cathodal stimulation through the microelectrode (skull screw anodal) was tested at subsequent experimental sessions. Monophasic pulse stimuli were delivered from a constant current source at 100/sec, 0.2 msec duration, beginning with at least 5 15-sec trials at 10μA and increasing the current by 5 to 10μA for subsequent tests until some behavioral effect was observed. Current was monitored continuously via a 10K resistor in series with the electrode. When escape was elicited, at least 15 confirmatory trials were given; intensities of up to 300μA were tested for escape effects. The entire procedure was repeated at several points along the track as the electrode was withdrawn.

## RESULTS

Figure 1 shows the anatomical location of sites at which escape was elicited at various levels of stimulating current; up to 300μA was tested at sites indicated as ineffective.

Stimulation at points near the midline frequently evokes stimulus-bound, unidirectional turning of the head and trunk. At the midline points shown as ineffective, the motor response may prevent escape since increasing current only augments tonic contraction of the involved muscles. Both during and after stimulation, however, there is no natural pain behavior or effort to escape. Similar results are obtained at ineffective sites in the lateral reticular nucleus and in the vestibular nuclear complex. Other behaviors elicited at ineffective sites include: jaw and tongue movement (near hypoglossal nucleus), vomiting (near tractus solitarius), urination and defecation (motor nucleus of vagus), and tonic contraction of facial muscles (facial nucleus).

The effect of escape-eliciting bulboreticular stimulation is immediately generalized, for each cat escapes when the first effective site is tested;

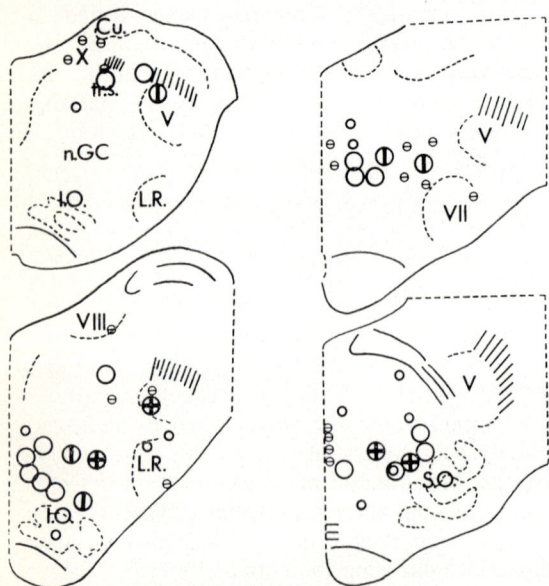

FIGURE 1 *Anatomical distribution of stimulation sites.* Small circles with horizontal bar: neither escape nor natural pain behavior elicited with stimuli up to 300μA. Small open circles: escape elicited with stimuli between 100 and 200μA. Large circles: escape elicited with stimuli below 100μA. Vertical bars indicate escape stimuli between 26 and 50μA; plus signs, 25μA. Cu: external cuneate nucleus; I.O.: inferior olivary nucleus; L.R.: lateral reticular nucleus; n.GC: nucleus gigantocellularis; S.O.: superior olivary nucleus; V: trigeminal nucleus; VII: facial nucleus; VIII: vesticular nucleus; X: motor nucleus of vagus.

no response shaping is required. When escape was elicited, stimulating current was maintained at or near the lowest initially effective value for all subsequent confirmatory trials at that site. At effective points near the midline, stimulus-bound turning is sometimes evoked at currents below that required to elicit escape, but the motor response is sufficiently weak to not interfere with escape execution. Escape latencies range from 3 to 27 seconds during the initial trials at the first point tested and decrease to an average of 8.5 seconds during the last 5 confirmatory trials at each site.

Two types of responses are observed at sites eliciting escape. In one case, stimulation elicits behavior which appears to localize some sensory event. In the nucleus of the solitary tract, for example, escape was accompanied by considerable licking, salivation, and wiping at the mouth. Escape-producing stimuli near the trigeminal nucleus were associated with wiping vigorously at the face. At sites within the reticular formation, however, such localizing responses are not observed. At current intensities just sufficient to elicit escape, ongoing behavior is arrested and the cat orients toward the barrier; escape is executed smoothly and without supervenient behavior (Figure 2). If, as is usually the case, the cat is sitting quietly, stimulus onset is signaled by raising the head and looking about; this is followed by standing, orientation toward the barrier, and escape. Vocalization frequently accompanies this behavior.

Additional observations were used to examine the possible aversive character of bulboreticular stimulation. Two cats, each with the stimulating electrode in the bulboreticular formation, were tested in a classical conditioning paradigm using brain stimulation as the unconditioned stimulus (UCS). The conditioning stimulus (CS) was a 1.0 KHz tone which preceded the UCS by 10 seconds and continued throughout escape execution. Following an initial training session of 10 pairings, 3 additional sessions of 20 pairings each were given to each cat. The training was limited to a total of 70 trials in order to avoid the possible interference of a well learned conditioned response with subsequent observation of stimulation effects at other sites in the same animal. In spite of these limitations, it is clear that both cats were learning to avoid bulboreticular stimulation by crossing the barrier within 10 seconds of CS onset (Figure 3). During the fourth training session, the tone elicited avoidance responses on 55% and 65% of the trials.

A form of operant conditioning was also used in 5 cats, taking advantage of their natural desire to

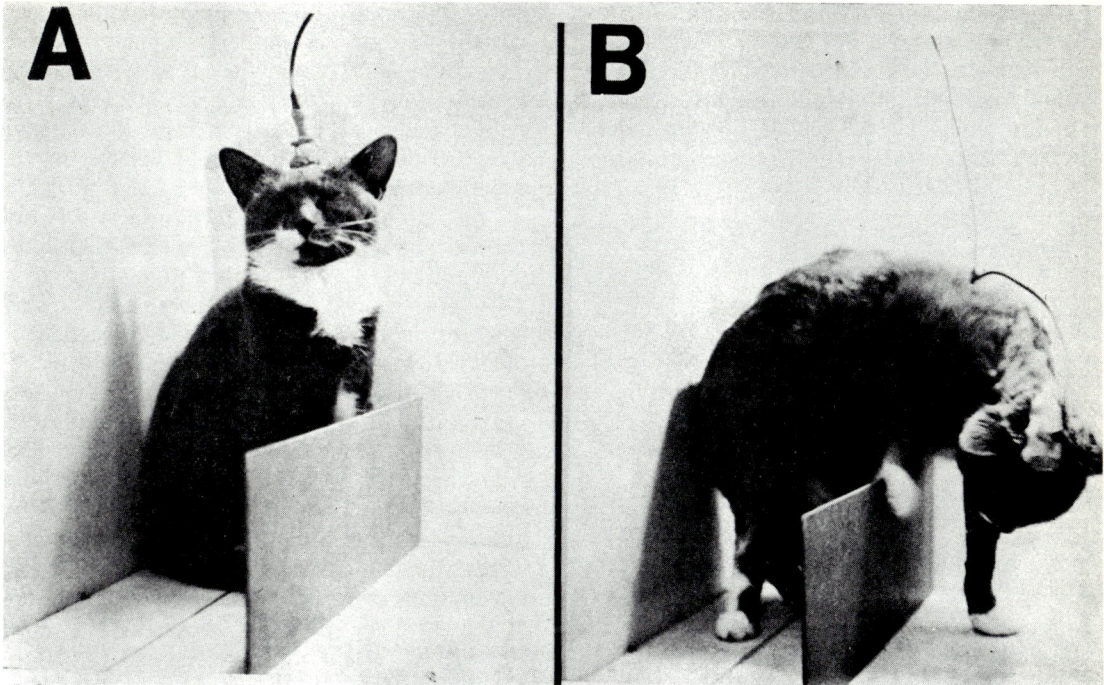

FIGURE 2 *Cat escaping NGC stimulation of* 100μA. A: Before stimulation. B: Six seconds after stimulus onset, cat crosses barrier smoothly and without evident motor impairment. The stimulus is still on in B.

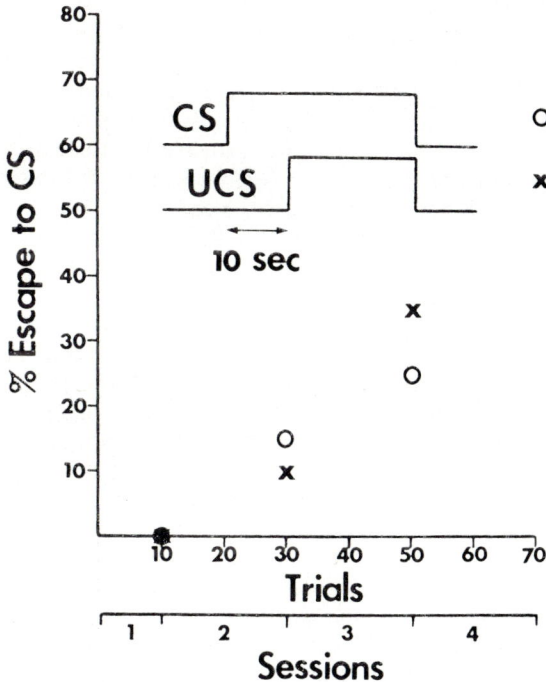

FIGURE 3 *Bulboreticular stimulation used as UCS in conditioning avoidance to 1.0 KHz tone.* UCS intensity of 120μA in one cat (X) and 75μA in the other (0). By the end of 70 trials, delivered in 4 sessions, both cats were avoiding the UCS on over 50% of the trials.

leave the chamber when the door is opened and the experimenter is seated well away from the opening. As each cat extended its head or a forelimb outside the chamber, brain stimulation was applied until the extended part was pulled back. Stimulation of three sites not eliciting escape were ineffective in training the cat to stay in the chamber even though currents up to 300μA were tested. Stimulation at escape-producing sites (5) and intensities, however, were effective in driving the cat back into the chamber within a few seconds of stimulus onset and, after less than 15 stimulus presentations, prevented further attempts to exit for the next 10 minutes.

Although stimulation parameters were maintained at 100 0.2 msec pulses/sec throughout these experiments, strength-duration curves were obtained at representative sites evoking a 'pure' escape response, an escape response including localizing behavior, and a stimulus-bound motor response without escape. These data were obtained in an attempt to characterize some of the electrical properties of the effective neural elements (BeMent and Ranck, 1969a, 1969b). Stimulation near the hypoglossal nucleus, for example, produced slight opening of the mouth and protrusion of the tongue without escape even at higher stimulus intensities.

The strength-duration curve for the onset of this motor response is shown in Figure 4 where it is to be compared with the curves derived from reticular formation points eliciting escape within 15 seconds of stimulus onset. The chronaxie for the motor response is 0.1 msec whereas both escape chronaxies are between 0.2 and 0.3 msec.

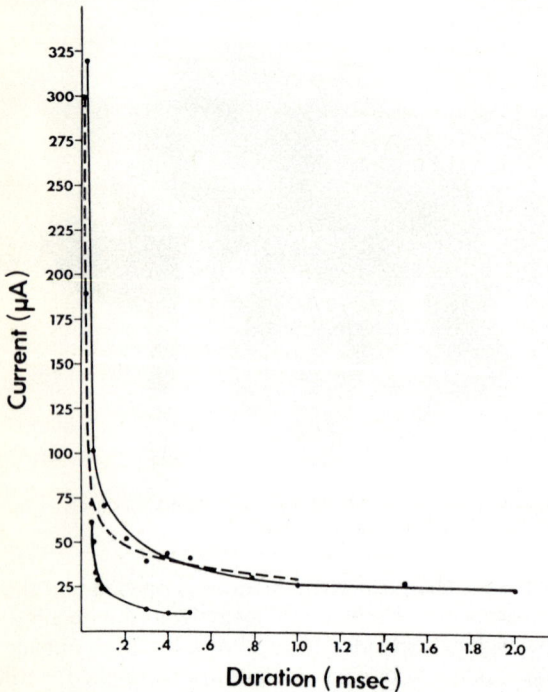

FIGURE 4　*Strength-duration curves using behavioral events.* Two upper curves obtained from different cats, using escape within 15 sec as the event. Cat represented by upper solid line showed 'pure' escape to stimulation in NGC at inferior olivary level. Cat represented by dashed line rubbed face during escape from stimulation in trigeminal nucleus. Lower curve from another cat, using a stimulus-bound motor effect (tongue protrusion), elicited by stimulation near hypoglossal nucleus, as the event.

## DISCUSSION

These results confirm and extend those of a previous study (Keene and Casey, 1970) showing that stimulation of NGC elicits escape in the rat. In the present study, however, monopolar stimulation was employed, so a comparison of the intensities required at each point is not valid.

Although the radial extent of effective stimulating current may be greater when monopolar, rather than bipolar, electrodes are used (Stark, Fazio and Boyd, 1962), it is unlikely that the NGC escape responses are attributable to stimulating currents

outside this area. Recent studies of the effective current radius with monopolar stimulation (Stoney, Thompson and Asanuma, 1968; BeMent and Ranck, 1969a, 1969b) lend support to this contention and provide a useful basis for estimating the distribution of stimulating current in these experiments. Stoney *et al.* (1968) derive an estimate of the strength-distance relationship for intracortical stimulation of cell bodies with 0.2 msec duration pulses. Their estimates, applied to the NGC area, indicate an effective radius of less than $200\mu$ for the lowest threshold ($25\mu$A) escape points obtained; less than $400\mu$ would apply to the sites requiring up to $100\mu$A. A similar estimate is obtained using the data acquired by BeMent and Ranck (1969a) from stimulation of dorsal column fibers. Allowing for the differences in pulse duration and stimulating conditions, stimulation at the lowest threshold escape sites should not excite even the largest, lowest threshold fibers $400\mu$ away; the higher threshold elements which predominate in gray matter are likely to be unaffected beyond $200\mu$.

Unfortunately, the strength-duration curves obtained in this study give no clear indication as to the type of stimulated element responsible for escape. The chronaxies for escape are between 0.2 and 0.3 msec, which, as BeMent and Ranck (1969b) point out, fails to distinguish between stimulation of fibers and cell bodies. The stimulus-bound motor effects, however, may be due to stimulation of myelinated fibers since the chronaxie for this effect obtained near the hypoglossal nucleus was 0.1 msec (BeMent and Ranck, 1969b). Although not specifically tested, the same is likely to be true for the motor effects obtained by stimulation near the medial longitudinal fasciculus.

These results also complement the findings of the previous study (Casey, 1971) for, as shown in Figure 5, there is considerable overlap among the sites eliciting escape and the location of units with escape-related activity. Indeed, the majority of the lowest threshold escape points are located within clusters of such NGC units and, in most instances, these sites are identical to the location of a unit with escape-related activity because the first stimulation trial was given at the final unit recording position in that animal. There are escape points in the trigeminal nucleus where units responding to light tactile stimuli are recorded, but this is not surprising because of the possibility of stimulating first-order somatic sensory afferents. In the context of the previous unit study, then, the findings lend essential support to the

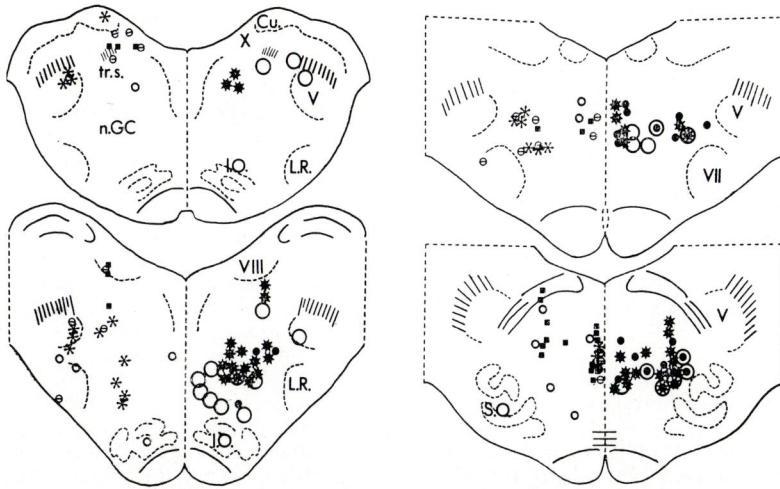

FIGURE 5 *Summary plot of combined stimulation and recording results.* Symbols for stimulation points are the same as in Figure 1 except that large plus signs and vertical bars are omitted. Unit location symbols are taken from the unit analysis of this region in the same cats (Casey, 1971). On the left, ineffective stimulus points and those requiring between 100 and 200μA for escape are shown together with units lacking escape-related activity (asterisks and filled squares). On the right, stimulation sites eliciting escape with 100μA or less are plotted with the location of units showing escape-related activity (stars and filled circles). Unit symbol enclosed by a circle indicates identity of stimulation and recording point. Tractus solitarius: tr.s.

hypothesis that the bulboreticular formation in the NGC area is an integral part of central pain sensory mechanisms.

Direct stimulation of the brain is, of course, a highly unnatural way of eliciting behavior, and the results must be interpreted with caution. In parcular, one can never be certain of the significance of such stimuli so far as sensory experience is concerned. This raises the question as to whether or not escape behavior was elicited by internal cues which were not aversive and in no way related to somatic pain. A number of observations bear on this point. First, it should be noted that a stimulus-bound motor effect alone fails to elicit either escape or natural pain behavior. The autonomic responses evoked at some points may be aversive, but this could not be tested because of the accompanying motor activity; the same is true of stimulation near the midline brainstem. However, when motor effects do not impede movement, there is no evidence of pain response at ineffective sites. Second, at points eliciting escape, there is often vocalization and, at some points, withdrawal, biting, or rubbing of some localized part of the body. This is consonant with the natural pain behavior observed during cutaneous nerve stimulation at intensities sufficient to elicit escape. Third, the effect of NGC stimulation is immediately generalized so that it results in a behavior which had previously been uniquely associated with somatic pain. There is little reason to suspect that a non-aversive experience would have this effect since there was no evidence that avoidance responses had developed at any time during the experiments. Finally, the limited trials of classical conditioning were sufficient to show that NGC stimulation could serve as an unconditioned stimulus. This is further supported by the effectiveness of such stimuli in shaping operant behavior.

Taken together, then, the results of both unit recording and brain stimulation indicate that the activity of a neural population in the NGC area generates behavior operationally identified with pain. The lack of localizing responses elicited by NGC stimulation, in agreement with the lack of somatotopic organization in this area, suggests a motivational, rather than a spatial discriminative, role in pain sensory mechanisms. In this regard, this bulboreticular area might be considered part of a system with somatic afferents (Casey, 1969) and thalamic projections (Bowsher *et al.*, 1968; Mitchell and Kaelber, 1966) together comprising the neural determinants of the motivational dimension of pain while the discriminative somesthetic functions are subserved by separate and parallel connections to the ventro-basal thalamus and primary somesthetic cortex (Melzack and Casey, 1968).

## REFERENCES

BeMent, S. L., and Ranck, J. B., Jr., 1969a, A quantitative study of central myelinated fibers. *Exp. Neurol.* **24:** 147–170.

BeMent, S. L., and Ranck, J. B., Jr., 1969b, A model for electrical stimulation of central myelinated fibers with monopolar electrodes. *Exp. Neurol.* **24:** 171–186.

Bowsher, D., Mallart, A., Petit, D., and Albe-Fessard, D., 1968, A bulbar relay to the centre median. *J. Neurophysiol.* **31:** 288–300.

Casey, K. L., 1971, Responses of bulboreticular units to somatic stimuli eliciting escape behavior in the cat. *Intern. J. Neuroscience.* **2:** 15–28

Casey, K. L., 1969, Somatic stimuli, spinal pathways, and size of cutaneous fibers influencing unit activity in the medial medullary reticular formation. *Exp. Neurol.* **25:** 35–56.

Halpern, B. P., and Halverson, J. D., 1967, Elevated escape latencies after hindbrain lesions. *Physiologist.* **10:** 193.

Keene, J. J., and Casey, K. L., 1970, Excitatory connection from lateral hypothalamic self-stimulation sites to escape sites in medullary reticular formation. *Exp. Neurol.* **28:** 155–166.

Melzack, R., and Casey, K. L., 1968, Sensory, motivational, and central control determinants of pain. In: The Skin Senses, Kenshald, D. R., ed. C. C. Thomas, Springfield.

Mitchell, C. L., and Kaelber, W. W., 1966, Effect of medial thalamic lesions on responses elicited by tooth pulp stimulation. *Am. J. Physiol.* **210:** 263–269.

Olds, M. E., and Olds, J., 1963, Approach-avoidance analysis of rat diencephalon, *J. Comp. Neurol* **120:** 259–295.

Stark, P., Fazio, G., and Boyd, E. S., 1962, Monopolar and bipolar stimulation of the brain. *Am. J. Physiol.* **203:** 371–373.

Stoney, S. D., Jr., Thompson, W. D., and Asanuma, H., 1968, Excitation of pyramidal tract cells by intracortical microstimulation: Effective extent of stimulating current. *J. Neurophysiol.* **31:** 659–669.

# MODIFICATIONS OF DORSAL ROOT POTENTIALS DURING RESPIRATORY ACIDOSIS

JEAN-MARIE BESSON, JEAN-MICHEL BENOIST and PIERRE ALEONARD

*Laboratoire de Physiologie des Centres Nerveux*

*Faculté des Sciences—4, avenue Gordon-Bennett Paris* (16°)

The depressant action of $CO_2$ on various spinal reflexes has been emphasized by many authors. The present study concerns the effects of respiratory acidosis on various types of lumbar dorsal root potentials (DRP).

Inhalation of a gas mixture containing 10–20% $CO_2$ is followed by: (a) a constant increase in the amplitude of DRPs of segmental and cortical origin; (b) variable changes (most frequently diminution) in the amplitude of heterosegmental DRPs.

The magnitude of the changes induced by a given concentration of inhaled $CO_2$ varies greatly from one preparation to another.

Various mechanisms are discussed which may explain the increase in the DRP during respiratory acidosis.

There have been many studies of the effect on the central nervous system of the inhalation of gas mixtures rich in carbon dioxide ($CO_2$) (see references in Woodbury and Karler, 1960, and Wyke, 1963). The general conclusion from these works is that the effects of $CO_2$ on the CNS depend on the $CO_2$ tension of the inhaled gas. On arousal mechanisms, first studied by Bremer and Thomas (1936), it has been shown that the inhalation of low concentrations of $CO_2$ results in arousal (Dell and Bonvallet, 1954; Bonvallet *et al.*, 1955), while greater concentrations have a depressant action (Gellhorn, 1953; Woodbury *et al.*, 1958). Different structures, however, show different sensitivity. These results have received some support from recent work on the effects of $CO_2$ on synaptic transmission in the cuneate nucleus (Morris, 1969a and b).

Inhalation of $CO_2$ in a concentration of 10%, or more, generally has a depressant action on the spinal cord, and depression of the flexor reflex (Glazer, 1929) and the patellar reflex (Brooks and Eccles, 1947; King *et al.*, 1932) have been recorded. King *et al.* (1932) showed that spinalisation reduced the sensitivity of the patellar reflex to depression by $CO_2$ and attributed the cord effects chiefly to inhibition arising from higher centres more sensitive to $CO_2$ action. In fact, the $CO_2$ induced

modification of spinal electrical activity is often minimal, for example Krnjevic *et al.* (1965) found no conclusive changes in the resting potentials of motoneurones nor in the activity of Renshaw cells even with inspired tensions of 20% $CO_2$. However, Papajewski *et al.* (1969) described a reduction in motoneurone excitability and changes in membrane polarization which varied in different cells. In addition, Bradley *et al.* (1950), Kirstein (1951) and Don Esplin and Rosenstein (1963) found that in the spinal cat $CO_2$ inhalation reduced the monosynaptic reflex considerably and the polysynaptic reflex to a lesser extent, which suggests that $CO_2$ has a purely spinal action.

We have studied the effects of inhalation of gas mixtures containing 10 to 20% $CO_2$ on dorsal root potentials (DRP). It is known from the work of Wall (1958) and Eccles (1964) that the DRP reflects changes in polarization of primary afferents and is an index of presynaptic inhibition. These phenomena are widespread at the spinal level where various types of presynaptic control of the afferent volley have been shown.

We recorded three types of DRP from the lumbar cord:

1) potentials of segmental origin (DRPS) from stimulation of the ipsilateral hindlimb; (Barron and Matthews, 1938)

2) heterosegmental potentials (DRPHseg) from stimulation of a forelimb. These have been

described by Bergmans *et al.* (1964) in the frog and in the cat under chloralose by Mallart (1965);

3) potentials induced by stimulation of certain cortical areas (DRPC) (Andersen *et al.*, 1964; Carpenter *et al.*, 1963; Abdelmoumène *et al.*, 1970).

## METHODS

Nineteen experiments were carried out on cats of 2 to 3 kg weight, fifteen under chloralose anaesthesia (80 mg/kg i.v.) and four using pentobarbitone (30 mg/kg i.v.). The preparations were immobilised with Gallamine and artificially ventilated. The surgical procedures, homeostatic control and recording techniques were as previously described (Besson and Abdelmoumène, 1970). Peripheral transcutaneous stimuli were delivered to the limbs between pairs of needles using 0.3 msec 7 V pulses and the sensorimotor cortex was stimulated with trains of 4 to 5 pulses of 0.3 msec duration at 200–300 c.p.s. and 5–6 V amplitude. The $CO_2$ tension of the expired gas was measured with a Godart $CO_2$ analyzer. The ventilation was adjusted to give initial levels of 4 to 4.5% $CO_2$. Mixtures of $CO_2$ with oxygen were prepared in a spirometer and delivered to the animal using a Palmer respiration pump. Concentrations of 10, 15 and 20% $CO_2$ were administered for an average of 8 minutes.

## RESULTS

### 1. General

Inhalation of $CO_2$-enriched gas was followed by increases in the amplitude of dorsal root potentials induced by segmental or cortical stimulation (Figure 1). This effect appeared to be fairly consistent and was absent in only one preparation of the nineteen. However, the action of $CO_2$ on heterosegmental potentials was less consistent; in about 75% of cases the amplitude was reduced as shown in Figure 1, while in the remainder it was increased. The occurrence of these changes under both chloralose and pentobarbitone suggests that they do not depend on the type of anaesthetic. Only cortical and segmental potentials were examined under barbiturate anaesthesia, since this is known to abolish heterosegmental potentials (Besson and Abdelmoumène, 1970; Besson *et al.*, 1971). The increase in amplitude of segmental potentials persists after spinal section at $C_1$.

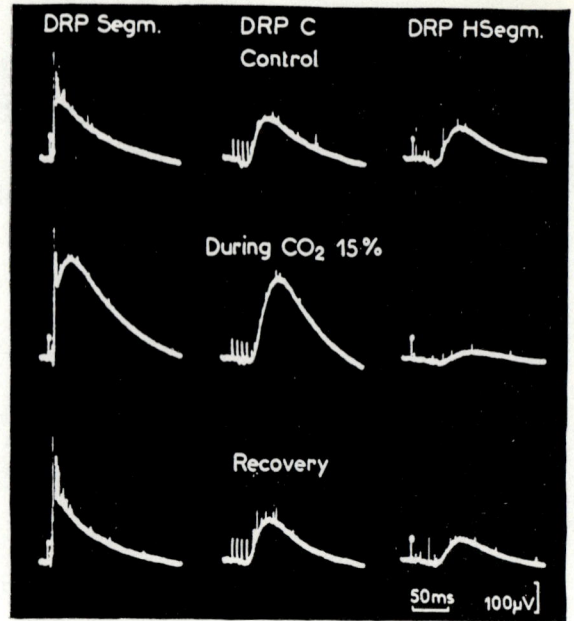

FIGURE 1    Effects of inhalation of a mixture of $CO_2$ 15% and $O_2$ 85% on DRPs of various origin recorded at L7: DRPS stimulation of ipsilateral hindlimb DRPC stimulation of contralateral sensorimotor cortex; DRPHseg stimulation of ipsilateral forelimb.

In contrast to the changes in amplitude, the time-course of the DRP was practically unaffected by $CO_2$.

Although the dorsal root reflex (DRR) was not studied systematically, it did not seem to undergo any significant change, except in two cases in which it was considerably reduced in amplitude and in one case where we observed a slight increase.

We found that the changes in DRP were not dependent on the alterations in arterial pressure which occur during $CO_2$ inhalation. This is in agreement with the results of Bradley *et al.* (1950) who showed that the $CO_2$ induced depression of mono- and polysynaptic reflexes was not related to arterial pressure changes.

### 2. Differences in sensitivity between animals

Although the increase of amplitude of the DRPS and DRPC during respiratory acidosis was a constant and repeatable finding there was considerable variation between animals in the magnitude of the changes. Figure 2 demonstrates this variation in 5 separate experiments (in different animals) during inhalation of an identical mixture

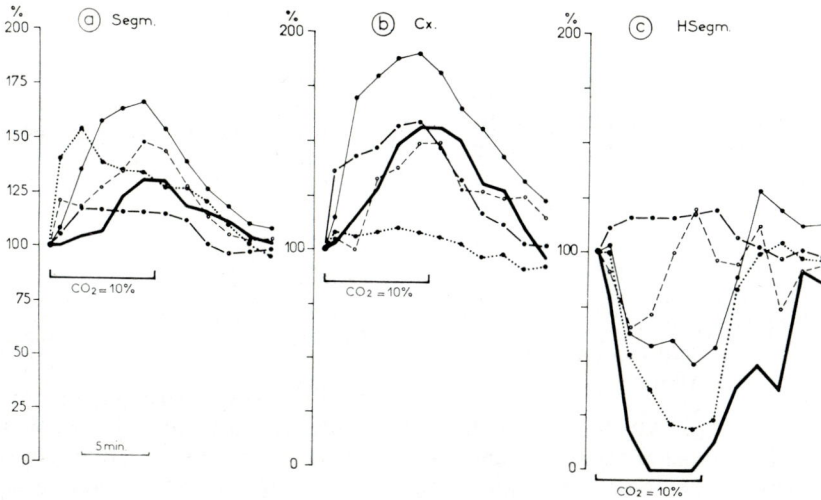

FIGURE 2  Variations in the amplitude of DRPs in different preparations during inhalation of a mixture of $CO_2$ 10% and $O_2$ 90%. DRP amplitude is expressed as a percentage of its initial value.

containing 10% $CO_2$ and 90% $O_2$. The amplitude of the DRP is shown as a percentage of its initial value and the time course of its alterations is plotted. The DRPS (Figure 2a) and DRPC (Figure 2b) increased in each case but the magnitude of the increase varies considerably between preparations. As mentioned above, the DRPHseg sometimes increased and sometimes diminished in amplitude, in one of the five cases shown there was an increase, while all the others showed decreases of varying magnitude. Sometimes the reduction in DRPHseg followed a brief increase in amplitude (Figure 3c) and might be followed by a rebound increase later during inhalation of $CO_2$ or after the end of this inhalation (Figure 2c and 3c).

3. *Effects of varying the concentration of inhaled $CO_2$*

In a given preparation the effects on the amplitude of the DRP depend on the $CO_2$ tension of the inhaled gas mixture. As the $CO_2$ tension increases from 10% to 20%, the following were usually observed:

(a) a parallel increase in DRPS (Figure 3a);
(b) changes in the DRPC which may increase further, or be reduced (Figure 3b), or remain unchanged (Figure 3b);
(c) a progressive reduction in the amplitude of DRPHseg, although sometimes this reduction appears only at the highest tensions of $CO_2$ used (20%) as in Figure 3c.

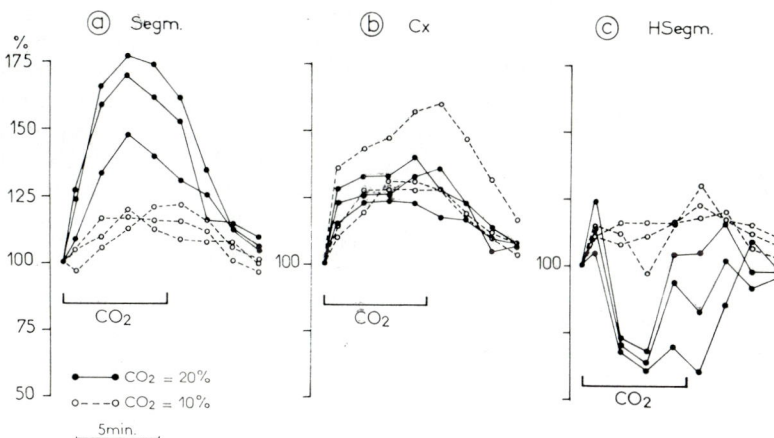

FIGURE 3  Changes in DRP amplitude as a function of the $CO_2$ tension of the inhaled gas mixture

### 4. Time course of the variations

There was considerable variation between animals inhaling the same $CO_2$ concentration in the latency of the maximum change in the DRP and in the time required after inhalation of $CO_2$ had ceased to regain the control state (Figure 2). Sometimes the time course was similar for different concentrations in the same animal (see Figure 3), but this was not always the case. However, during any single inhalation period, the latencies of the maximum effects on the three types of potential were comparable no matter what the direction of the effect. This is shown in the two examples of Figure 4.

1. (Figures 4a and b). All three potentials, segmental, cortical and heterosegmental are increased. The slopes of the three curves and the latencies of the maxima are superposable for each

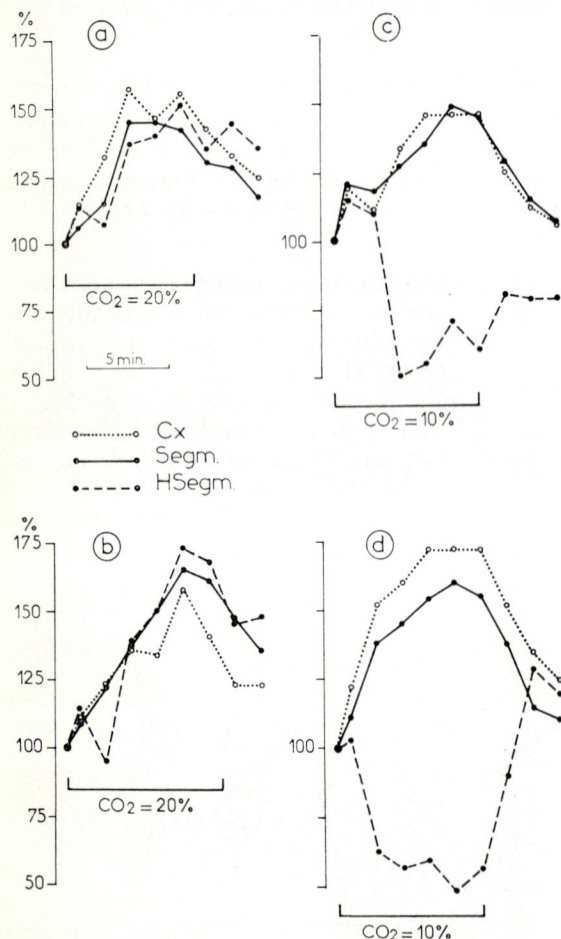

FIGURE 4  Time course of the changes in amplitude of DRPs during $CO_2$ inhalation (see explanation in text).

inhalation period; however, they vary from one inhalation to another in the same animal.

2. (Figures 4c and d). The segmental and cortical potentials increase, while the heterosegmental diminish. The latencies of the maxima are comparable for the three types, but they vary from one inhalation to another.

### DISCUSSION

Inhalation of a gas mixture containing 10–20% $CO_2$ increases the amplitude of the DRPS and DRPC in intact animals under chloralose or pentobarbitone anaesthesia. In addition, the increase in the DRPS is found in the spinal animal, suggesting that some modification has occurred of activity at the segmental level, independently of any change in descending control systems.

The variability of effects of $CO_2$ inhalation on the heterosegmental DRP does not conflict with the constant increase found in the DRPS and DRPC.

Besson and Abdelmoumène (1970), Besson et al. (1971) have shown that the DRPHseg probably depends in part on activation of a supraspinal loop and dependence on this more complex mechanism probably accounts for the divergent results obtained during different experiments using various concentrations of inspired $CO_2$. It seems probable that, in the concentrations which we used, $CO_2$ depresses transmission in the supraspinal circuit, or circuits, and thus leads to a reduction in the descending activity which is essential to the production of the DRPHseg. The following observations lend support to this hypothesis: diminution in the DRPHseg occurred particularly with high $CO_2$ concentrations and in preparations with a low-voltage electrocorticogram and infrequent spikes while increases were usually seen in animals with many spikes in the electrocorticogram. Similarly, the early increase in DRPHseg which often occurred before depression or after the end of $CO_2$ inhalation may well be due to the known excitatory effects of low concentrations of $CO_2$ (Dell and Bonvallet, 1956; Morris, 1969a and b).

In contrast, we never found reduction of the DRPC. As the cortical stimuli were supramaximal and presumably recruited activity in the descending paths (which are to a large extent pyramidal, Carpenter et al., 1962), it appears that this recruitment is not markedly affected by relatively large inspired $CO_2$ concentrations in spite of the diminished cortical excitability reported by several workers (Pollock, 1949; Gellhorn, 1953; Woodbury *et al.*, 1958). We conclude, therefore, that the increase which is found in the DRPC depends upon a spinal mechanism which is influenced by $CO_2$ in the same way as the mechanism of the segmental DRP.

We shall now consider the various hypotheses which we believe may explain the segmental mechanisms concerned in the increase of DRP. The negative wave of the DRP is believed to be caused by depolarization of the primary afferent fibres (Wall, 1958; Eccles, 1964), which, in turn, is due to activity in a network of interneurones situated perhaps in the substantia gelatinosa of Rolando (Wall, 1962, 1964) or perhaps at a deeper level (Eccles *et al.*, 1962). According to the hypothesis of Mendell and Wall (1964), this interneuronal network is in a state of 'tonic' activity produced by afferents, especially those of cutaneous origin, and thus the primary afferents are kept constantly depolarized. The positive waves which are found in the DRP under certain conditions are explained by a relaxation of this 'tonic' activity which results in hyperpolarization of the afferents.

The increases in amplitude of the DRPS and DRPC which we found under $CO_2$ might be explained in various ways:

1. $CO_2$ might increase the activity of the interneuronal network brought into action by segmental or cortical stimulation and thus increase the depolarization generated by these interneurones. This hypothesis seems to be untenable for the following reasons:

(a) we used supramaximal stimuli and it is probable that recruitment of interneurons was already maximal before $CO_2$ was administered;

(b) in the concentrations which we used, $CO_2$ is generally believed to depress neuronal activity, (Brooks and Eccles, 1947; Lorente de No, 1947; Shanes, 1948; Washizu, 1960; Monnier, 1962; Carpenter, 1963; Krnjevic *et al.*, 1965; Morris, 1969; Papajewski *et al.*, 1969).

2. $CO_2$ might cause permanent hyperpolarization of the primary afferent terminals.

The second hypothesis is more compatible with the generally accepted effects of $CO_2$ on the membrane potential. If the second hypothesis proposed is correct, two mechanisms, which are not, however, mutually exclusive, might be involved in the hyperlarization of the primary afferent terminals; a

direct effect due to a permanent change in the membrane properties of the primary afferents or an indirect effect resulting from blocking of 'tonic' activity in the interneuronal network involved in the production of the DRP which would in turn in relative hyperpolarization of the terminals. A point against such an indirect effect is the absence of any significant change in the duration of the DRPS during $CO_2$ inhalation, as this suggests that activity of the interneuronal network which produces PAD is not involved in the changes which we found.

In this connection, the observations of Morris (1969) on the activity of the terminals of dorsal column fibres in the cuneate nucleus are interesting. Inhalation of $CO_2$ (10–20%) reduced the amplitude of the antidromic potential recorded from a peripheral nerve following stimulation of the cuneate nucleus through a microelectrode. The author attributed this reduction to decreased excitability of the primary afferent terminals of the cuneate fasciculus. However, although Morris stressed the importance of these presynaptic phenomena, she did not say whether she attributed the changes to hyperpolarization.

We had thought that observations on the dorsal root reflex (DRR) might provide information on the time course of the changes in membrane potential and in the excitability of primary afferents which occur during $CO_2$ inhalation. If $CO_2$ hyperpolarizes afferent terminals and reduces their excitability we ought to find a reduction in the DRR as well as an increase in the amplitude of the DRP. Indeed, these were the findings in several experiments but, in most of the others, there was no apparent change in the amplitude of the DRR.

While we have no direct evidence of hyperpolarization of primary afferents during $CO_2$ inhalation, the increases in the amplitude of the DRPS and DRPC do suggest that such hyperpolarization may occur. This would also be consistent with the results of Morris (1969a and b) who found diminished excitability of afferents in the cuneate nucleus during $CO_2$ inhalation, and those of Lorente de No (1947) who found an increase in the membrane potential of peripheral fibres. Krnjevic *et al.* (1965) have reported similar findings on cortical cells. Conclusive evidence on the presence of hyperpolarization will require intrafibre recording from the primary afferents during $CO_2$ inhalation.

We thank Dr. Robert Burke from N.I.M.H. for some suggestions.

## SUMMARY

1) The depressant action of $CO_2$ on various spinal reflexes has been emphasized by many authors. The present study concerns the effects of respiratory acidosis on various types of lumbar dorsal root potentials (DRP).

2) Inhalation of a gas mixture containing $10-20\%$ $CO_2$ is followed by:
 (a) a constant increase in the amplitude of DRPs of segmental and cortical origin;
 (b) variable changes (most frequently diminution) in the amplitude of heterosegmental DRPs.

3) The magnitude of the changes induced by a given concentration of inhaled $CO_2$ varies greatly from one preparation to another.

4) In any one preparation increasing inspired $CO_2$ concentration is associated with further increase in amplitude of the segmental DRP and diminution of the heterosegmental DRP, while the DRP of cortical origin may increase further or diminish.

5) The depressant effect which is common on the heterosegmental DRP is not inconsistent with the augmenting actions on segmental and cortical DRPs, but is probably due to a depressant action on the supraspinal mechanisms of the heterosegmental DRP.

6. Various mechanisms are discussed which may explain the increase in the DRP during respiratory acidosis.

## REFERENCES

Abdelmoumène, M., Besson, J. M. and Aléonard, P., 1970, Cortical areas exerting presynaptic inhibitory action on the spinal cord in cat and monkey. Brain Res., 20: 327–329.

Andersen, P., Eccles, J. C. and Sears, T. A., 1964, Cortically evoked depolarization of primary afferent fibers in the spinal cord. J. Neurophysiol., 27: 63–77.

Barron, D. H., and Matthews, B. H. C., 1938, The interpretation of potential changes in the spinal cord. J. Physiol., 92: 276–321.

Bergmans, J., Colle, J. et Lafere, Y., 1964, Etude électrophysiologique des relations entre les membres antérieurs et postérieurs chez la grenouille. J. Physiol. (Paris), 56: 290–291.

Besson, J. M., and Abdelmoumène, M., 1970, Modifications of dorsal root potentials during cortical seizures. Electroenceph. clin. Neurophysiol., 29: 166–172.

Besson, J. M., Rivot, J. P., and Aléonard, P., 1971, Action of picrotoxin on presynaptic inhibition of various origins in the cat's spinal cord. Brain Res., 26: 212–216.

Bonvallet, M., Hugelin, A., et Dell, P., 1955, Sensibilité comparée du système réticulé activateur ascendant et du centre respiratoire aux gaz du sang et à l'adrénaline. J. Physiol. (Paris) 47: 651–654.

Bradley, K., Schlapp, W., and Spaccarelli, G., 1950, The effect of carbon dioxide on the spinal reflexes in decapited cats. J. Physiol. 111: 62P.

Bremer, F., and Thomas, J., 1936, Action de l'anoxémie, de l'hypercapnie et de l'acapnie sur l'activité électrique du cortex cérébral, C.R. Soc. Biol. 123: 1256–1261.

Brooks, C. Mc-C., and Eccles, J. C., 1947, A study of the effects of anaesthesia and asphyxia on the monosynaptic pathway through the spinal cord. J. Neurophysiol. 10: 349–360.

Carpenter, F. G., 1963, The stabilizing action of carbon dioxide on peripheral nerve fibers and on the neurons of the medullary reticular formation in the rat. Ann. N.Y. Acad. Sci. 109: 480–493.

Carpenter, D., Lundberg, A., and Norrsell, U., 1962, Effects from the pyramidal tract on primary afferents and on spinal reflex actions to primary afferents. Experientia (Basel) 18: 337–338.

Carpenter, D., Lundberg, A., and Norrsell, U., 1963, Primary afferent depolarization evoked from the sensory motor cortex, Acta Physiol. Scandinav. 59: 126–142.

Dell, P., et Bonvallet, M., 1954, Contrôle direct et réflexe de l'activité du système réticulé activateur ascendant du tronc cérébral par l'oxygène et le gaz carbonique du sang, C.R. Soc. Biol. 148: 855–858.

Dell, P., et Bonvallet, M., 1956, Mise en jeu des effets de l'activité réticulaire par le milieu extérieur et le milieu intérieur. In XXst International Physiological Congress (Brussels), Abstracts of Reviews, pp. 286–306.

Eccles, J. C., 1964, Presynaptic inhibition in the spinal cord. In: Progress in Brain Research, Vol. XII, Physiology of spinal neurons, Eccles, J. C., and Schadé, J. P. eds., Elsevier, Amsterdam, pp. 66–91.

Eccles, J. C., Kostyuk, P. G., and Schmidt, R. F., 1962, Central pathways responsible for depolarization of primary afferent fibres. J. Physiol (Lond.) 161: 237–257.

Esplin, D. W., and Rosenstein, R., 1963, Analysis of spinal depressant actions of carbon dioxide and acetazolamide. Arch. int. Pharmacodyn. 143: 498–513.

Gellhorn, E., 1953, On the physiological action of carbon dioxide on cortex and hypothalamus. Electroenceph. clin. Neurophysiol., 5: 401–413.

Glazer, W., 1929, Regulation of respiration. The effects of mechanical asphyxia and administration of carbon dioxide, sodium carbonate, sodium bicarbonate and sodium cyanide on the reflex response of the anterior tibial muscle of the dog. Amer. J. Physiol. 88: 562–569.

King, C. E., Garrey, W. E., and Bryan, W. R., 1932, The effect of carbon dioxide, hyperventilation and anoxemia on the knee jerk. Amer. J. Physiol. 102: 305–318.

Kirstein, L., 1951, Early effects of oxygen lack and carbon dioxide excess on spinal reflexes. Acta Physiol. Scandinav., 23 (suppl. 80): pp. 54.

Krnjevic, K., Randic, M., and Siesjö, B. K., 1965, Cortical $CO_2$ tension and neuronal excitability. J. Physiol., 176: 105–122.

Lorente de No, R., 1947, A study of nerve physiology, Chapter III, Carbon dioxide and nerve function, Stud. Rockefeller Inst. med. Res. 131: 148–193.

Mallart, A., 1965, Heterosegmental and heterosensory presynaptic inhibition, Nature 206: 719–720.

Mendell, L. M., and Wall, P. D., 1964, Presynaptic hyperpolarization: A role for fine afferent fibers. *J. Physiol. (London)* **172**: 274–294.

Monnier, A. M., 1962, L'anhydride carbonique régulateur spécifique de l'activité nerveuse. *Arch. Int. Pharmacodyn.* **65**: 189–198.

Morris, M. E., 1969, The effects of hypercarbia on synaptic transmission in the cuneate nucleus. *J. Physiol. (London)* **205**: 27P.

Morris, M. E., 1969, The effects of respiratory acidosis on a sensory relay system, *Can. Anaes. Soc. J.* **16**: 494–507.

Papajewski, W., Klee, M. R., and Wagner, R. A., 1969, The action of raised $CO_2$ pressure on the excitability of spinal motoneurones. *Electroenceph. clin. Neurophysiol.* **27**: 618.

Pollock, G. H., 1949, Central inhibitory effects of carbon dioxide (Felis domesticus) *J. Neurophysiol.*, **12**: 315–324.

Shanes, A. M., 1948, Metabolic changes of the resting potential in relation to the action of carbon dioxide. *Amer. J. Physiol.* **153**: 93–108.

Wall, P. D., 1958, Excitability changes of primary afferents in relation with slow cord potentials. *J. Physiol. (London)* **142**: 1–22.

Wall, P. D., 1962, The origin of a spinal cord slow potential. *J. Physiol. (London)* **164**: 508–526.

Wall, P. D., 1964, Presynaptic control of impulses at the first central synapse in the cutaneous pathway. In: *Progress in Brain Research, Vol. XII*, Physiology of spinal neurons, Eccles, J. C., and Schadé, J. P. eds., *Elsevier, Amsterdam*, pp. 92–118.

Washizu, Y., 1960, Effect of $CO_2$ and pH on the responses of spinal motoneurons. *Brain and Nerve* **12**: 757–766.

Woodbury, D. M., and Karler, R., 1960, The role of carbon dioxide in the nervous system. *Anesthesiology*, **21**: 686–703.

Woodbury, D. M., Rollins, L. T., Gardner, M. D., Hirshi, W. L., Hogan, J. R., Rallison, M. L., Tanner, G. S., and Brodie, D. A., 1958, Effects of carbon dioxide on brain excitability and electrolytes, *Amer. J. Physiol.* **192**: 79–90.

Wyke, B., 1963, Brain function and metabolic disorders: the neurological effects of changes in Hydrogen ion concentration. Butterworths ed., London, pp. 242.

# CORTICAL PROJECTIONS OF CAT MEDIAL THALAMIC CELLS

D. ALBE-FESSARD, A. LEVANTE and R. ROKYTA

*Physiologie des Centres Nerveux*
*Faculté des Sciences—4, av. Gordon-Bennett Paris 16°*

In order to clear up the long debated question of the medial thalamic projections to the cortex, it was necessary to explore the structures at the unitary level and to make use of antidromic cell activations. The present work has made it obvious.

The principal fact that has been acquired by this technique is the demonstration of a direct connection between the medial thalamus and certain cortical areas. However, this connection seems to arise only from the anterior part of the CM-Pf complex, with the majority of its cells of origin situated in the nucleus centralis lateralis. Such a connection can explain the relatively short latency observed when stimulations are applied to the CM-Pf nucleus and evoked potentials are recorded at the cortical level from the same sites at which stimulating electrodes have been placed in the present experimental series (Albe-Fessard & Rougeul, 1957). However, in this former work, such short latencies (0.5 ms) as those in some antidromic experiments could never be found. We suggest that it is because well localized bipolar electrodes were then applied at Ant 7.5, in the most posterior medial thalamic zones, and that from there connections to cortex are made through a final relay situated in anterior CM-Pf or n. centralis lateralis.

The cells of the thalamic region that in the cat include the nuclei parafascicularis, centrum medianum, centralis lateralis and paracentralis can be driven by somatic impulses coming from broadly extensive zones of the body teguments. Moreover, the recovery time of evoked activities, when a double shock procedure is applied to the periphery, shows great similarities at the thalamic level to the ones obtained in bulbar and mesencephalic regions as well as in cortical convergent areas (Albe-Fessard et Rougeul, 1957; Buser et Bignall, 1967). The existence of these similarities, together with the fact that stimulation applied at the medial thalamic level gives rise to responses with relatively short latencies in the different convergent cortical areas, induced us to propose the hypothesis that this thalamic region may play the role of a relay between reticular regions and these cortical areas.

Furthermore, we had found that a block applied at this level with cooling probes was able to suppress the convergent cortical responses without greatly affecting the primary ones (Albe-Fessard et Fessard, 1963). However, these coolings were efficacious only at relatively low temperatures $(-5°)$, which means that a wide thalamic region needs to be cooled in order to block these pathways.

This effect was denied recently by investigators using either local destructive lesions (Bignall, 1967; Buser et Bignall, 1967) or localized cooling (Condé et al., 1968; Bénita et Condé, 1971).

On the other hand, anatomical techniques had failed to demonstrate cortical projections of the CM-Pf nuclei (see ref. in Rose and Woolsey 1949; Albe-Fessard et Rougeul, 1957; Bowsher, 1966; Mehler, 1966). For this reason, it has been claimed by anatomists that, if a cortical projection of the medial thalamus does exist, it must be through a relayed thalamo-cortical pathway. However the latency of the premotor cortical responses after stimulation of CM-Pf is so short in the cat as to be compatible with the hypothesis of a direct pathway. Recent anatomical data by Totibadzé and Moniava (1969) support this assumption.

The present state of the question, uncertain as it was, induced us to ask with more refined techniques whether this thalamic region had direct connections with cortical areas or if intermediate subcortical nuclei are necessarily involved.

We then found it appropriate to stimulate convergent cortical areas and Putamen in the same preparation (anatomical relations of CM and Putamen are well known) and to record from the thalamic region that comprizes CM and Pf in order to investigate the eventuality of an antidromic activation. After a short series of preliminary experiments made with macroelectrodes, we realized how difficult it was to detect with this technique responses of a small group of fibers or cells driven antidromically. We then decided to use microelectrodes and made all efforts to identify antidromically provoked activities within medial thalamic structures.

## METHODS

These experiments have utilized 30 cats. In 6 of them, in order to get a better idea of the characteristics of the antidromic cell invasion in a well-known direct pathway, we have stimulated the SI cortex and recorded in the nucleus ventralis posterior.

In the other 24 animals, recording electrodes were placed in the medial thalamus and intralaminar region. In the first period, all animals were anesthetized, either with a volatile agent (ether or Halothane), or with a short-acting anesthetic (Ketamine—intra-muscular injection of 10 mg/kg). All the operative procedures were made under these conditions of anaesthesia; later on, the anesthesia was suppressed, localized analgesics were applied or other anesthetics, like Chloralose, Nembutal, or Brevital (Methohexitone), were injected. 24 cats were recorded under local analgesia, 5 under chloralose, 1 under nembutal. In 6, repeated doses of Brevital and Ketamine (an Arylcycloalkylamine) were injected. All the animals were intubated and maintained under artificial respiratory conditions after injection of a muscular blocking agent (Gallamine). End-tidal $CO_2$ was always monitored and maintained between 3 and 4 per cent. The femoral pressure was measured.

Glass micropipettes of 8 to 12 megohms filled with KCl were introduced into the brain of the cat through a metallic tube stereotaxically implanted and left in place at level +4 or +5 depending on the experiments. DC together with AC coupled amplifiers were used to make intracellular derivations recognizable. As a general rule, all the recordings presented here were extra-juxtacellular.

Cortical stimulations were applied through small silver balls resting on the dura mater. As shown on the photographic representation of the anterior cortex of the cat's brain (Figure 1), 5 of these electrodes were implanted into the region of the pericruciate cortex, 2 on the anterior marginal cortex (a) and 2 on the suprasylvian (b). As we

rapidly realized that stimulations applied between electrodes a and b were always without effect, in the last 20 experiments we only implanted the pericruciate stimulating electrodes. The distance between stimulating electrodes was of the order of 4–5 mm. Closer proximity of the points seems to have to be avoided (see discussion). The silver balls were applied in close contact with the dura mater and the holes made in the bone for their passage closed with dental cement. The introduction and fixation of stimulating electrodes were done one after the other to avoid cerebral pulsations that can be due to the loss of cerebrospinal fluid and correlative entrance of air under the bone. For the same reason, the metallic tube that gives passage to the microelectrode through the bone is implanted last. Finally, in the experiments where Putamen was stimulated, a bipolar electrode was implanted into this structure under electrophysiological control (coordinates: Ant 15–17, Lat 8, H O). Stimulations were applied through the cortical electrodes grouped by pairs (1–1, 2–3, 3–4; 4–5 and 1–5 were used only in a few cases). They were made of single shocks or trains of identical shocks distributed by a Digitimer. The shock duration was 0.2 ms (exceptionally 0.5 msec). The threshold stimulation necessary to elicit antidromic responses was always determined. It was generally of the order of 2.5 to 3 volts; exceptionally intensities up to 9 volts were used. The frequencies of the trains to shocks varied from 100 to 500 per sec.

At the end of all experiments, two electrodes parallel to the recording tracks were placed 2 mm posterior to the most posterior one; and after an increase of the anesthetic level, an intracarotid perfusion was made with formalin (10 per cent). Frozen sections were made and Nissl staining was applied.

The locus of the recorded cell was generally recognized by searching for the deepest level where the tip of the microelectrode could be perceived on a magnified image of the sections. Then the coordinates which had been read on the microdriver during the experiment were corrected and placed on the track. In three cases, before starting the perfusion, the microelectrode was replaced by a fine bipolar electrode of which the depth of the tip is more accurately recognizable on the sections (case of figure 8).

## RESULTS

Successful experiments have shown that it is possible to obtain antidromic cellular activations of

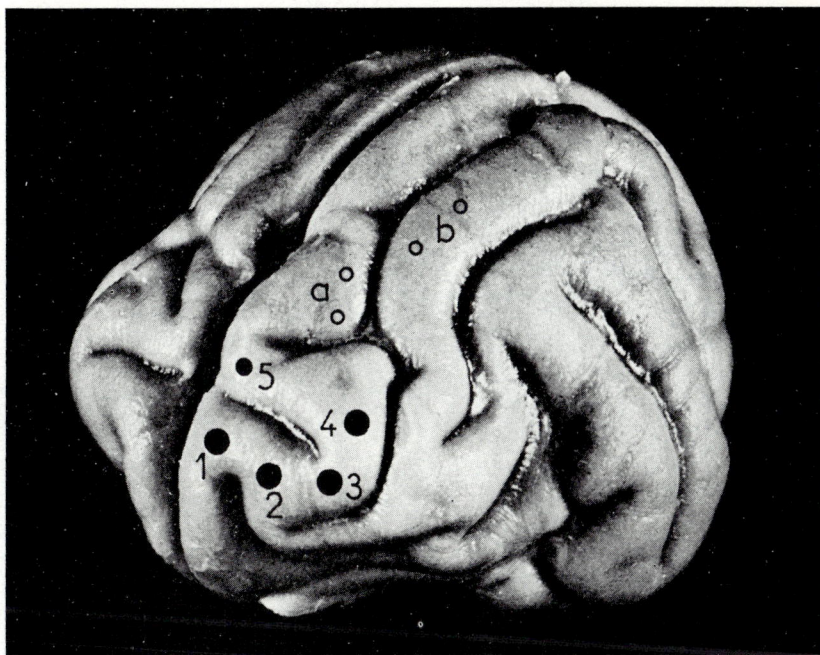

FIGURE 1 On a photographic anterolateral view of a Cat cortex the positions of the stimulation points are presented: in pericruciate regions (1 to 5), anteromarginal (a), suprasylvian (b), gyri (pairs of stimulation points). Points 1 to 4 were utilized in all experiments; point 5 and couples a and b only in some.

the medial thalamus by cortical stimulations. We are going to give here the criteria that permit one to recognize indisputably the antidromic origin of these activations, followed by a description of the localization of these cells in the medial thalamus and their physiological properties.

## 1. Characteristics of an antidromic activation of thalamic cells

The conditions of stimulation that have to be fulfilled to evoke antidromic responses in cells of the medial thalamus and of nucleus ventralis posterior (VP) were compared. They proved to be very similar. This preliminary evidence being given, we shall describe these properties on cells recorded at the level of the limits between anterior CM and CL nuclei (see later for their exact localization), a place where they are more easily detected.

At this level, in an animal under local analgesia, the cells generally have a permanent spontaneous activity with a relatively slow frequency rate (10–20/sec). A stimulation made of one shock applied to certain cortical areas can drive these cells in three different ways. It may give rise either to a spike with a very short latency, to a train of spikes of longer latency, or, in some cases, to a

long slow wave that in juxta- or intracellular recording reveals itself to be the sign of hyper-polarization.

a) A cell which responds with a short latency spike can also respond with long latencies and, in the latter case, be the site of an excitatory or an inhibitory process. The former association is the most frequent. In this case, repetition of stimulations at high rates respects the spike with the shortest latency but makes the long-latency one disappear (Figure 2).

b) When stimulating shocks are delivered at low frequency (1 per sec for example), the short-latency response frequently does not appear with its whole development. Instead of the spike resembling that in the spontaneous activity, there appears a small diphasic wave (Figure 3) or nothing at all (Figure 4, 1). Stimulations at higher rates, 100 up to 500/sec, set up the normal spike shape and the one-to-one responses (Figure 4).

The appearance of a stable response at high frequencies of stimulation is one more property which differs in antidromic and orthodromic responses. The orthodromic ones, in general, do not follow frequencies higher than 50–100 per sec.

c) In order to recognize safely the antidromic

nature of the responses, we used in addition in most of the cells presenting the above mentioned characteristics, the collision test illustrated by Paintal in his analysis of muscle innervation (1959),

and applied to the central nervous system by Darian-Smith et al. (1963), Gordon and Miller (1966) and many others.

If an impulse of synaptic origin fires a nerve cell, an antidromic activation of the same cell body is impossible during a time which is slightly more than twice the latency of the antidromic impulse (time for the orthodromic impulse to reach the axon terminal + delay for the spike elicited at the level of the axon terminal to reach the cell body + recovery period of the axon terminals). This test of collision was easily applied thanks to the occurrence

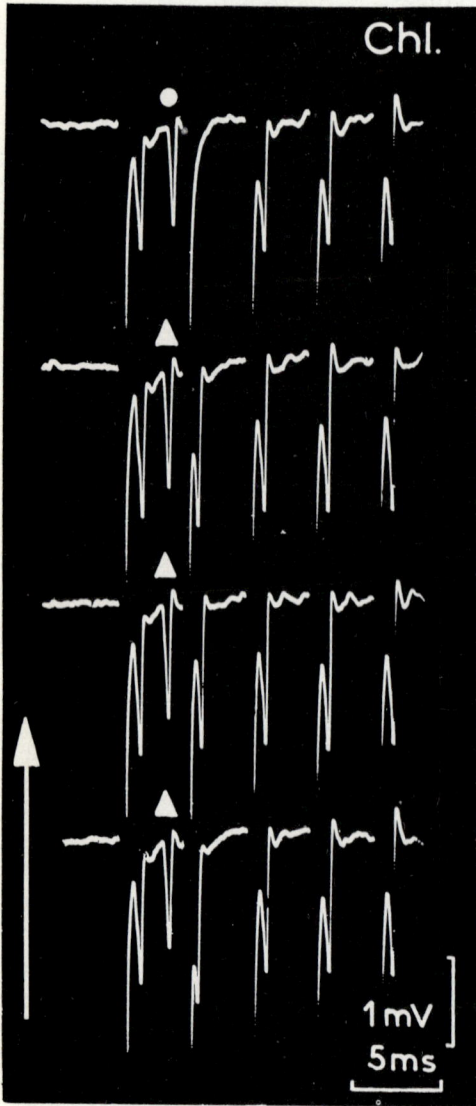

FIGURE 2   Cat under chloralose anesthesia.
Cell in medial thalamic structure responding to repetitive cortical stimulation (270/sec). A long latency response (dot and triangles) is visible just after the short latency one that follows the initial shock artefact. Repetition of stimulation suppresses the long latency (orthodromic) response while the short latency one, admittedly antidromic, follows the high rate stimulation. Upper tracing shows a case in which the orthodromic spike (dot) has occluded the nearest antidromic response (collision test, see fig. 5). In this and the following figures, positivity of the microelectrode corresponds to a downward deflection on the tracings.

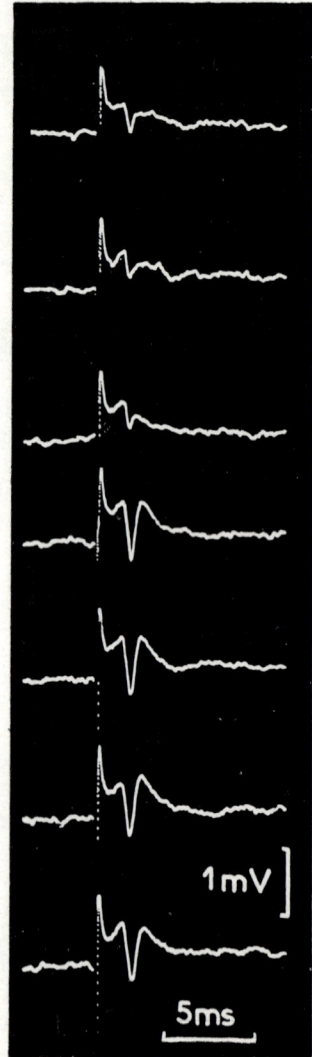

FIGURE 3   Cat under local analgesia. Cortical stimulation. Antidromically generated effects in a cell of medial thalamus. Upper three tracings: only field effects are visible, probably due to abortive conduction or neighbouring "fibres de passage".

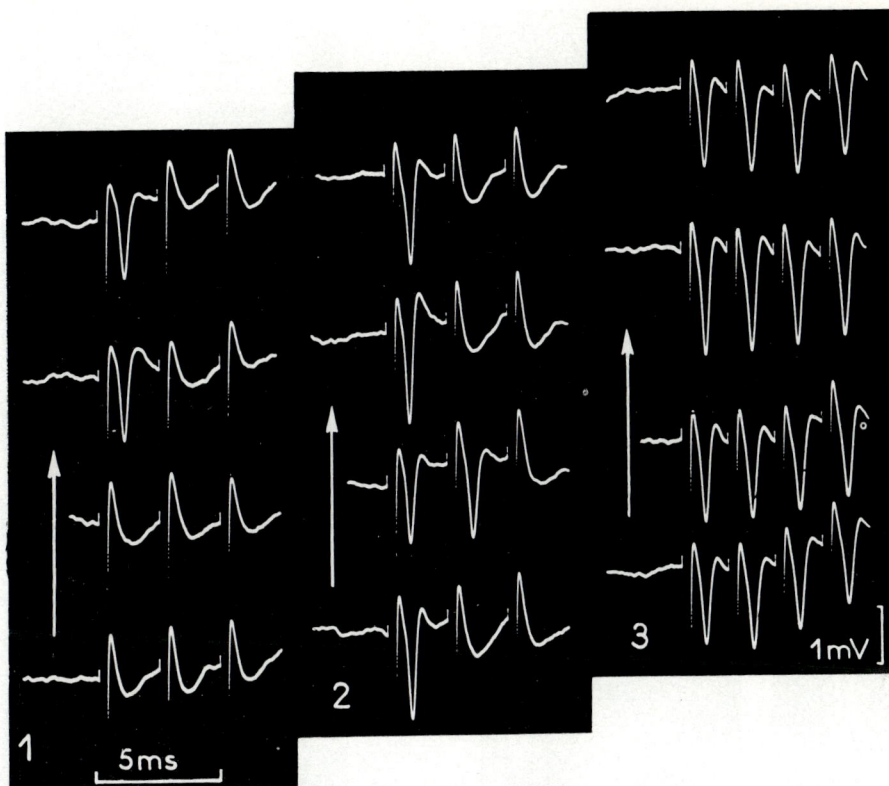

FIGURE 4   Cat under local analgesia.
The electrode was close to a cell recognized by its spontaneous activity (not illustrated here). Trains of shocks of
.2 ms – 3 V, at frequency of 400/sec (1, 2), then 500/sec (3) were applied to premotor cortex (see artefacts). Series
of 4 tracings (bottom to top) follow each other in the order 1, 2, 3. Note that the first two trains fail to elicit any
response. Repetition of trains makes one or two responses appear. Then the small frequency increase that follows
results in a one-to-one response.

of spontaneous orthodromic impulses that appear regularly in awake animals (Figure 5), but we had to make use of responses provoked by peripheral electrical stimulation in animals under chloralose (Figure 6). The occlusion time is thus necessarily longer than the refractory period, a superior limit to which can be deduced in two ways: shortest interval existing between two spontaneous impulses (Figure 5), or between impulses provoked by stimulations of increasing frequencies (Figure 5, 6).

Two difficulties may arise with this test:

a) When the latency of the antidromic spike is very short (around .5 sec) the occlusion time is only slightly different from the refractory period (Figure 6); this case, frequent in VP is rare in medial thalamus. Anyhow, the shortness of latency in this case is certainly another proof of the antidromic origin of the response. A latency of 0.5 msec

could not be interpreted by the addition of a conduction time and a synaptic delay.

b) In a few cases, spikes which present short and stable latencies cannot follow high frequencies except during a short time (see an example Figure 6). There it can be supposed that the train of cortical stimulations had provoked a somewhat delayed orthodromic inhibition (Albe-Fessard and Gillett, 1961).

2.   *Localization in the medial thalamus of cells responding antidromically to cortical stimulation*

We have in this group of experiments limited our exploration to the thalamic region comprised between the planes 7 and 9 at laterality 3. In the first animals, we have often made two explorations on one thalamic side Ant 7–Ant 8,5. In these cases, it is always in the anterior electrode track that we

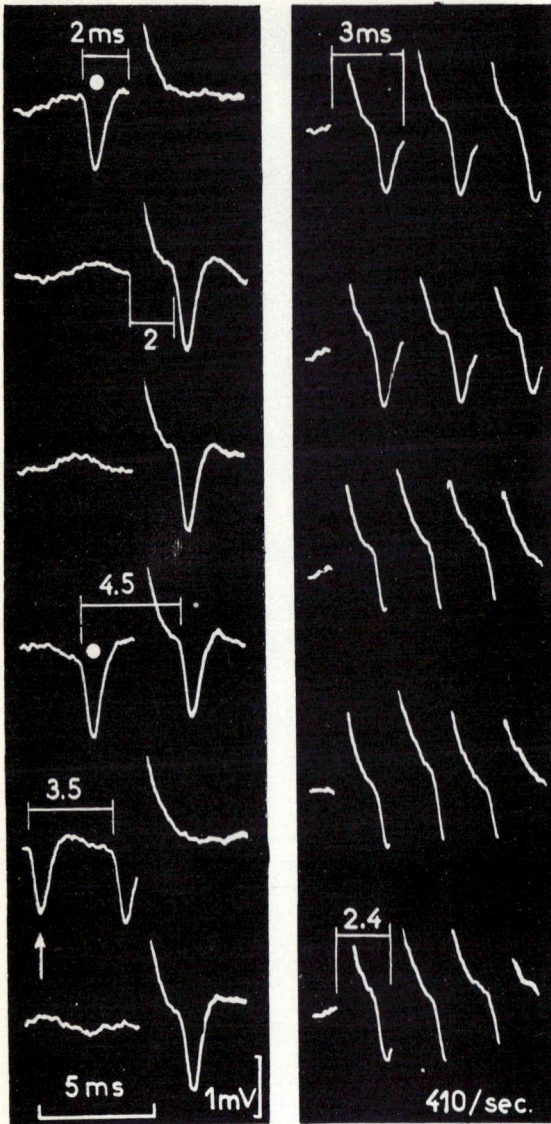

FIGURE 5 Cat under local analgesia. Example of collision test.
Left, from top to bottom: thalamic cell activated antidromically by stimulation of premotor cortex (latency slightly inferior to 2 msec). Response fails to appear in the 1st and 5th tracings, due to the interference with a spontaneous orthodromic response. The orthodromic conduction suppresses the antidromic activation if the beginning of the orthodromic response precedes the stimulation by 2 msec (1st tracing) or less (5th tracing). It leaves intact the antidromic response in the 4th tracing. Thus the collision time is comprised between 4 and 4.5 msec.
Right; repetitive stimulation at increasing frequencies (top to bottom). Response still present at 410/sec, showing that refractory period (< 2.4 msec) is greatly inferior to collision time. Moreover, the 5th tracing at left (arrow) shows a 3.5 msec interval between two spontaneous spikes, a value illustrating again the superiority of collision time over the refractory period of the fiber.

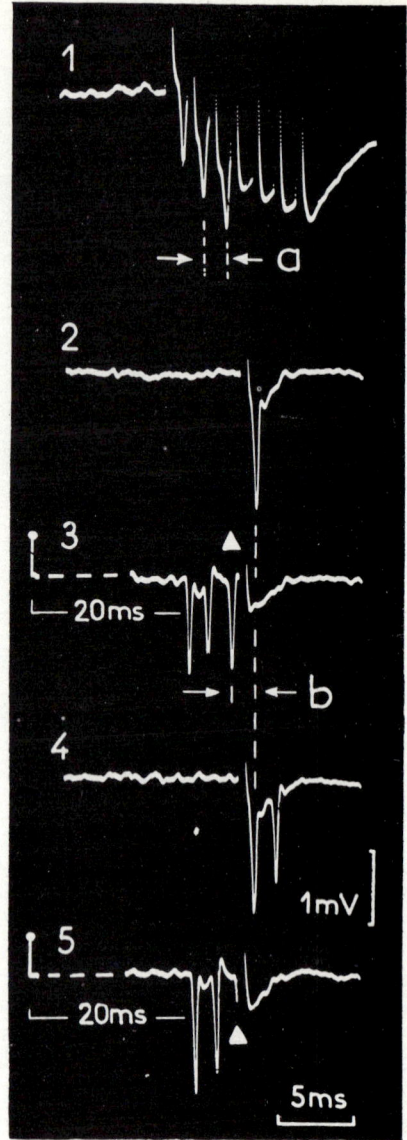

FIGURE 6 Cat under chloralose anesthesia.
All tracings display "antidromic" spikes generated by stimulation of premotor cortex. Orthodromic spikes appear in 3 and 5 (stimulation of contralateral anterior paw) and in 4 (corticofugal activation). Tracings at slower sweep on a separate channel not appearing here enabled us to measure the orthodromic activation delay (here around 20 ms).
1 Cortical stimulation at high frequency (a = 1.5 msec). Responses do not follow from the 4th stimulation on, a failure that may be due to a corticofugal inhibitory control.
2 to 5 Same cell showing exceptional brevity of antidromic spike latency (around .5 ms). Note that in the 3rd record, the interval b between an orthodromic spike (triangle) and the expected antidromic activation represents an inferior value to the collision time; and that it is about equal to a, which has a value clearly above the refractory period.
In 5, the last orthodromic spike (triangle) has arrived too late so that refractory period rather than collision can be made responsible for the occlusion (see text).

have encountered antidromic spikes. Therefore, in the other animals, we have systematically explored this anterior region of CM (Ant 8–Ant 9). After verification of the anatomical location, many tracks were found to be more anterior than it was predicted by the stereotaxic coordinates employed.

It is the reason why, in our general results, a part of the exploration is reported to be in the anterior plane 9,5 and even 10. After histological verification, the laterality revealed to be comprised between 2.5 and 3.5, the majority of recordings being at Lat 3.

In the totality of the cats (24) in which medial thalamus was explored between +4 and −1, 781 cells were studied; 62 of these cells were recognized to be activated antidromically by cortical stimulation. In addition, in the 3 experiments in which Putamen stimulation was used, we found 5 cells activated by putamen stimulation, and two of these activated both by putamen and cortex. However the number of experiments where cells have been studied with putamen stimulation is too small and further experiments at this point have to be performed.

As shown in Figure 7, the cells activated antidromically were distributed between the planes 8 and 9.5. If we refer to the limits of structures as given by the atlas of Ajmone-Marsan and Jasper, they are thus localized at the extreme anterior part of CM-Pf and at the level of nucleus centralis lateralis. (see Figure 8).

### 3. Latencies of antidromic responses

The latencies of antidromic spikes evoked by stimulation of Putamen were always very short (<0.5).

Antidromic cellular spikes elicited by cortical stimulations, had the shortest latencies around 0.5 ms, the largest between 4 and 5 ms (see Figure 9), while for the majority of the cells latencies were comprised between 0.5 and 2.5 ms. These latencies crudely correspond to a propagation of impulses along an axonal distance of about 25 mm; the velocity in the thalamo-cortical fibers is thus comprised between 10 up to 50 m/sec.

### 4. Cortical localization of the effective stimulation

In all the experiments in which antidromic responses could be recorded, the electrodes had been placed as schematically represented in Figure 1. The distribution of the couple of stimulating electrodes that gave antidromic responses with the lowest threshold is the following one: 29 cells for couple of points 2–3, 5 cells (1–2), 17 cells (3–4); 7 cells were indifferently driven by couples 2–3 and 3–4.

Thus the axon terminals of the studied cells seem to be scattered at the level of the pericruciate cortex, the majority being in the lateral anterior part.

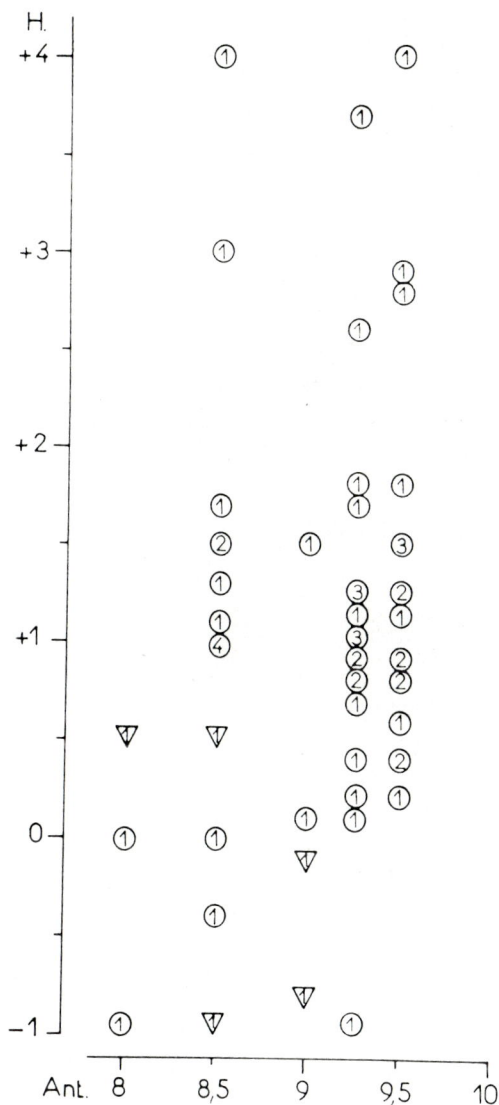

FIGURE 7 Sites of the majority of the antidromic cells encountered. This diagram represents the stereotaxic anteriority (abscissae) and depth (ordinates) corrected with the help of histologic sections. The laterality of these tracks was in all cases near 3 mm from the mid-line. The numbers in the circles indicate the number of all the cells encountered at this particular level. The triangles represent the positions of the cells activated by putaminal stimulation.

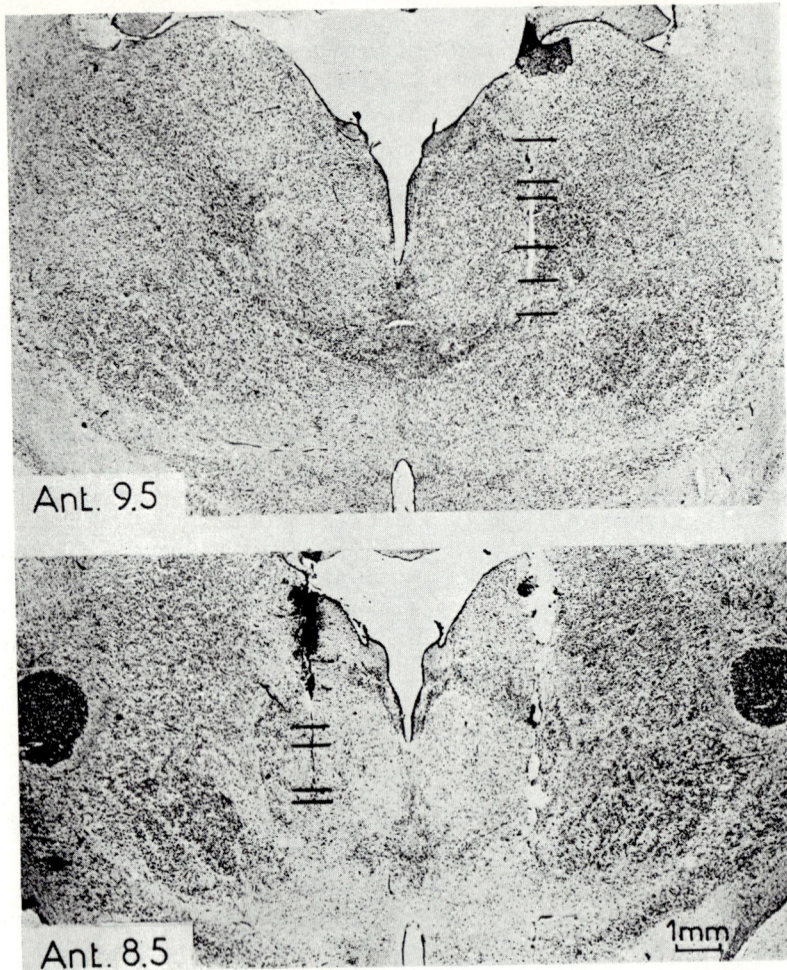

FIGURE 8   Two histological sections presenting two traces of electrode tracks along which positions of anti-dromic activated cells are represented by small horizontal bars. At anteriority 8.5, on the last part of the experiment, the glass microelectrode was replaced by a metallic one in order to improve depth determination. The thick trace appearing at right has been caused by the penetration of the driving metallic tube into the thalamic structures.

## 5.   *Natural stimulations activating orthodromically the 'antidromic' cells*

In a few experiments, localization of the peripheral somatic stimulations that activated the 'antidromic' cells was saught. This was done for all the animals which were anesthetized under chloralose, a case in which it was essential to obtain orthodromic evoked responses to be able to use the collision test (Figure 6).

The peripheral field was always large, often bilateral, and natural effective stimulations were in this case small taps applied to one or the other body side. These responses were clearly of the non-specific activation type. Only in a few cases, peripheral field activations were explored in awake animals.

They were again widely distributed either on one side of the body or on two symmetrical limbs. Often passive movements of limbs or pressure on muscles were as efficacious stimuli as tactile ones.

## 6.   *Effect of anesthesia on antidromic responses*

The fact that, in animals prepared under nembutal or which had received nembutal later on, we did not find antidromic cells, led us to verify if the type of anesthesia had an effect on antidromic activation.

In three cats, we searched for an antidromic response and, after having verified the nature of the responses by using the tests described in section 1, we injected a short acting barbiturate (Brevital) into the systemic circulation. The antidromic responses

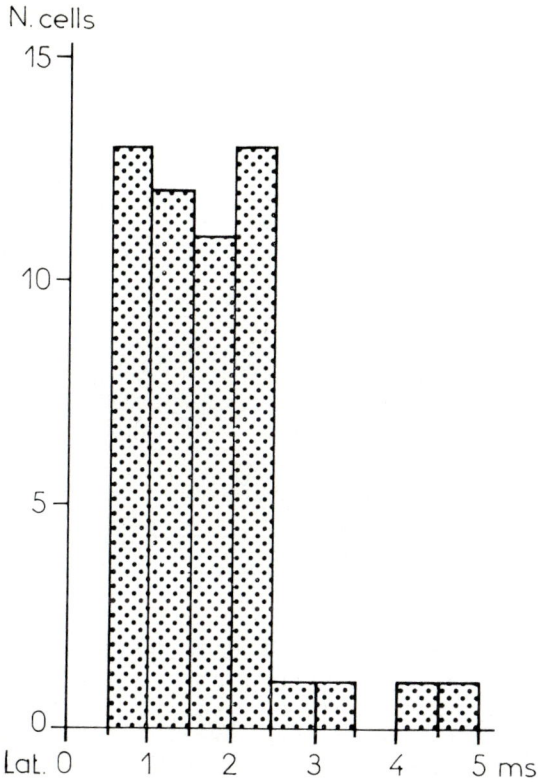

FIGURE 9   Histogram of the latencies corresponding to responses from cells activated antidromically as testified by the collision test.

were in general, either entirely suppressed, or made more variable during 10 to 20 minutes after the injection (see Figure 10 and 11).

As a verification of this fact, we have made exactly the same experiment with cells of the n. ventralis posterior recognized by their small receptive field and antidromically activated by SI stimulation. The antidromic responses were here again suppressed or had their probability of apparition reduced (see Figure 11). However, during the blocking phase, a residual field effect due to some fiber activity is often more apparent in VP than in CM. During this period, an increase of the stimulation did not reestablish the antidromic invasion of the cells, neither in CM-CL, nor in VP, if it had been suppressed.

In some cases, injection of Brevital was accompanied by an important decrease of blood pressure from 110–120 mm Hg to 70–50 mm/Hg. To verify that the modification of the antidromic responses was not due to these blood pressure changes, we have in one case produced a similar drop of pressure by a perfusion of Arfonad (Methioplegium). No change has appeared under these conditions in the antidromic responses. Similar tests, made with Ketamine and Chloralose anesthesia have proved that antidromic responses can be recorded quite unchanged under these conditions.

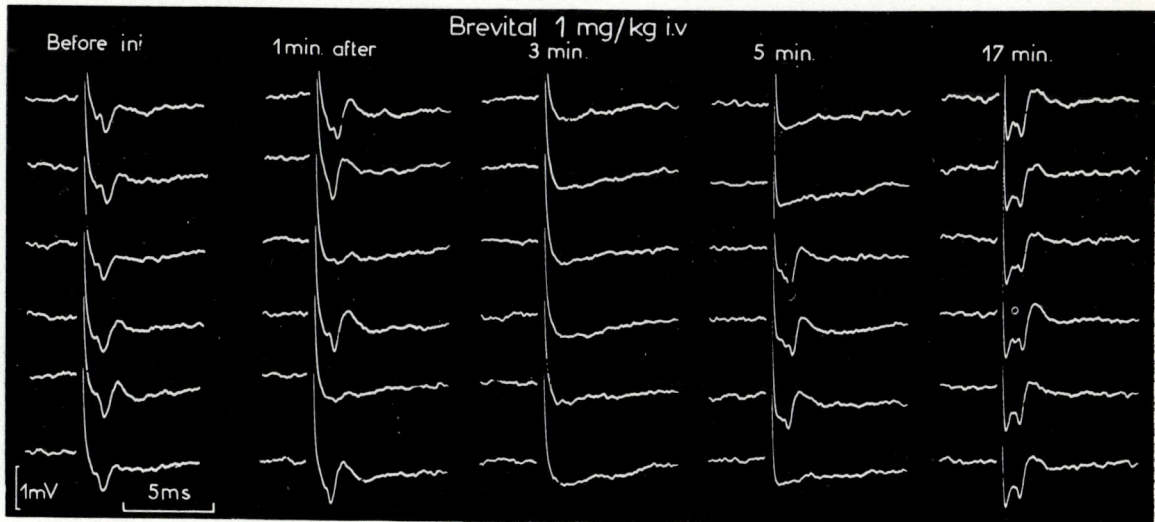

FIGURE 10   An example of phases of suppression concerning antidromic activities after an injection of 1mg/kg of Brevital in the general circulation. The antidromic origin of the spike had been verified beforehand. At the 17th minute, same type of recording after recovery of the response, but the direction of the artefact had been changed.

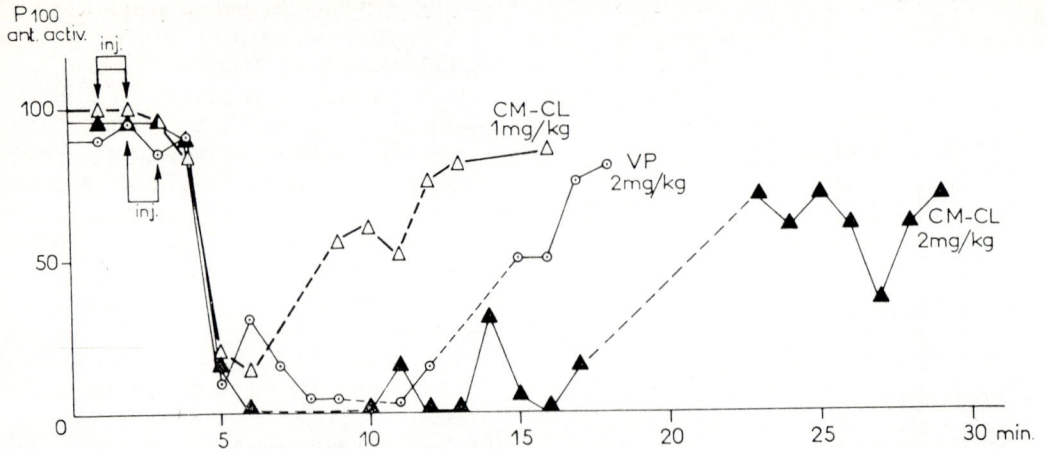

FIGURE 11   Time course of suppression and recovery of antidromic responses following an injection of the same short acting barbiturate (Brevital). The·responses were counted on continuous recordings that were taken during the injections and then afterwards from time to time during recovery period. The doted parts of the curves correspond to the periods without recordings. Two cells, at the limit of CM-CL, had their activities recorded. The injections were respectively of 1mg/kg and 2 mg/kg. A third cell recorded in VP and submitted to the same action shows a similar evolution of its antidromic response to SI cortex stimulation.

## DISCUSSION

In order to clear up the long debated question of the medial thalamic projections to the cortex, it was necessary to explore the structures at the unitary level and to make use of antidromic cell activations. The present work has made it obvious.

1.) The principal fact that has been acquired by this technique is the demonstration of a direct connection between the medial thalamus and certain cortical areas. However, this connection seems to arise only from the anterior part of the CM-Pf complex, with the majority of its cells of origin situated in the nucleus centralis lateralis. Such a connection can explain the relatively short latency observed when stimulations are applied to the CM-Pf nucleus and evoked potentials are recorded at the cortical level from the same sites at which stimulating electrodes have been placed in the present experimental series (Albe-Fessard and Rougeul, 1957). However, in this former work, such short latencies (0.5 ms) as those in some antidromic experiments could never be found. We suggest that it is because well localized bipolar electrodes were then applied at Ant 7.5, in the most posterior medial thalamic zones, and that from there connections to cortex are made through a final relay situated in anterior CM-Pf or n. centralis lateralis.

Results of the group of experiments reported here do not offer an explanation of the responses observed at other cortical levels after the same thalamic stimulations; in particular in marginalis anterior and suprasylvian cortical gyri. Assuming that these responses might be relayed through cortico-cortical pathways, we have repeated former experiments and made a large ablation of peri-cruciate area 1, without greatly affecting the responses in other non-specific cortical areas. Another pathway to these regions has thus to be searched for. However, we have also to consider that our experiments have only dealt with cells placed in the thalamus at 3 mm from the mid-line. Further experimental series, where more medial regions will be explored, have to complete the present work. A longer thalamo-cortical pathway involving a putaminal-pallidal detour has also to be considered (Drooglever-Fortuyn et Stefens, 1951; Hassler, 1955, 1970).

2.) In some experiments, we were totally unable to find antidromically activated cells. In the majority of these cases, we had to deal with experiments in which we had tried to make a more acurate localization and for this reason had utilized shorter distances between stimulating electrodes. We suggest that it explains the failure because, the electrodes lying on the dura mater, a certain distance between them must be maintained so that a sufficient intensity of stimulating current can arrive at the underlying cortical layers.

3.) The suppressing effect of an injection of barbiturates on antidromic activation is not easily explained. The same phenomenon was found at the spinal level (Shapovalov, 1964). We can invoke the impossibility of an invasion of the cell body due to a block at the axon hillock. In this case, we have to await a residual field effect as a sign of an invasion of the fibers. In fact, this effect is well observed in the case of VP cells activated by SI and recorded in animals under Brevital. On the contrary, in medial thalamus, we have not often observed this residual field effect, possibly because the thalamo-cortical fibers are too scattered to enable visualization of their proper activity.

The lack of antidromic cell body invasion observed in this case can be explained by the hyper-polarization of the membrane as produced by injection of barbiturates (Albe-Fessard *et al.*, 1970). This is a fact which is in good agreement with the observation we have made that repetitive stimulations are at times necessary to elicit an anti-dromic cellular response. In the same way it appears that, when a cell is not giving a full antidromic response for each stimulus, it is easy to obtain it by sending natural orthodromic excitations towards this cell.

However, we cannot explain why anesthetics like Chloralose that also have an increasing effect on the membrane potential do not block the anti-dromic activation as well as barbiturates.

4.) The sort of experiment we have made shows no more than the existence of a projection from the medial thalamus to the precruciate cortex. It does not give an idea of the role that this projection can play, either on the sensory or the motor side. It also does not exclude that 'de passage' fibers can go through the CM to other thalamic nuclei like VL and from there relayed messages arrive at area 4 and area 6 (Zilber et Buser, 1963; Bénita *et al.*, 1971).

As CM stimulation can have an excitatory action on precruciate cells (Lux and Klee 1952) and an inhibitory one on SI cells (Albe-Fessard, 1961) (including the region just lateral to the tip of cruciate), we also have to think either that the pathways that we have demonstrated are respon-sible for both effects, or that one of the pathways is excitatory and the other inhibitory.

## SUMMARY

1.) The existence in the cat's brain of a direct connection between medial thalamic nuclei (centrum medianum, parafascicularis, centralis lateralis) and certain cortical areas is still a con-troversial subject, in spite of the short latencies of the premotor cortical responses following CM-Pf stimulation. In order to clear up the question, we thought it appropriate to make use of antidromic activations. This was done by stimulating the pericruciate gyrus and recording unitary spikes with microelectrodes in the medial thalamic nuclei. Cats were either under chloralose anesthesia or maintained under local analgesia after preparation under short-acting anesthetics.

2.) The features characterizing the responses to antidromic activation have been studied by taking preliminary records in the nucleus ventralis posterior (VP), stimulation being applied to SI. Responses at medial thalamic level have similar properties: very short latencies of the elicited spikes, capacity to follow stimulations at relatively high frequencies, successful collision test. The collision was realized between antidromic spikes and spontaneous orthodromic ones in animals under local analgesia. Under chloralose, ortho-dromic spikes had to be generated by peripheral somatic stimulations.

3.) Thalamic antidromically activated cells can be found scattered between planes Ant 8 and Ant 9.5. Consequently, the direct thalamo-cortical pathway that has thus been safely disclosed takes its origin from the head of the CM-Pf nuclei and from the CL nucleus.

4.) Most of the successful antidromic activa-tions have been obtained when stimulation was applied to the precruciate cortex and to the part of SI cortex lateral to the tip of cruciate fissure.

5.) In a few experiments, the effects of different anesthetics on the antidromic activation have been tested. Short-acting barbiturates (Brevital) sup-pressed it or made it more fluctuant. Same effects were produced on antidromically activated cells of the VP nucleus, cortical stimulation being then in SI. By contrast, antidromic activations can be obtained in animals anesthetized with Chloralose or Ketamine.

## ACKNOWLEDGMENT

We are grateful to Prof. A. FESSARD for his valuable help in the preparation of our manuscript.

## REFERENCES

Albe-Fessard, D., 1961, Nouvelles données sur l'origine des composantes des potentiel évoqués somesthesiques. *Actualités Neurophysiologiques, Masson, Paris*, pp. 24–60.

Albe-Fessard, D., Besson, J. M., and Abdelmoumène, M., 1970, Action of anesthetics on somatic evoked activities. In: *International Anesthesiology Clinics—Anesthesia and Neurophysiology, Vol. VIII*, No 1, H. Yamamura, ed.; Little, Brown and Company, pp. 129–166.

Albe-Fessard, D., et Fessard, A., 1963, Thalamic integrations and their consequences at the telencephalic level. Specific and unspecific mechanisms of sensory-motor integration. In: *Progress in Brain Research, Vol. I*, Brain Mechanisms: pp. 115–148.

Albe-Fessard, D., and Gillett, E., 1961, Convergences d'afférences d'origines corticale et périphérique vers le centre médian du chat anesthésié ou éveillé. *Electroenceph. clin. Neurophysiol.*, 13: 257–269.

Albe-Fessard, D., and Rougeul, A., 1958, Activités d'origine somesthésique évoquées sur le cortex non-spécifique du Chat anesthésié au chloralose: Rôle du centre médian du thalamus. *Electroenceph. clin. Neurophysiol.*, 10: 131–152'

Bénita, M., et Condé, H., 1971, Etude des efférences du noyau centre médian du thalamus du chat vers le cortex et les structures strio-pallidales. *Exp. Brain Research* (sous presse).

Bignall, K. E., 1967, Effects of subcortical ablations on polysensory cortical responses and interactions in the cat. *Exp. Neurol.* 18: 56–67.

Boivie, J., 1971, The termination of the spinothalamic tract in the cat. An experimental study with silver impregnation method. *Exp. Brain Res.* (sous presse).

Bowsher, D., 1966, Some afferent and efferent connections of the Parafascicular—Centre Median complex. In: *The Thalamus*, Purpura, D. P., and Yahr, M. D., eds., Columbia Univ. Press, N. Y. & London, pp. 99–108.

Buser, P., and Bignall, K. P., 1967, Nonprimary sensory projections on the cat neocortex. Int. *Rev. Neurobiol.*, 10: 111–165.

Condé, H., Schmied, A., et Bénita, M., 1968, Quelques données électrophysiologiques sur les efférences du centre médian du thalamus. *J. Physiol.* (Paris) 60: 420.

Darian-Smith, I., Philips, G., and Ryan, R. D., 1963, Functional organization in the trigeminal main sensory and rostral spinal nuclei of the cat. *J. Physiol.* (London) 108: 120–146P.

Drooglever-Fortuyn, J., and Stefens, R., 1951, On the anatomical relations of the intralaminar and midline cells of the thalamus. *Electroenceph. clin. Neurophysiol.*, 3: 393–400.

Gordon, G., and Miller, R., 1966, Identification of corticofugal cells projecting to the gracile and cuneate nuclei of the cat. *J. Physiol.* (London) 186: 34–35P.

Hassler, R., 1955, Functional anatomy of the thalamus. Congress latino-amer., *Neurocir., VI* (Montevideo): 754–787.

Hassler, R., 1970, Dichotomy of facial pain conduction in the diencephalon. In: *Trigeminal Neuralgia, Pathogenesis and Pathophysiology*, Hassler, R., and Walker, A. E., eds., Georg Thieme Verlag, Stuttgart, pp. 123–138.

Lux, H. D., and Klee, M. R., 1962, Intracelluläre Untersuchungen über Einfluß hemmender Potentiale im motorischen Cortex. I. Die Wirkung elektrischer Reizung unsperzifischer Thalamuskerne. Archiv für Psychiatrie und Zeitscrift f. d. ges. *Neurologie* 203: 648–666.

Mehler, W. R., 1966, Further notes on the Center Median Nucleus of Luys. In: *The Thalamus*, Purpura, D. P., and Yahr, M. D., eds., Columbia Univ. Press, N.Y. & London, pp. 109–127.

Paintal, A. S., 1959, Intramuscular propagation of sensory impulses. *J. Physiol.* (London) 148: 240–251.

Rose, J. E., and Woolsey, C. N., 1949, Organization of the mammalian thalamus and its relationships to the cerebral cortex. *Electroenceph. clin. Neurophysiol.*, 1: 391–404.

Shapovalov, A. I., 1964, Intracellular microelectrode investigation of effect of anesthetics on transmission of excitation in the spinal cord. *Fed. Proc.* (Translation Supplement), 1, part II: 113–116.

Totibadze, N. K., and Moniava, E. S., 1969, On the direct cortical connections of the nucleus centrum medianum thalami. *J. Comp. Neurol.*, 137: 347–360.

Zilber, N., et Buser, P., 1963, Structures acheminant les influx sensoriels vers le cortex moteur du chat. *J. Physiol.* (Paris) 55: 358.

# ELECTROPHYSIOLOGY OF CONTRALATERAL AND IPSILATERAL VISUAL PROJECTIONS TO THE WULST IN PIGEON (COLUMBA LIVIA)†

M. PERIŠIĆ, J. MIHAILOVIĆ, and M. CUÉNOD

*Brain Research Institute, University of Zürich, Switzerland*

Evoked potentials and single unit discharges in the wulst of the pigeon were observed in response to contralateral and ipsilateral eye stimulation. Mean latencies of the first component of the ipsilateral evoked responses following electrical stimulation of the optic nerve papilla were 14.3 msec; those from contralateral eye were 10.1 msec. Photically evoked responses showed similarly longer latencies to stimulation of the ipsilateral eye. Electrolytical lesion or reversible cooling of the rostral part of the supraoptic decussation (DSO) induced a significant decrease of ipsilaterally evoked potential amplitudes; these procedures had no effect on contralaterally evoked responses. 125 wulst units activated by electrical stimulation of the optic nerve papilla were studied. 52% of the wulst neurones responded to contralateral stimulation with first spike latencies of 7–39 msec; these were localized dorsally in the wulst. Units responding exclusively to ipsilateral stimulation (26.4%) were located ventrally and had latencies of 10–86 msec. A population of neurones found in the intermediate zone (21.2%) were activated by both contralateral and ipsilateral stimulation. This group exhibited longer discharge latencies to ipsilateral than to contralateral eye stimulation. Despite total crossing of the optic nerve fibres, integration of contralateral and ipsilateral visual input occurs in the wulst. The DSO plays an important role in mediating the ipsilateral input.

Recent anatomical investigations have shown two different visual pathways, which attain avian telencephalon (see Karten, 1969). They are termed (1) the retino-tecto-rotundo-ectostriatal path and (2) the retino-thalamo-hyperstriatal path. The first pathway has been the subject of numerous investigations utilizing anatomical, physiological and behavioural approaches. Electrophysiological data indicate a direct retinal projection to the tectum (Hamdi and Whitteridge, 1954; Holden, 1968a, 1968b; Robert and Cuénod, 1969). After synapsing in the tectum the pathway continues to the nucleus rotundus and further to the ectostriatum (Revzin and Karten, 1966; Revzin, 1967).

The second, phylogenetically younger, visual pathway was experimentally investigated by Cowan *et al.* (1961). Physiological studies of this pathway showed that retinal stimulation elicits a response in the hyperstriatum accessorium and hyperstriatum intercalatum superius in the pigeon (Revzin, 1969). No electrophysiological data about relay stations in this pathway are presently available. Behavioural investigations of the functions performed by the structures constituting this pathway have exclusively involved lesions at the telencephalic level. These studies demonstrated some visual discrimination deficits after ablations (see Ten Cate, 1936; Zeigler, 1963; Pritz *et al.*, 1970). Perlia's original study (1889) indicated and those of subsequent investigators confirmed that all avian optic fibres cross in the chiasm and attain the contralateral brain hemisphere. Consequently any monocularly obtained visual information, which is available to the ipsilateral hemisphere must traverse commissural connexions (see Cuénod, 1971).

The present study was undertaken to clarify the input characteristics of the wulst (hyperstriatum accessorium, hyperstriatum intercalatum superius and hyperstriatum dorsale) neurones in response to contralateral and ipsilateral retinal stimulation and to identify a possible interhemispheric pathway, whereby the ipsilateral input is mediated.

## MATERIAL AND METHOD

The experiments were performed on 65 adult pigeons (Columba livia) of either sex.

†This work was supported by the Swiss National Foundation for Scientific Research (Grant Nr. 4806, 3.329.70, 3.133.69 and by the Slack-Gyr Foundation).

119

a. *Preparation*

The animals were anaesthetized by intramuscular application of Equithesine ® (0.25 ml/100 g of body weight). The trachea was canulated; the animals paralyzed by Flaxédil ® (0.20 ml/100 g of body weight) and artificially ventilated. The $CO_2$ level in the expired air was periodically measured by a Harvard $CO_2$ analyzer; the cloacal temperature was also monitored. The animals were fixed in a head holder allowing stereotaxic manipulations according to Karten and Hodos (1967). The telencephalon was exposed and the dura opened above the hyperstriatum accessorium. The surface of the brain was protected with a saline-agar solution. In 24 experiments employing the electrical stimulation of the optic nerve papilla the eye lids and the nictitating membrane were removed, the cornea, the lens and the crystalline body aspirated. Photic stimulation was employed in six animals which were monocularly enucleated.

b. *Stimulation*

For electrical stimulation bipolar stainless steel electrodes were placed on the optic nerve papilla and an array of four electrodes in the supraoptic decussation. Single square wave pulses of 0.05–0.75 mA, 0.1–0.3 msec duration and 0.3/sec frequency, were delivered by a Grass S–8 stimulator. Diffuse light flashes were used for photic stimulation (Robert and Cuénod, 1969).

c. *Recordings*

Evoked potentials were recorded with tungsten wire electrodes with tip diameters of 5–10 $\mu$ and 5–25 pF capacitance. The electrodes were placed in the central part of the wulst (coordinates: AP 10.0–11.0, lat. 2.0–3.0 and depth from the surface 0–3.5 mm. according to Karten and Hodos, 1967); maximal amplitude responses were observed in this locus. Glass micropipettes filled with 2M NaCl or 3M KCl saturated with fast green† were used for single unit recordings. The indifferent electrode was placed in the neck muscles. Recording electrodes were connected to a standard cathode follower (BAK). Signals were amplified 1,000 times with Tektronix 122 preamplifiers and displayed on Tektronix 556 cathode ray oscilloscope and on the Tektronix storage oscilloscope type R564B. The Computer of Average Transients CAT 400B was used to average photically evoked potentials using 50

† F.C.F. Nat. Aniline Div., All. Chem. Corp., U.S.A.

single sweeps and for post-stimulus time histograms of single unit recordings.

d. *Reversible Cooling of the DSO*

Reversible cryogenic blockade of the supraoptic decussation (DSO) fibres was performed with a closed system freon cryoprobe, 1.1 mm tip diameter (Bénita, 1970). The probe was stereotaxically located in the interhemispheric space. Fibre conduction blockade was achieved by cooling the tip up to 5°C. At the end of an experiment the temperature was reduced to −25°C so as to produce cryocoagulation to facilitate histologic localization.

e. *Histological Control*

A current of 100–200 $\mu$A for 2 min was passed through stimulating electrodes in DSO in order to localize their tips. Marking the tip position of the glass micropipettes was performed by passing a current of 12 $\mu$A for 15 min, the electrode tip negative with respect to the indifferent (Thomas and Wilson, 1965). Because of the low percentage of marked loci found spatial analysis of activated neurones was performed only on the basis of micromanipulator positions. At the end of each experiment the animal was first perfused with 200 ml of Ringer solution and then with formalin solution composed of 4% phosphate buffered formaldehyde (pH 7.2) and 0.32 M sucrose. The brains were frozen or embedded in celloidin or paraffin and frontally sectioned at 15–50 $\mu$. Alternative sections were stained following the Niss or Weil technique.

RESULTS

A. *Wulst responses to electrical stimulation of the optic nerve papilla*

Two microelectrodes were placed at symmetrical points in the wulst and the evoked potentials recorded to stimulation of the contralateral or ipsilateral optic nerve papilla. Responses were observed in 24 experimental animals.

1. *Contralaterally Evoked Responses*

In the contralateral hemisphere a biphasic positive-negative potential was evoked by stimulating the optic nerve papilla. The positive component had an average peak latency of 10.1 ($\pm$1.7 SD) msec and peak amplitude 68.3 ($\pm$27.7) $\mu$V, while the negative wave appeared with a latency of 22.8

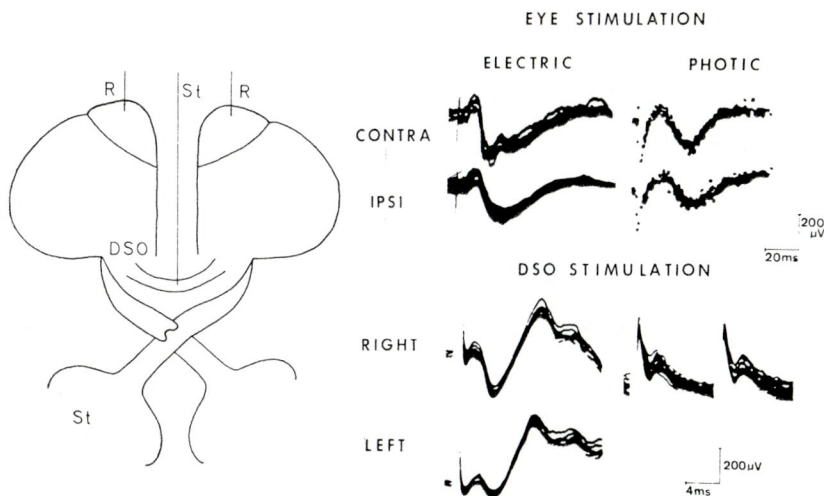

FIGURE 1 Wulst responses evoked by electrical and photic stimulation of the eye and of DSO. Responses to electrical stimulation are presented as superpositions of 5 single sweeps. Responses to photic stimulation are averages of 50 single sweeps displayed by CAT 400B. DSO stimulation: stimulation frequency 0.3/sec (left) and 110/sec (right). Legend: DSO, supraoptic decussation; St, stimulation; R, recording.

($\pm 2.9$) msec and amplitude of 184.7 ($\pm 51.1$) $\mu$V (Figure 1).

### 2. Ipsilaterally Evoked Responses

The response evoked in the ipsilateral hemisphere had a wave shape virtually identical to that seen in the contralateral hemisphere, i.e., biphasic positive-negative. The positive component appeared with an average latency of 14.3 ($\pm 2.5$) msec and an average amplitude of 61.2 ($\pm 17.0$) $\mu$V. The peak latency of the negative component was 31.5 ($\pm 5.9$) msec, its amplitude 136.4 ($\pm 41.5$) $\mu$V (Figure 1).

Neither contralaterally nor ipsilaterally evoked potentials followed stimulation frequencies higher than 10/sec.

### B. Wulst Responses to Photic Stimulation

The wave shapes of photically evoked potentials showed a more variable pattern. Consequently all the responses were averaged. The averaged responses appeared very similar to those evoked by electrical stimulation but with longer latency. In six experiments the positive component appeared with an average peak latency of 18.4 ($\pm 1.8$) msec to stimulation of the contralateral eye and 22.0 ($\pm 3.0$) msec with ipsilateral eye stimulation. The negative component of the photically evoked responses was seen with a latency of 39.2 ($\pm 6.4$) msec in the contralateral hemisphere and 47.0 ($\pm 9.2$) msec in the ipsilateral (Figure 1).

The amplitude of the potentials recorded in the wulst varied in size as a function of the depth of

the recording electrode. Thus, in the contralateral hemisphere the greatest amplitudes were seen at depths of 500–1000 micra, while in the ipsilateral hemisphere the evoked potential amplitudes attained their maxima at 2000–2500 micra (Figure 2, right).

### C. Wulst Responses to Stimulation of the DSO

In fifteen experiments the stimulating electrodes were placed in DSO. Wulst potentials were recorded from both left and right hemispheres. The earliest positive response was always present with an average latency of 2.5 msec. Later responses, having latencies of 4 and 6 msec were not observed in all experiments. Only the first, positive wave followed stimulation frequencies over 100/sec. Short latency responses were taken to indicate correct DSO placement, a condition necessarily preceding electrolytical lesion (Figure 1).

### D. Effects of Lesion in DSO on Wulst Responses

Wulst responses to electrical stimulation of the contralateral or ipsilateral optic nerve papilla after electrolytical lesion in DSO were analyzed in eleven animals, those to photic stimulation in four animals.

### 1. Contralaterally Evoked Responses

Neither the latencies of the positive or negative components of the contralaterally evoked responses nor their amplitudes showed any significant changes after the lesion in DSO (Figure 3).

## SINGLE UNITS

## EVOKED RESPONSES

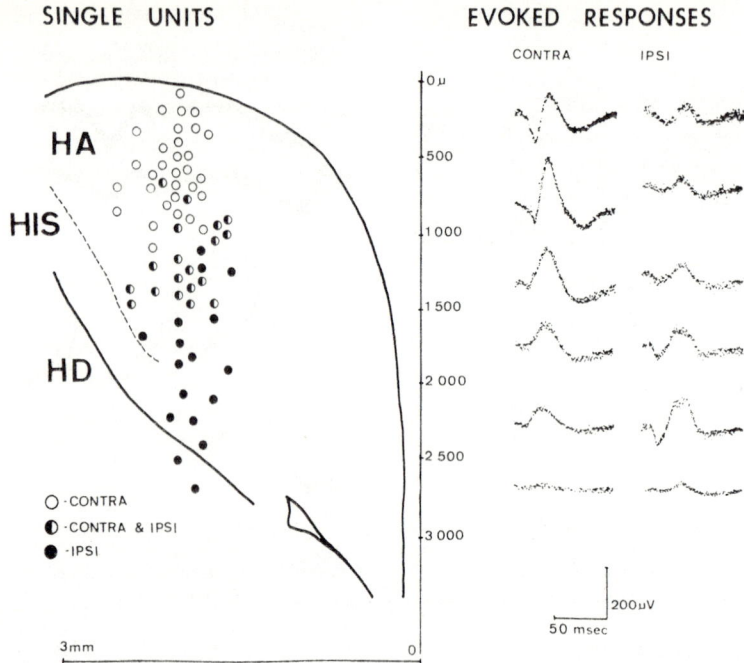

FIGURE 2  Spatial distribution of single units and evoked potential elicited by electrical stimulation of contralateral or ipsilateral optic nerve papilla. Topographical distribution of single units (left). The number of symbols is not a function of the number of units recorded at any given locus. Averaged evoked responses, using 20 sweeps (right). Calibration: Negativity up. Legend: HA, hyperstriatum accessorium; HIS, hyperstriatum intercalatum superius; HD, hyperstriatum dorsale.

### 2. Ipsilaterally Evoked Responses

Noticeable changes were observed in both latency and amplitude of the positive and negative components of the electrically evoked responses after DSO lesion. Latency of the positive wave increased from 14.3 ($\pm 2.5$) msec in control animals to 15.6 ($\pm 3.4$) msec after DSO lesion ($0.3 > p > 0.2$). The amplitude of this wave decreased from 68.3 ($\pm 27.7$) $\mu$V to 17.7 ($\pm 12.1$) $\mu$V ($p < 0.01$) after the DSO lesion. The latency of the negative component increased from 31.5 ($\pm 5.9$) msec before the lesion to 35.7 ($\pm 6.0$) msec after the DSO lesion ($0.1 > p > 0.05$). Its amplitude decreased from 136.4 ($\pm 41.5$) $\mu$V to 50.9 ($\pm 19.3$) $\mu$V ($p < 0.01$). In some experiments the amplitude decrease after DSO lesion was so profound that it was very difficult to distinguish the wave from the baseline. A similar effect was also recorded in response to photic stimulation (Figure 3). Lesions occasionally placed in the anterior or pallial commissure or in the medial hypothalamus were not followed by these effects. Figure 3 illustrates these data.

Electrolytical lesions never completely severed DSO. The rostralmost lesions yielded the greatest amplitude decrease for both positive and negative

components of the ipsilaterally evoked responses. Figure 4 is a plot of the relationship between evoked potential amplitude decrease and anatomic localization of the electrolytic lesion. A correlation coefficient of 0.68 exists between the amplitude decrease and the approximate percentage of the rostral portion of DSO fibres lesioned.

### E. Reversible Cryogenic Blockade of DSO

To verify these data with minimal, controlled loss of DSO function this region was progressively cooled in four animals. Cooling the rostral part of DSO to 5°C also resulted in a decrease of the amplitude of the ipsilaterally evoked response. The responses returned to control amplitudes as soon as the temperature of the cryoprobe increased above 15°C. Contralaterally evoked potentials were unchanged by cooling (Figure 5).

### F. Single Unit Recordings in Wulst Associated with Electrical Stimulation of the Optic Nerve Papilla

125 single units observed in the central wulst area in 31 pigeons responded to optic nerve stimulation. The overwhelming majority displayed single spikes or repetitive firing to currents of 0.5 mA or less. Afferent inhibition of the background activity was

## PHOTIC STIMULATION

LESION

IPSI

DSO

CONTRA

## ELECTRIC STIMULATION

LESION

IPSI

DSO

CONTRA

CONTROL

LESION

IPSI

CONTRA

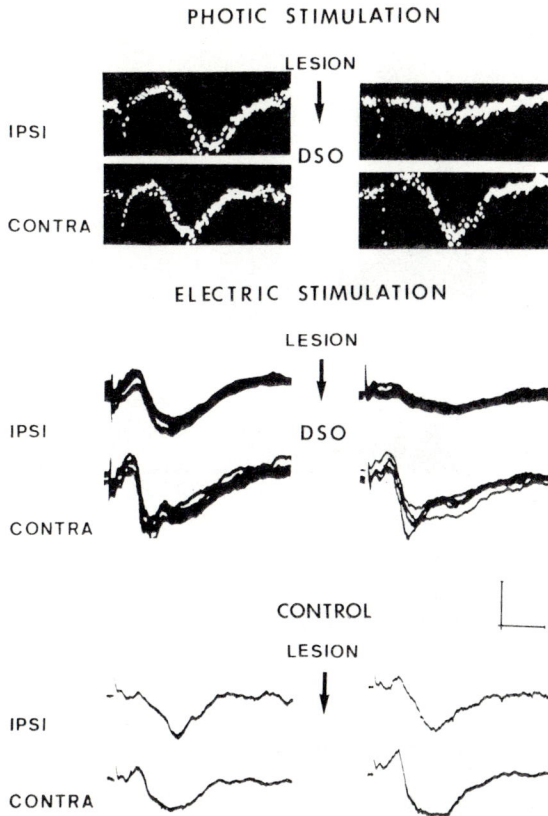

FIGURE 3   Effect of electrolytical lesion in DSO on wulst responses evoked by electrical and photic stimulation of the eye. Left column: Ipsilaterally and contralaterally evoked potentials before the DSO lesion. Right column: Responses after the DSO lesion. Note the decrease of ipsilaterally evoked responses after lesion in the DSO. Control: Lesion was placed in the anterior commissure and in the pallial commissure; DSO intact. Calibration: 20 msec–200 µV, positivity up.

$$y = 0.25x + 23.7$$
$$r = 0.68$$

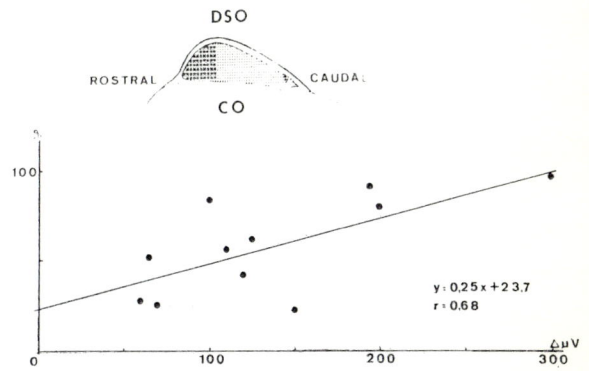

FIGURE 4   Correlation between the decrease of ipsilaterally evoked potential amplitudes and the percentage of destruction of the rostral part of the DSO. Abscissa: Difference between peak-to-peak amplitude before and after lesion. Ordinate: Approximate percentage of the DSO fibres lesioned. Above: Schematic representation of a sagittal section of DSO. Dark dotted area is the rostral part of the DSO where the size of the lesion was measured and correlated with the evoked potential decrease in amplitude. CO: Optic chiasm.

seen with suprathresholded stimuli. Almost all the spikes had positive-negative conformation; their amplitudes varied from 0.2 mV – 5 mV (Figure 6). Additional units exhibited only spontaneous activity, independent of optic nerve stimulation.

Sixty-eight units (52.4%) responded exclusively to contralateral stimulation and had earliest spike latencies of 7–39 msec. They could not be activated by stimulation of the ipsilateral papilla even when the stimulation current was increased to 3 mA (Figure 6).

Thirty-three (26.4%) wulst units responded exclusively to ipsilateral optic nerve stimulation. The latencies measured to the earliest spike of this unit population varied from 10–86 msec (Figure 6).

A third group of units responded to both contra-lateral and ipsilateral optic nerve stimulation. This group, composed of 24 single units (21.2%), responded to contralateral stimuli with latencies of 8–40 msec and to ipsilateral stimulation with latencies of 12–50 msec (Figure 6). This group of neurones fired with obviously longer latencies to ipsilateral than to contralateral optic nerve papilla stimulation. The correlation between these two latency values is shown on Figure 7. The correlation coefficient is 0.88.

Single units responding exclusively to contra-lateral stimulation, were located in the dorsalmost parts of the wulst area, whereas units found in the ventral layers responded exclusively to ipsilateral stimulation. Neurones responding to both contra-lateral and ipsilateral optic nerve stimulation were localized in the intermediate zone of the wulst. The topographic distinctness of these three regions was significant (p < 0.01) (Figure 2, left).

## DISCUSSION

After monocular stimulation responses could be recorded not only in the contralateral but also in the ipsilateral wulst of the avian telencephalon. The greater latency of the ipsilateral evoked responses (4 msec) strongly suggests a longer pathway possibly containing slow conducting fibres and/or one or more relays of the ipsilateral retino-wulst projections. No responses could be evoked

LATENCY HISTOGRAMS OF WULST UNITS RESPONDING
TO OPTIC NERVE STIMULATION

REVERSIBLE BLOCKADE
OF DSO REGION

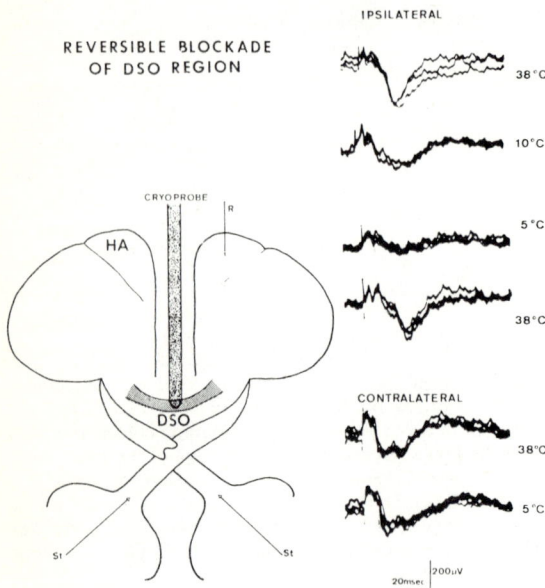

FIGURE 5 Effects of reversible cryogenic blockade of DSO region on wulst responses evoked by electrical stimulation of the optic nerve papilla.

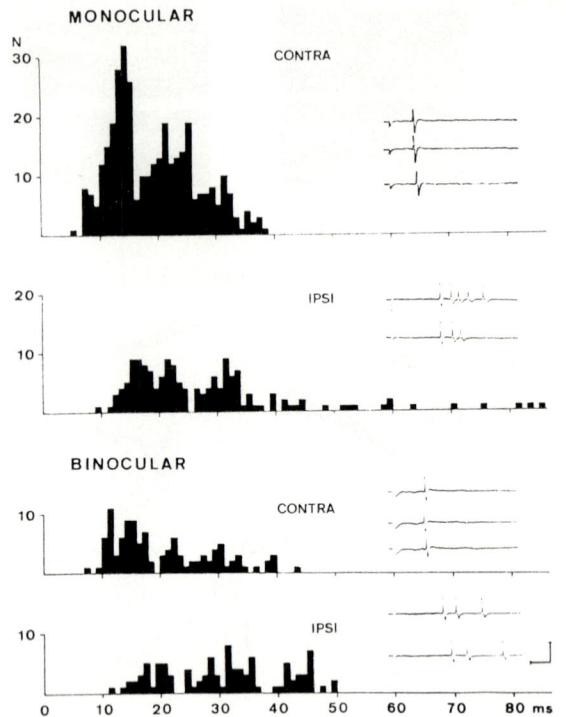

FIGURE 6 Monocular: (Contra) histogram of the first spike latencies of single units activated exclusively by contralateral optic nerve stimulation. (Ipsi) histogram presents those units activated exclusively by stimulation of the ipsilateral eye. Binocular: (Contra and ipsi) latency histograms of units responding to both contralateral and ipsilateral stimulation. Inserts: Representative samples of unit discharges to stimulation of the designated eye (right). Calibration: 10 msec–1 mV, positivity up. First spike latency to each of five successive stimuli is plotted.

in either contralateral or ipsilateral wulst which followed stimulation frequencies higher than 10/sec. This pathway, therefore, appears to be trans-synaptic.

A topographical distribution of contralateral and ipsilateral visual projections could be delineated. Three different cell populations were distinguished in the wulst region in single unit recordings. One group of neurones, located dorsally and a second group located ventrally, receive exclusively monocular input; the dorsal group from the contralateral and the ventral from the ipsilateral eye. The third intermediately located cell population yielded responses from both eyes. This intermediate area very likely represents a region where the telencephalic integration of binocular inputs occurs.

Since the optic nerve fibres appear to be totally crossed in the chiasm of the pigeon, the ipsilateral response must derive from recrossing interhemispheric pathways. From our data it can be seen that the rostral part of the supraoptic decussation (DSO) contains most of the fibres mediating the ipsilateral visual projections to the wulst. Acute electrolytical lesions or reversible cooling of the DSO fibres remarkably impaired the ipsilaterally evoked wulst responses. The amplitudes of these

responses decreased significantly. However, as the ipsilaterally evoked responses were not entirely eliminated by these procedures, it seems very likely that the DSO contains the overwhelming majority but not all of the interhemispheric visual afferents to the wulst.

No evidence can be presently found regarding the exact anatomical route(s) from the retina to the ipsilateral wulst. Moreover, electrophysiological studies have only examined responses induced in the contralateral telencephalon to stimulation of they eye. Bremer et al. (1939) first showed that retinal illumination provokes an initial 'on' wave on the dorsal surface of the telencephalon of the pigeon. The single unit recordings of Revzin (1969) from wulst neurones in pigeons presumably contra-

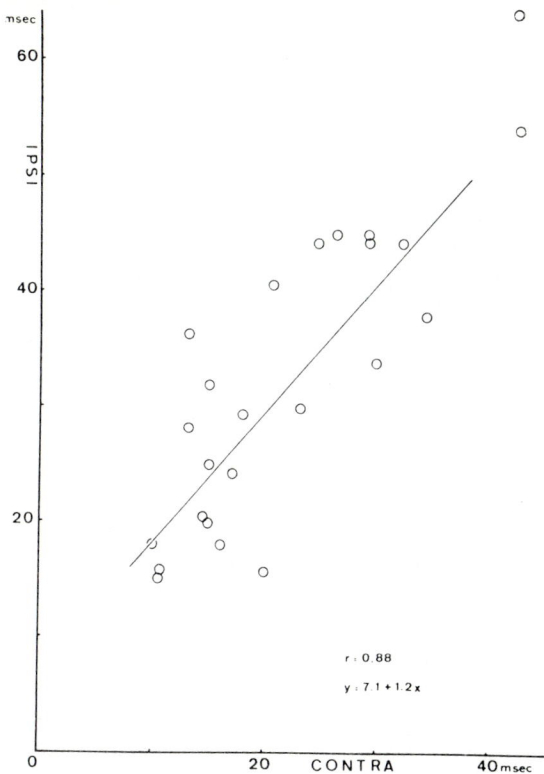

FIGURE 7 Correlation of first spike latencies to ipsilateral versus contralateral stimulation, for units responding to binocular stimulation. Mean first spike latencies for 5 successive stimuli for each neurone were calculated, those to contralateral stimulation are plotted on the abscissa and to ipsilateral on the ordinate.

lateral to the stimulated eye, showed that these neurones had very small receptive fields (2–15˚). The units were topographically organized. Studies conducted on reptiles showed potentials in the dorsal areas of the telencephalon as well as in the dorsal thalamus evoked by stimulating the contralateral eye (Karamian et al., 1965).

The contralateral retinal projections to the wulst have been anatomically studied by a number of investigators. The data presently available indicate: (1) a retino-thalamic pathway, and (2) a thalamo-wulst pathway. It has been suggested that these two pathways are continuous (Karten, 1969).

The retinal afferents of the dorsal thalamic nuclear group, the lateral geniculate nucleus, posteroventral nucleus and ectomammillary nucleus have been detailed following anterograde degeneration (Cowan et al., 1961; Karten and Revzin, 1966; Hirschberger, 1967; Karten and Nauta, 1968).

Dorsal thalamic projections to the avian telencephalon were described by Edinger et al. (1903)

and later investigated by Powell and Cowan (1961) utilizing the retrograde degeneration method. After lesion of the wulst, degeneration was observed in nucleus dorsolateralis anterior[†] and nucleus parvocellularis superficialis in the pigeon. Karten and Nauta (1968) have examined the dorsal thalamic projections to the wulst in the owl. These investigations shed no light, however, on anatomical pathways which might mediate ipsilateral retino-wulst evoked electrical activity.

Since Bellonci's study in the pigeon (1888), tracing DSO fibres to different thalamic nuclei, this fibre bundle has been the subject of extensive investigations. These investigations show the DSO to contain fibres of diverse origins and equally diverse terminations. Among the nuclei sending or receiving fibres constituting the avian DSO are the optic tectum, lateral geniculate nucleus, bed nucleus of the DSO, posteroventral nucleus, nucleus ovoidalis (Karten, 1967), nucleus decussationis supraopticae (Kappers et al., 1960), nucleus commissurae transversae Gudden (Frey, 1933), lateral hypothalamic nucleus (Kuhlenbeck, 1937), nucleus opticus suprarotundus (Hirschberger, 1967). In consideration of these connexions, which include the primary meso-diencephalic avian visual centres, one might have predicted a visual role for this bundle. The visual function of DSO first suggested by Chichinadze (1940) was recently substantiated by Meier (1971) who described a severe impairment in interhemispheric transfer of colour and pattern discrimination following DSO lesion. This observation is in good agreement with the data presented here.

[†]nucleus opticus suprarotundus.

ACKNOWLEDGMENTS

We would like to thank Prof. K. Akert for his valuable advice and continuous support, Drs. M. C. Trachtenberg and R. E. Meier for reading the manuscript, Mr. M. Felder, Mr. A. Fäh and Dr. M. Bénita for experimental assistance, and Mr. A. Fidéler, Miss L. Decoppet and Mrs. M. Müller for technical assistance.

REFERENCES

Bellonci, J., 1888, Ueber die centrale Endigung des Nervus opticus bei den Vertebraten, Z'schr. wissenschaftl. Zoologie 47: 1–46.
Bénita, M., 1970, Démonstration d'un appareil et de sondes cryogéniques, isolées thermiquement par gaine de vide, et fonctionnant en circuit fermé, J. Physiol. (Lille) 62: 333–334.
Bremer, F., Dow, R. S., and Moruzzi, G., 1939, Physiological analysis of the general cortex in reptiles and birds, J. Neurophysiol. 2: 473–487.

Chichinadze, N., 1940, Das Problem der Lokalisation kortikaler Prozesse, welche durch optische Reize hervorgerufen werden, *Mitt. Georg. Abt. Akad. Wiss. USSR* **1**: 609–614.

Cowan, W. M., Adamson, L., and Powell, T. P. S., 1961, An experimental study of the avian visual system, *J. Anat. (Lond.)* **95**: 545–563.

Cuénod, M., 1971, Split-brain studies. Functional interaction between bilateral central nervous structures. In: G. H. Bourne (Ed.) *The structure and function of the nervous tissue.* New York: Acad. Press (in press).

Edinger, L., Wallenberg, A., and Holmes, G. M., 1903, Untersuchungen über die vergleichende Anatomie des Gehirnes, 5, Untersuchungen über das Vorderhirn der Vögel, *Abh. Senckenberg nat. Ges., Frankfurt a.M.* **20**: 343–426.

Frey, E., 1933, Ueber die basale Opticuswurzel und die caudalen Verbindungen der Commissura transversa Gudden der Vögel, *Proc. kon. ned. Akad. Wet.* **36**: 351–359.

Hamdi, F. A., and Whitteridge, D., 1954, The representation of the retina on the optic tectum of the pigeon, *Quart. J. exp. Physiol.* **39**: 111–119.

Hirschberger, W., 1967, Histologische Untersuchungen an den primären visuellen Zentren des Eulengehirns und der retinalen Repräsentation in ihnen, *J. Orn.* **108**: 187–202.

Holden, A. L., 1968, The field potential profile during activation of the avian optic tectum, *J. Physiol.* **194**: 75–90.

Holden, A. L., 1968, Types of unitary response and correlation with the field potential profile during activation of the avian optic tectum, *J. Physiol.* **194**: 91–104.

Kappers, A. C. U., Huber, G. C., and Crosby, E. C., 1960, The comparative anatomy of the nervous system of vertebrates, including man, Vol. II. New York: Hafner Publ. Co., pp. 1046–1048.

Karamian, A. I., Vesselkin, N. P., Belekhova, M. G., and Zagorulko, T. M., 1966, Electrophysiological characteristics of tectal and thalamocortical divisions of the visual system in lower vertebrates, *J. comp. Neurol.* **127**: 559–576.

Karten, H. J., 1967, The organization of the ascending auditory pathway in the pigeon (Columba livia), *Brain Res.* **6**: 409–427.

Karten, H. J., 1969, The organization of the avian telencephalon and some speculations on the phylogeny of the amniote telencephalon, *Ann. N.Y. Acad. Sci.* **167**: 164–179.

Karten, H. J., and Revzin, A. M., 1966, The afferent connections of the nucleus rotundus in the pigeon, *Brain Res.* **2**: 368–377.

Karten, H. J., and Hodos, W., 1967, A stereotaxic atlas of the brain of the pigeon (Columba livia). Baltimore: Johns-Hopkins Univ.

Karten, H. J., and Nauta, W. J. H., 1968, Organization of retinothalamic projections in the pigeon and owl, *Anat. Record.* **160**: 373 (abstract).

Kuhlenbeck, H., 1937, The ontogenic development of the diencephalic centers in a bird's brain (chick) and comparison with the reptilian and mammalian diencephalon, *J. comp. Neurol.* **66**: 23–75.

Meier, R. E., 1971, Interhemisphärischer Transfer visueller Zweifachwahlen bei kommissurotomierten Tauben, *Psychol. Forsch.* **34**: 220–245.

Perlia, R., 1889, Ueber ein neues Opticuscentrum beim Huhne, *v. Graefe Arch. Ophthal.* **35**: 20–24.

Powell, T. P. S., and Cowan, W. M., 1961, The thalamic projection upon the telencephalon in the pigeon (Columba livia), *J. Anat. (Lond.)* **95**: 78–109.

Pritz, M. B., Mead, W. R., and Northcutt, G. R., 1970, The effects of wulst ablations on color, brightness and pattern discrimination in pigeon (Columba livia), *J. comp. Neurol.* **140**: 81–100.

Revzin, A. M., 1967, Unit responses to visual stimuli in the nucleus rotundus in the pigeon, *Fed. Proc.* **26**: 2238 (abstract).

Revzin, A. M., 1969, A specific visual projection area in the hyperstriatum of the pigeon (Columba livia), *Brain Res.* **15**: 246–249.

Revzin, A. M., and Karten, H. J., 1966, Rostral projections of the optic tectum and nucleus rotundus in the pigeon, *Brain Res.* **3**: 264–276.

Robert, F., and Cuénod, M., 1969, Electrophysiology of the intertectal commissures in the pigeon, I. Analysis of the pathway, *Exp. Brain Res.* **9**: 116–122.

Ten Cate, J., 1936, Physiologie des Zentralnervensystems der Vögel, *Ergebn. Biol.* **13**: 93–173.

Thomas, R. C., and Wilson, V. J., 1965, Precise localization of renshaw cells with a new marking technique, *Nature (Lond.)* **206**: 211–213.

Zeigler, H. P., 1963, Effects of endbrain lesions upon visual discrimination learning in pigeons, *J. comp. Neurol.* **120**: 161–182.

# RECURRENT EXCITATION IN THE CA3 REGION
# OF CAT HIPPOCAMPUS†

R. M. LEBOVITZ,‡ M. DICHTER,§ and W. A. SPENCER

*Department of Physiology, New York University Medical School,*
*550 First Avenue, New York, N.Y. 10016*
*and*
*The Department of Neurobiology and Behavior, The Public Health Research Institute,*
*455 First Avenue, New York, N.Y. 10016*

The deafferented fornix preparation was utilized to detect recurrent excitatory connections of hippocampal pyramidal cells. The intensity and axosomatic locus of the well-known recurrent inhibitory action posed special difficulties for detecting the more subtle recurrent excitation. Therefore, initially negative extra-cellular unitary recordings, in which cellular damage should be minimal, were employed to study the recurrent excitatory actions. Such recordings from identified pyramidal cells of the CA3 region of cat hippocampus revealed synaptic types of activation as well as antidromic invasion upon stimulation of the deafferented fornix. These data thus demonstrate the existence of a recurrent excitatory pathway in this region. This is a previously undisclosed circuit required by recent hypotheses concerning the origin of experimentally induced interictal spike discharges in the hippocampus.

## INTRODUCTION

Among Wade Marshall's long-standing interests are the intrinsic mechanisms of cortical function. This paper is presented to honor his many contributions in this area and to acknowledge his encouragement and inspiration to those who have pursued this field following his leads.

The hippocampus is a particularly advantageous structure for studying intrinsic feedback mechanisms in cortical tissue. A large percentage of the regularly arranged pyramidal cells sends axons into the fimbria and fornix (Cajal (1911), Daitz and Powell (1954), Lorente de Nó (1934) Raisman *et al.* (1966)). These axons can be stimulated in isolation following chronic fornix deafferentation (Andersen *et al.* (1964a), Kandel *et al.* (1961), Spencer and Kandel (1961)), thereby allowing study of the recurrent pathways that provide pyramidal cell feedback.

A recent study of 'interictal spike' epileptiform activity utilized the deafferented fornix preparation to reveal the participation of recurrent pathways

(Dichter and Spencer (1969)) in the genesis of this abnormal response. Findings suggested that both recurrent excitation and recurrent inhibition play a critical role in the genesis of this special type of seizure activity.

A number of studies have now documented the existence of recurrent inhibitory action following stimulation of the deafferented fornix (Andersen *et al.* (1964a), Dichter and Spencer (1969), Kandel *et al.* (1961), Kandel and Spencer (1961)). By contrast, there is as yet no direct evidence for recurrent excitation. Occasional signs of the orthodromic (i.e., trans-synaptic) mode of activation of CA3 pyramidal cells upon stimulation of the deafferented fornix were noted by Andersen *et al.* (1964b), but were assigned to possible direct stimulus spread to the hippocampus. Further examination of recurrent excitation seemed important because the recently advanced hypothesis concerning the genesis of hippocampal epileptiform activity specifically postulates the existence of local recurrent pyramidal cell excitation in the hippocampus (Dichter and Spencer (1969)).

Because of the immense power of the pyramidal cell recurrent inhibition, and the bias of intracellular recordings for axosomatic synaptic actions, it seemed possible that more subtle and probably remote recurrent excitatory actions, if they exist,

† Work supported by USPHS Grants: BN 05980, NS 09361, MSTG 5 T05 GM 01668, and K3-NS-19485.

‡ Current Address: Department of Physiology, University of Texas Southwestern Medical School, Dallas, Texas.
§ Current Address: EEG Laboratory, NINDS, National Institute of Health, Bethesda, Maryland.

might be obscured by the powerful inhibition evident in intracellular recordings. We therefore have adopted a sensitive extracellular unitary recording approach similar to that which Eccles *et al.* (1966) employed to detect the recurrent actions of Purkinje cells. To minimize cell damage we have relied primarily on recordings of initially negative extracellular action potentials (Mountcastle *et al.* (1957)). Such recordings have revealed clear evidence of significant local recurrent excitatory actions on hippocampal pyramidal cells in the CA3 region, as indicated in a brief preliminary publication (Lebovitz *et al.* (1969)).

## METHODS

The data in this study were derived from 25 cats anesthetized with thiopental or pentobarbital (30 mg/kg initial dose, IP) and prepared as described in previous publications (Dichter and Spencer (1969), Kandel *et al.* (1961) Spencer and Kandel (1961)). At least twenty days prior to the acute experiments, the fornix on one side and the hippocampal commissure were sectioned with an electric cautery under direct vision; this procedure results in a preparation having no hippocampal afferents in the fornix or fimbria (Spencer and Kandel (1961)). The experimental setup for the final acute experiment is diagrammed in Figure 1. A four-pole platinum wire electrode was gently placed upon the proximal end of the transected fornix-fimbria outflow tract as far as possible away from the hippocampus. The array was connected to a switch that could select any one of the four serial pairs of wires as the stimulating dipole. The stimuli were monophasic pulses of 0.01 duration. Unit activity was recorded from the CA3 zone of the hippocampus with stainless steel or tungsten micro-electrodes, or with micropipettes—usually filled with NaCl, but occasionally with K-acetate when data were collected during other studies based on intracellular recording. The reference electrode was a platinum wire firmly inserted into the body of the temporalis muscle. We chose to use NaCl extracellular recording microelectrodes so that neuronal activity could be detected with minimal trauma to the responding cell; the spike potentials were therefore generally small and initially negative, although initially positive spikes were also occasionally studied. Frequently, in order to display the recorded unit activity more clearly, the micro-electrode signal was passed through an electronic high pass filter (low frequency cut off at 200 to 600 Hz). Timing and stimuli were monitored on one

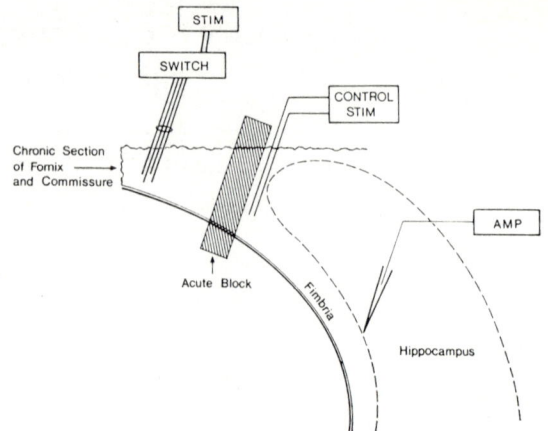

FIGURE 1  Schematic diagram of stimulating and recording setup. Fornix and hippocampal commissure had previously been sectioned several weeks previously to destroy the afferent fibers (to the hippocampus) and their terminals. The four wire stimulating electrode covered an area approximately 1 mm square. Selected serial pairs of wires were connected to the stimulator via a rotary switch yielding a choice of four slightly displaced dipoles. Control for stimulus spread was accomplished by local anesthetization or section of the fornix-fimbria between the stimulating and recording electrodes (cross-hatched rectangle). The disappearance of the antidromic field response and its reappearance when stimulating proximal to the acute block ('Control Stim') verified that there was no significant stimulus spread.

beam of a multi-beam oscilloscope and the records presented are of photographs taken directly from the tube face.

Following the recording of spike activity, impulse conduction in the fornix-fimbria was blocked between stimulating and recording electrodes either by placing a piece of filter paper saturated with 2% xylocaine across the fimbria, or by carefully sectioning this structure at the same intermediate point. The disappearance of the hippocampal field potentials over the range of stimulus strengths utilized indicated that the results were not contaminated by electrical spread to hippocampal cells. That the hippocampus itself was not anesthetized or damaged was verified by stimulating again on the other (hippocampal) side of the block. At the conclusion of the experiment the brain was removed and hardened in 10% Formalin for later inspection of the extent and contour of the original fornix section. During some experiments, the electrical activity of the exposed contralateral hippocampus was recorded as an additional check on the completeness of the original deafferentation by noting the absence of responses relayed through the hippocampal commissure.

## RESULTS

*Antidromic Responses of Pyramidal Cells Recorded Extracellularly*

It has been repeatedly demonstrated (Andersen *et al.* (1964a), Dichter and Spencer (1969), Kandel *et al.* (1961), Kandel and Spencer (1961)) that stimulation of the deafferented fornix generates antridromic spikes followed by large IPSPs in pyramidal cells: thus, as first recognized by Kandel *et al.* (1961), these IPSPs are recurrent. An example of these IPSPs is shown in Figure 2B. Andersen, Eccles and

FIGURE 2  Antidromically evoked field potential response (upper trace) and intracellularly recorded IPSP (lower trace) on same time scale. Single shock to fornix-fimbria in deafferented fornix preparation. In this and all figures following, extracellular as well as intracellular unit recordings will be displayed positive up. 2 mV and 10 mV labels for voltage calibration refer to upper and lower traces respectively.

Loyning (1964b) have suggested that the basket cells with their widespread axosomatic pyramidal cell connections (Lorente de Nó (1934), Blackstad and Flood (1963)) are the inhibitory interneurons in this recurrent pathway. The extracellular field potential associated with the recurrent inhibition generated by stimulation of the deafferented fornix, shown in Figure 2A at higher gain, exhibits two components: an initial, downward (negative) component related, at least in part, to the antidromic invasion of pyramidal cells; and an upward (positive) component associated with the initial phases of the IPSP. These field potentials formed useful reference points for studying the timing of extracellularly recorded unitary responses.

Antidromic responses of pyramidal cells were easily recognized, definitively identifying the responding unit as a pyramidal cell. The specific criteria employed for identifying antidromic responses in pyramidal cells were similar to those

used in extracellular analyses of neocortical pyramidal tract cell responses (Phillips (1959), Towe *et al.* (1963)): (1) the spike latency was brief, regular and independent of stimulus strength; and (2) spike generation followed rapid stimuli (up to 100/sec). In the vicinity of the cell body layer the negative antidromic field potential was prominent and defined the approximate latency of antidromic invasion. It is significant, however, that units which satisfied these criteria† could discharge after the negative wave, and that not all units firing during the negative wave satisfied these criteria, i.e., some of these appeared to be initiated trans-synaptically.

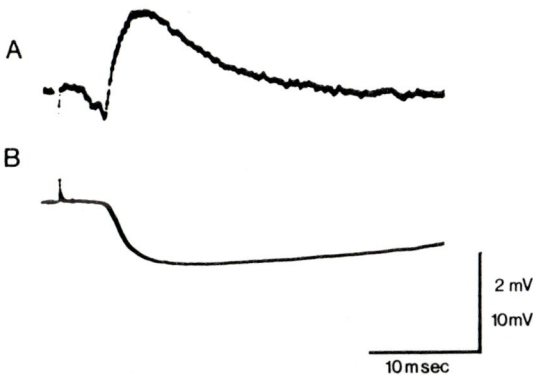

*Trans-synaptic Activation*

In a certain sense, one might be obliged to label as 'orthodromic' or 'trans-synaptic' any unitary response which fails to satisfy one or more of the criteria for antidromic activation listed above. It was common, however, to observe action potentials generated in response to fornix stimulation which satisfied the *latency* criterion, but which followed stimuli at no more than 10 or 20 per second, even at the maximum stimulus intensities. While there may be some hesitation in labeling such a response 'antidromic', there would obviously be little justification for concluding it was 'orthodromic'. In fact, it seems most probable that these are cases in which antidromic invasion is blocked by an unusually effective recurrent IPSP. Injury or other factors would perhaps account for the poor high-frequency following. We therefore took as our primary criterion for orthodromic activation that the spike latency show either 1) obvious jitter (that is, spontaneous and irregular latency variation at a given stimulus intensity), or 2) a marked dependence of latency upon stimulus strength. There is, of course, nothing to preclude a fortuitous early orthodromic spike appearing late in the antidromic field, and this was frequently observed at high stimulus intensities. However, the first suggestion of an orthodromic type of activation as stimulus intensity was raised gradually was usually that the single spike fell a millisecond or more after the antidromic field potential. In the deafferented fornix preparation it was common to observe such single-

---

† On occasion, units were sampled which satisfied all of the above criteria but which also exhibited a slight but definite decrease or increase in latency at high stimulus rates. This behavior was not studied in detail although it could be demonstrated easily with paired stimuli that the magnitude of the latency shift was a function of the inter-stimulus interval and spike-dependent. Such shifts may be related to afterpotentials.

discharge, orthodromically activated units in the vicinity of the cell body layer.

The intervals during which reliably identified single-discharge orthodromic types of activation were observed may be indicated best by representative spike latency histograms (Figure 3). The ordinates represent the fraction of unitary responses

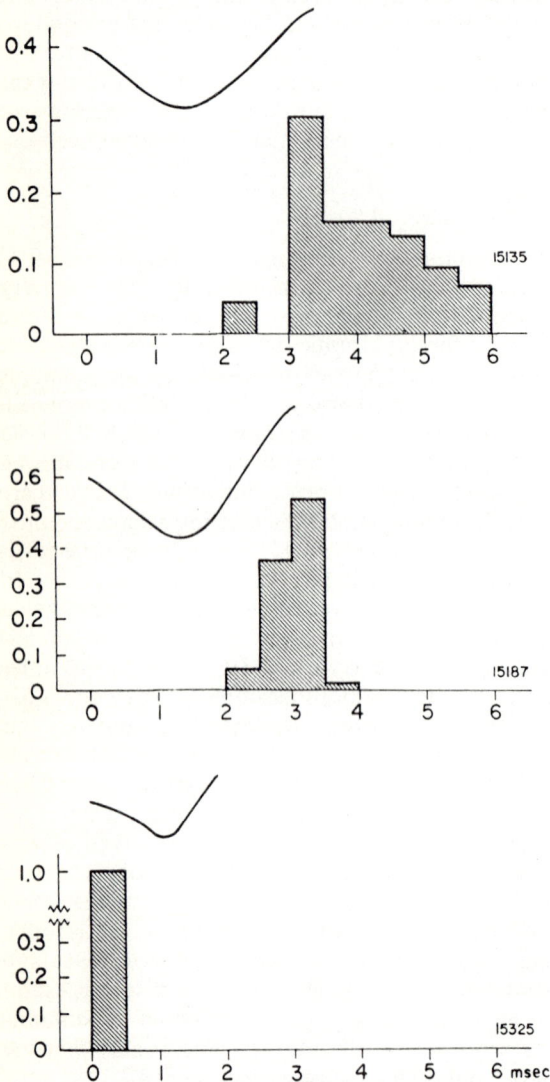

occurring within the stated time bin. In general the orthodromic spikes occurred during the early phase of the IPSP, judged by their timing with respect to the field potentials as shown for two separate units in the top two graphs of Figure 3. Antidromically activated units discharged earlier and at constant latency, as shown for another cell in the bottom graph. The firing latency decreased at high stimulus intensities such that some responses at the shortest latencies occurred while the antidromic field potential was still prominent. Figure 4 is derived from responses of a single unit activated at different

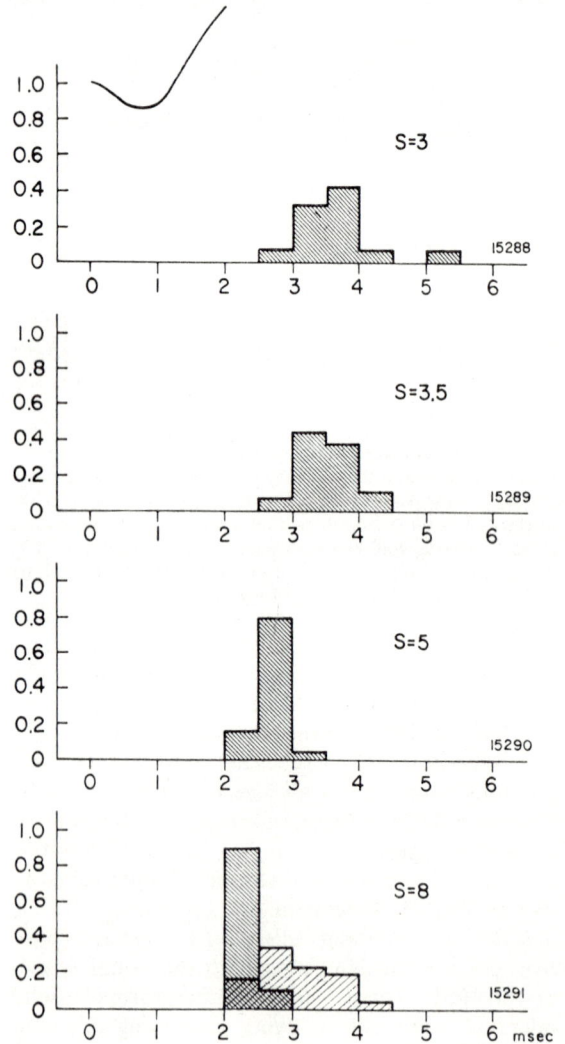

FIGURE 3  Spike latency histograms of orthodromically evoked unit discharge and antidromic (identified pyramidal cell) action potentials. Spike latencies are relative to the onset of the antidromic field potential. Antidromic field potential plus initial portion of IPSP field is sketched above each histogram. Bin width is 1/2 msec. Ordinate indicates fraction of spikes in stated bin.
Top two graphs: Two orthodromically driven units (50 spikes each histogram).
Bottom graph: Antidromically activated unit (100 spikes).

FIGURE 4  Spike latency histograms of a single orthodromically driven unit at 4 different stimulus intensities (S) indicated in arbitrary units. A total of 155 spikes were counted; summary histogram (i.e., for all stimulus intensities) shown in coarse crosshatching superimposed on bottom histogram for S = 8.

intensities (S = 3, 3.5, etc.) and illustrates latency shortening. We would emphasize the necessity of examining unitary firing *pattern* as well as time of firing before judging that a unitary response is an antidromic, hence pyramidal cell, spike. Indeed, it seems probable that the presumed antidromic field potential contains substantial contributions from units driven in an orthodromic manner.

Occasionally a much different pattern characterized by multiple firing in response to the fornix stimuli was observed (Figure 5). Units showing

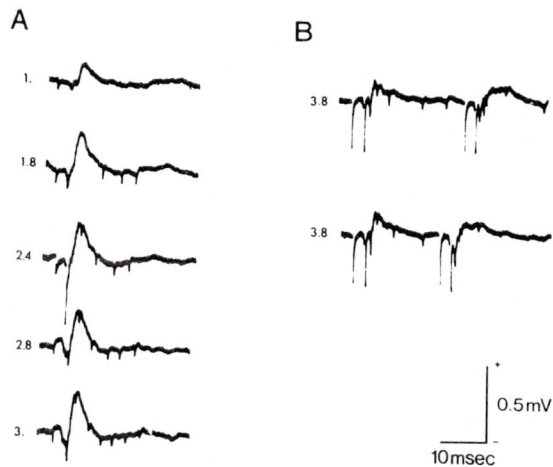

FIGURE 5   Responses of tentatively identified inhibitory interneuron, extracellularly recorded. Stimulus strength noted in arbitrary units at left of each sweep.
*A*. Response (as measured by burst duration, frequency and onset latency) is enhanced by stronger stimuli. Antidromic firing of a nearby pyramidal cell occurred intermittently.
*B*. Double stimuli at short intervals greatly decreased interspike interval of first few spikes in each burst. These records are from the same unit as in *A*, but recorded somewhat later when a large antidromic spike was present.

such rapid firing were extremely sensitive to stimulating electrode position, and were often only barely detectable above the noise level. When such units could be studied, we often noted an increased number of spikes per response, enhanced firing frequency, and shortened initial spike latency when progressively higher stimulus strengths were employed (Fig. 5A). Circumstantial evidence suggests that the cells giving these responses are the same as those postulated by Andersen *et al.* (1964b) to be the inhibitory basket cells of the statum oriens. The evidence is as follows: (1) such recordings were rare and highly sensitive to electrode position, suggesting a small and infrequent cell; (2) the onset time and length of the burst response corresponded to the time course of at least the early phase of the

recurrent pyramidal cell IPSP; (3) augmentation of firing occurred with higher stimulus intensities; (4) such recordings were frequently obtained at levels somewhat superficial (100 $\mu$ or so) to the pyramidal cell layer. It was often noted that the firing frequency of the response to the second of paired stimuli was enhanced at short interstimulus intervals (Figure 5B), suggesting potentiation.

*Definitive Identification of Pyramidal Cells*

The above observations have shown that an orthodromic type of activation of hippocampal cells can follow stimulation of the deafferented fornix. Definitive identification of these units, however, has not yet been discussed. The field potentials and the measured depth corresponding to the position of the orthodromically driven units indicated that they were in the vicinity of the pyramidal cell body layer of the hippocampus. This, in addition to the large number of such units and their spatial admixture with clearly antidromically invaded pyramidal units, made it most probable that the orthodromic discharges arose from hippocampal pyramidal cells. However suggestive, these were still an inconclusive means of identification and it remained to demonstrate unambiguously that some of the synaptically activated units were indeed pyramidal cells. The most direct identification of a pyramidal cell would be to show that it is antidromically invaded following stimulation of the deafferented fornix (Kandel *et al.* (1961)). Accordingly, we attempted several different stimulating procedures with each stable unit, and were able to locate 16 units, out of approximately 120 examined, which responded both antidromically and orthodromically to deafferented fornix stimuli.

One such procedure was to slightly alter the stimulus locus, once a stable unit had been identified, by reversing the stimulus polarity and/or selecting another dipole from the four available. (Attendant changes in the field potentials confirmed that different populations of cells were being activated by the fornix stimuli.) This slight, but often critical, change shifted the center of the most intense activity relative to the monitored unit; this evidently altered the balance of antidromic and orthodromic input since it resulted in a clear shift of firing pattern from one mode to the other (Figure 6).

Simply changing the stimulus strength was also often effective in demonstrating two distinct modes of firing. Figure 7 shows a unit whose latency and response mode was dependent upon stimulus strength. The firing pattern was characteristically antidromic with strong stimuli (Figure 7A), and orthodromic with weak (Figure 7B).

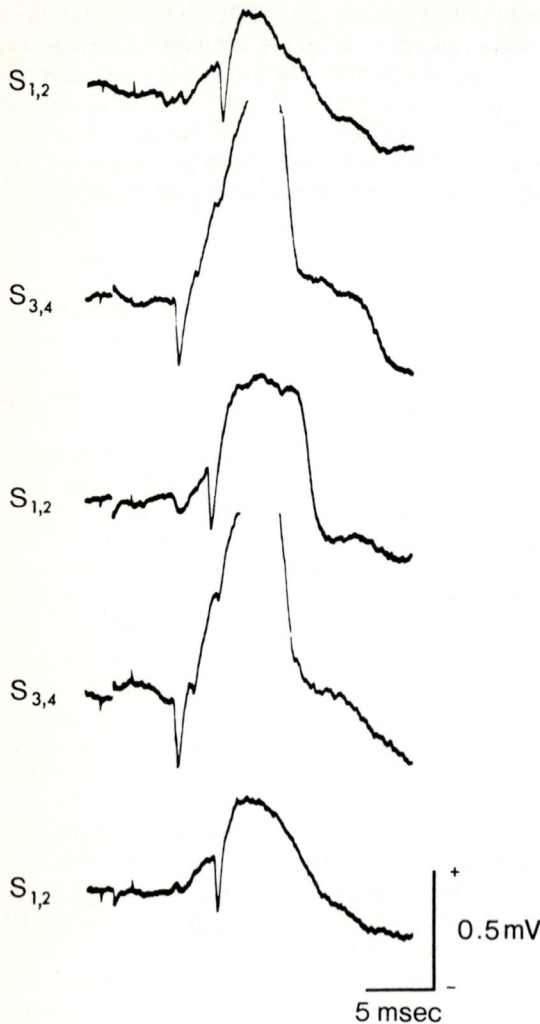

$S_{1,2}$

$S_{3,4}$

$S_{1,2}$

$S_{3,4}$

$S_{1,2}$

0.5mV

5 msec

FIGURE 6 Effect of change of position of the stimulus dipole on mode of activation. $S_{1,2}$ (dipole 1–2): evoked orthodromic type response which showed obvious jitter. $S_{3,4}$ (dipole 3–4): same stimulus of same strength evoked unit discharge which satisfied criteria for antidromic invasion; therefore, this can be identified as a pyramidal cell.

With slightly less intense stimuli the firing latency increased, showed marked jitter and inability to follow rapid stimuli for more than a few repetitions (Figure 7B). It may be presumed that the recorded cell's axon was no longer being activated and that these orthodromic types of responses represent trans-synaptic recurrent excitation from activation of other pyramidal cell axons. A schematic summary of the dual mode of activation is shown in Figure 8. The range of onset times for orthodromic activation is shown by the bidirectional arrow. Note that its briefest latency is sufficiently short to suggest media-tion with only a single synaptic delay. Even the

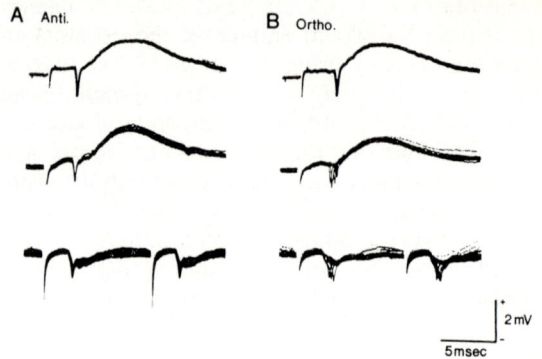

A Anti.      B Ortho.

2 mV

5msec

FIGURE 7 Orthodromic type activation of identified pyramidel cell.
A. Single trace (upper), superimposed multiple traces at 2/sec (middle) and superimposed multiple traces at 100/sec (lower). The stability of response latency indicates antidromic invasion, hence that this is a pyramidal cell response.
B. Same unit at slightly decreased stimulus intensity. Firing latency is increased and jitter is obvious in the superimposed traces at 2/sec (middle). Unit now unable to regularly follow high frequency (100/sec) stimuli (lower).

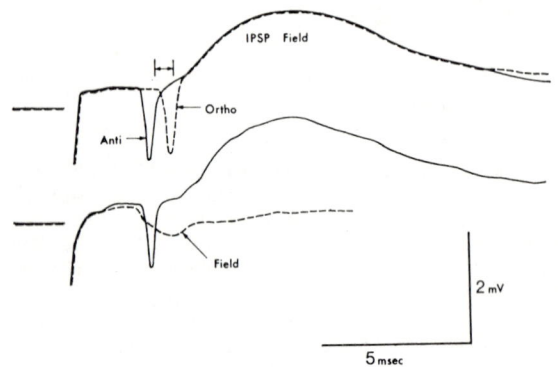

IPSP Field

Ortho

Anti

Field

2 mV

5 msec

FIGURE 8 Line tracings of the antidromic and orthodromic type firing modes for unit illustrated in Figure 7. The range of onset latencies for the orthodromic spike is given by the vertical bars bracketing double-ended arrow and is 0.5 to 1.2 msec after the time of onset of the antidromic spike. 'Field' tracing refers to slow wave evident during 100/sec stimulation.

discharges with longest latency examined here would still not represent 'rebound' discharges following the IPSP (Andersen et al. (1964c), Kandel et al. (1961), Spencer and Kandel (1961)).

A few units showed a double response over a limited range of stimulus strengths (Figure 9). The earlier response fell temporally within the anti-dromic field potential and could be shown fully to satisfy the antidromic criteria. A second spike generally followed the first by 3–5 msec and was clearly orthodromic. It was our impression that

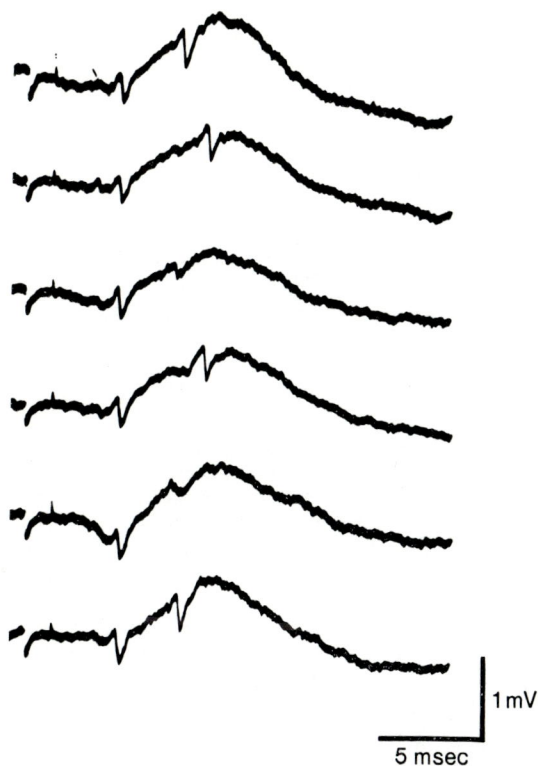

FIGURE 9 Double firing of identified pyramidal cell at threshold for the second response. Earlier response is antidromic. Later response from the same cell is intermittent and clearly orthodromic. The small negative deflection when the second invasion fails (the third and fifth trace) may represent partial invasion of the cell.

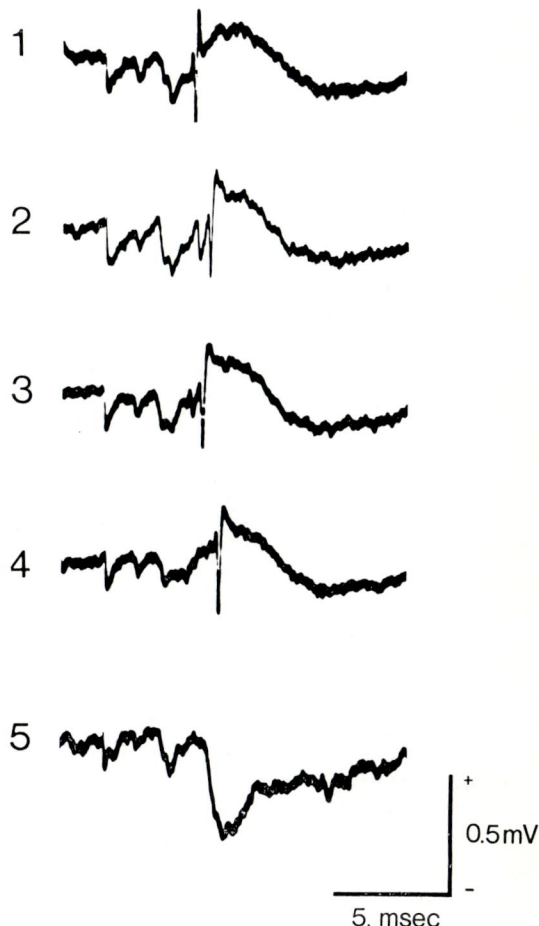

FIGURE 10 Orthodromic mode of firing. Unit tentatively identified as a pyramidal cell since its impalement (with loss of action potential) revealed an IPSP in response to the fornix stimulus (Line 5). Small amplitude of IPSP is due to large diameter of electrode tip used to obtain initially negative extracellular spikes.

high frequency driving promoted this double response in some units.

In figure 10 is illustrated the responses of an orthodromically activated cell, showing pronounced spike latency jitter (traces 1–4), which was subsequently impaled by the microelectrode. The impalement rapidly destroyed the cell but not before a few small IPSPs were identified in response to the fornix stimuli (trace 5). Since these intracellular recordings had a configuration characteristic of hippocampal pyramidal cells (Kandel *et al.* (1961)), this provides further, though less direct, evidence to suggest that pyramidal cells may be driven in an orthodromic manner following stimulation of the chronically deafferented fornix. In such cases the impalements were usually not of high quality since, as indicated before, fairly large electrodes were chosen to record initially negative extracellular spikes to minimize cell damage, and these are not optimal for cell impalement. Also, the EPSPs, which

may be presumed as the agent responsible for the orthodromic type of firing in pyramidal cells, are probably remote and hence not easily discerned in intrasomatic recordings (see Andersen *et al.* (1966)).

## DISCUSSION

Recurrent excitation in the neocortex has been well documented (Eccles (1968), Phillips (1959), Stefanis and Jasper (1964 a and b), Takahashi *et al.* (1957)). Although the organization of hippocampal cortex lacks the structural complexity of neocortex, it possesses an unusually powerful recurrent inhibitory system, believed to be axosomatic in site of action, which makes the analysis of recurrent excitation a difficult task. In intracellular recordings from hippocampal pyramidal cells the recurrent IPSP generated by fornix stimulation usually dominates the soma membrane potential, whether or not a spike has been elicited. One would expect that, if excitatory PSPs are elicited on the extensive pyramidal dendrite system, they might be hidden very effectively by the 10–15 mV recurrent IPSP.

As demonstrated by Eccles *et al.* (1966) extracellular recording procedures are of great value in gathering evidence of rather subtle synaptic actions such as the recurrent hippocampal pyramidal cell excitation we have postulated. In the work reported here particular care was taken to avoid stimulus spread to the hippocampus by stimulating as close to the scar of the fornix section as was practical (that is, as far from the hippocampus as possible), and by using very brief (0.01 msec) stimuli. The magnitude of any residual current spread was checked by blocking the fornix with locally applied anesthetic or a second fornix section.

Because of the unique microanatomy of the hippocampal cortex (single layer of tightly packed pyramidal cell bodies) it was generally quite obvious, from the form of the field potentials, where the exploring microelectrode was located in relation to the pyramidal cell layer (stratum pyramidale). In the vicinity of the cell body layer we have observed cells which, to all appearances, are trans-synaptically driven by stimulation of the deafferented fornix. We have interpreted this as presumptive evidence of recurrent excitation through pyramidal axon collaterals since there exists no other conventional anatomical substrate for such orthodromic activation at the short latencies observed. The crucial observation that some of these orthodromically fired cells would also demonstrate antidromic activation by fornix stimuli clearly identified them as pyramidal cells.

Thus our data indicate the existence of a recurrent fornix axon collateral excitatory system impinging upon pyramidal cells of the CA3 region of the cat hippocampus. The timing of unit discharges initiated by this pathway is contrasted with antidromic spikes and presumed inhibitory interneuron discharges in Figure 11. Although the shortest latency trans-synaptic unitary responses are probably monosynaptic, somewhat later discharges are undoubtedly generated by more circuitous pathways containing interneurons.

The anatomical descriptions of Cajal (1911) and Lorente de Nó (1934) provide several examples of

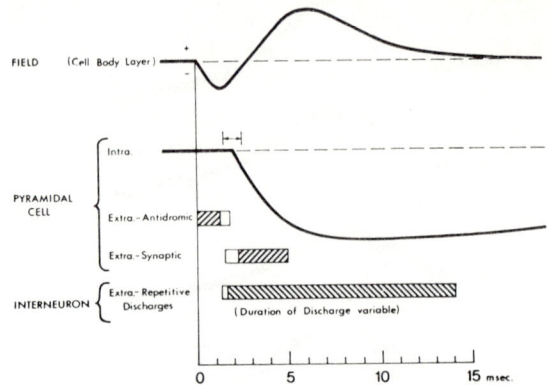

FIGURE 11 Summary diagram of time relations of responses of different cellular elements following single shock to deafferented fornix. Cross hatched portion of bars indicate periods of greatest firing probability; clear zones are periods of infrequent firing. This diagram presents the usual time relations we encountered.

local recurrent pathways which could produce these effects: most notably the longitudinal association system and the local, superficial pyramidal cell axon collaterals from the Shaeffer system which terminates in CA1 and CA2. These pathways should be considered as candidates for the recurrent excitatory actions our data have disclosed; further experiments are required to document which of these are actually involved.

The recurrent excitatory actions we have described are rather weak in comparison with the well known recurrent inhibitory action. However, this should not be taken to indicate that the excitation would be insignificant physiologically; it should be recalled that the recurrent inhibition in the hippocampus is perhaps the most powerful in the nervous system and other actions would thus appear weak by comparison. In addition, all our experiments were carried out under fairly deep anesthesia and with much of the hippocampal input eliminated; recurrent excitation may become much more effective under more physiological conditions.

Studies of the simplest type of hippocampal epileptiform activity, the so-called penicillin 'interictal spike', have revealed large and prolonged pyramidal cell depolarizations which have the characteristics of summated recurrent EPSPs (Dichter and Spencer (1969)). It seems likely that these large pathological depolarizations are mediated by the same recurrent excitatory pathways giving the rather subtle effects disclosed in more normal preparations by the experiments we have described.

It is important to note that the pathways mediating

the effects we have outlined deliver synaptic excitation close to regions of antidromic activation, hence could provide positive feedback actions. In contrast, the Shaeffer collateral system (Lorente de Nó (1934)), a pathway which might seem capable of mediating interictal spike genesis, is primarily to be regarded as an association bundle linking different zones of the hippocampus. Its participation in local positive feedback operations such as we've described would be confined to effects exerted in CA3 through the proximally branching collaterals of the longer collateral projection to CA1 and CA2. These proximal branches would constitute only one of several anatomical groups of axon collaterals which should be considered as candidates for mediation of the recurrent excitatory action we have shown here and, by inference, for genesis of the large pyramidal cell depolarizations associated with interictal spikes.

## REFERENCES

Andersen, P., Blackstad, T. W., and Lømo, T., 1966, Location and identification of excitatory synapses on hippocampal pyramidal cells, *Exp. Brain Res.* 1: 236–248.

Andersen, P., Eccles, J. C., and Løyning, Y., 1964, Location of postsynaptic inhibitory synapses on hippocampal pyramids, *J. Neurophysiology* 27: 592–607.

Andersen, P., Eccles, J. C., and Løyning, Y., 1964, Pathway of postsynaptic inhibition in the hippocampus, *J. Neurophysiology* 27: 608–619.

Andersen, P., Eccles, J. C., and Sears, T. A., 1964, The ventrobasal complex of the thalamus: types of cells, their responses and their functional organization, *J. Physiology* 174: 370–399.

Blackstad, T. W., and Flood, P. R., 1963, Ultrastructure of hippocampal axosomatic synapses, *Nature* 198: 542–543.

Cajal, Ramon y S., 1911, *Histologie du Systeme Nerveaux de l'Homme et des vertebres*, Paris: Maloine, Vol. II.

Daitz, H. M., and Powell, T. P. S., 1954, Studies of the connexions of the fornix system, *J. Neurol. Neurosurg Psychiat.* 17: 75–82.

Dichter, M., and Spencer, W. A., 1969, Penicillin-induced interictal discharges from the cat hippocampus. II. Mechanisms underlying origin and restriction, *J. Neurophysiology* 32: 663–687.

Eccles, J. C., 1965, Cerebral synaptic mechanisms. In: *Brain and Conscious Experience*; Eccles, J. C., ed. New York: Springer-Verlag, pp. 24–58.

Eccles, J. C., Llinas, R., and Sasaki, K., 1966, The action of antidromic impulses on the cerebellar Purkinje cells, *J. Physiol.* 182: 316–345.

Kandel, E. R., Spencer, W. A., and Brinley, F. J., 1961, Electrophysiology of hippocampal neurons. I. Sequential invasion and synaptic organization, *J. Neurophysiology* 24: 225–242.

Lebovitz, R., Dichter, M., and Spencer, W. A., 1969, Recurrent excitation in hippocampus, *Fed Proc.* 28: 455.

Lorente de Nó, R., 1934, Studies on the structure of the cerebral cortex. II. Continuation of the study of the ammonic system, *J. Psychol. Neurol.* 46: 113–177.

Mountcastle, V., Davies, P., and Berman, A., 1957, Response properties of neurons of cat's somatic sensory cortex to peripheral stimuli, *J. Neurophysiol.* 20: 374–407.

Phillips, C. G., 1959, Actions of antidromic pyramidal volleys on single Betz cells in the cat, *Quart. J. Exp. Physiol.* 44: 1–25.

Raisman, G., Cowan, W. M., and Powell, T. P. S., 1966, An experimental analysis of the efferent projection of the hippocampus, *Brain* 89: 83–108.

Spencer, W. A., and Kandel, E. R., 1961, Hippocampal neuron responses to selective activation of recurrent collaterals of hippocampofugal axons, *Exptl. Neurol.* 4: 149–161.

Stefanis, C., and Jasper, H., 1964, Intracellular microelectrode studies of antidromic responses in cortical pyramidal tract neurons, *J. Neurophysiol.* 27: 828–854.

Stefanis, C., and Jasper, H., 1964, Recurrent collateral inhibition in pyramidal tract neurons, *J. Neurophysiol.* 27: 855–877.

Takahashi, K., Kubota, K., and Uno, M., 1967, Recurrent facilitation in cat pyramidal tract cells, *J. Neurophysiol.* 30: 22–34.

Towe, A. L., Patton, H. D., and Kennedy, T. T., 1963, Properties of the pyramidal system in the cat, *Exptl. Neurol.* 8: 220–238.

# AN ANALYSIS OF DISHABITUATION AND SENSITIZATION OF THE GILL-WITHDRAWAL REFLEX IN *APLYSIA*†

THOMAS J. CAREW, VINCENT F. CASTELLUCCI and ERIC R. KANDEL

*Departments of Physiology and Psychiatry, New York University School of Medicine,*
*and*
*The Department of Neurobiology and Behavior,*
*The Public Health Research Institute of the City of New York, Inc.*

We have used a combined behavioral and cellular neurophysiological analysis to examine the relationship of sensitization to dishabituation of the gill-withdrawal reflex in *Aplysia*. The reflex withdrawal of the gill to tactile stimulation of the siphon or the purple gland (at the edge of the mantle shelf) shows habituation, dishabituation and sensitization. We have found that the purple gland and siphon provide independent afferent pathways each capable of eliciting the gill-withdrawal reflex. Habituation of one pathway did not affect the other, but a common 'dishabituatory' stimulus produced dishabituation of the habituated pathway as well as sensitization of the non-habituated pathway. These findings support the idea that dishabituation is not due to the removal of habituation but is an independent facilitation superimposed upon habituation. Our neurophysiological analysis showed that, on the cellular level, the neural correlates of sensitization and dishabituation are different reflections of a common heterosynaptic facilitatory process involving an increased effectiveness of excitatory synaptic transmission at the synapse between sensory and motor neurons.

## INTRODUCTION

A novel stimulus characteristically produces both somatic and autonomic reflex responses. The

† We join in dedicating this paper to Wade H. Marshall with the thought that several features of this paper parallel Wade's several careers. First, we have used natural somatosensory stimulation in this study to examine habituation and dishabituation in *Aplysia*. Wade's first career helped open up the neurophysiological exploration of the somatosensory (as well as other sensory) systems using natural stimulation. Second, each of us comes to the study of the cellular mechanisms of behavior having first been exposed to the capriciousness of behavioral analyses as a result of earlier studies of DC shifts and spreading depression (Brinley, Kandel and Marshall, 1960; Carew, Crow and Petrinovich, 1970; Castellucci and Goldring, 1970). Wade's second career was in part directed toward using spreading depression to analyze avalanche conduction and other intrinsic cortical mechanisms. Finally, one of us (ERK) would probably not be participating in these studies had it not been for a postdoctoral fellowship with Wade Marshall in the Laboratory of Neurophysiology. Even a casual reading of the literature and in particular this volume indicates the influence of Wade's third career. Wade's ability to sensitize (dishabituate) the young (and unknowing) to the mysteries of neural science has often been effective and has led to research not only on *Aplysia* but on lower forms as well, including man. We await with curiosity the outlines and impact of Wade's fourth career.

stimulus causes an animal to orient towards its source and also initiates changes in heart and respiratory rate. This set of responses, known collectively as the 'orienting reflex', was first described by Pavlov (1927) who also found that an animal gradually ceased responding to the novel stimulus if it was presented repeatedly. Pavlov referred to this decreased responsiveness as 'extinction' and attributed it to the building up of an inhibitory process. Pavlov also observed that a decremented orienting reflex could be restored by a period of rest and he attributed this restoration to disinhibition, or the removal of the inhibitory process.

After other reflex responses had been comparably analyzed (see Dodge, 1923; Humphrey, 1930a,b, 1933; Wendt, 1936; and Harris, 1943) it became evident that the waning of the orienting reflex was a special case of *habituation*, the general tendency of defensive as well as other reflex responses to decrease with repeated stimulation. Although this insight enlarged the scope of the problem, Pavlov's theoretical notions have continued to influence thinking about habituation and dishabituation. Thus, habituation is still often ascribed to an inhibitory process and dishabituation (the immediate restoration of a decremented response following the presentation of a novel stimulus) is thought to be a disinhibitory process that leads to the removal or abolition of habituation. For example, in his classic study, Humphrey (1930a,b) treats 'dehabituation' as synonymous with spontaneous *recovery* and attributes both phenomena to disinhibition. More recent examples of one or both of these theoretical positions are found in Konorski (1948, 1967), Sokolov (1963), Wickelgren (1967a,b) and Wall (1970).

There are now, however, two detailed studies that provide evidence that habituation is not an inhibitory process and that dishabituation is not due to disinhibition. Spencer, Thompson and Nielson (1966a,b,c) found that habituation of the flexion reflex of the spinal cat was best explained by a decreased excitatory drive and not by an increased inhibition. A similar conclusion was reached by Castellucci, Kupfermann, Pinsker and Kandel (1970) studying habituation of the gill-withdrawal reflex in *Aplysia*.

That dishabituation might not be a disinhibitory process, but a special case of sensitization, was suggested by Sharpless and Jasper (1956) in the course of studying the habituation of the EEG arousal response. Behavioral evidence in support of this notion was first provided by Spencer, Thompson and Nielson (1966a) while studying dishabituation of the flexion reflex. Spencer *et al.* (1966) found that non-habituated responses were also facilitated (sensitized) by a 'dishabituatory' stimulus and that habituated responses could be facilitated *beyond* their control values. Neither of these findings can be explained by disinhibition because that process could only restore decremented responses to their *initial* level. Thompson and Spencer (1966) and Groves and Thompson (1970) have recently reviewed experiments in other behavioral systems that also support this notion.

A more fundamental understanding of the relationship of habituation to dishabituation, and of dishabituation to sensitization requires a preparation in which these several behavioral modifications can be examined concurrently and analyzed on both behavioral and cellular neurophysiological levels. The gill-withdrawal response in *Aplysia* is a particularly useful system for examining these problems. The reflex withdrawal of the gill to tactile stimulation of the siphon or the purple gland (at the edge of the mantle shelf) shows both habituation and dishabituation. The purple gland and siphon provide independent afferent pathways each capable of eliciting the gill-withdrawal reflex. After one pathway has been habituated the effects of a common 'dishabituatory' stimulus can be examined concurrently on both the habituated and the non-habituated pathway. Moreover, part of the neural circuit of this reflex has been worked out on a cellular basis (Kupfermann and Kandel, 1969; Castellucci *et al.*, 1970). One can therefore examine the neural mechanisms of habituation, dishabituation and sensitization.

In this study we have used a combined behavioral and cellular neurophysiological analysis to examine the relationship of sensitization to dishabituation of the gill-withdrawal reflex in *Aplysia*. Our findings provide further support for the notion that dishabituation is not due to the removal of habituation but is an independent facilitation superimposed upon habituation. We have found that sensitization and dishabituation are different reflections of a common heterosynaptic facilitatory process involving an increased effectiveness of excitatory synaptic transmission at the synapse between sensory and motor neurons.

A preliminary description of some of the cellular analysis presented here has previously been reported (Castellucci *et al.*, 1970).

## METHODS

### I. *Behavioral Studies*

The receptive field of the gill-withdrawal reflex includes the mantle shelf and its edge, the purple gland, and the siphon (Kupfermann and Kandel, 1969). In these experiments we have restricted our analysis to two points at the extreme margins of the receptive field: (1) the anterior one third of the mantle shelf and purple gland (purple gland), and (2) the siphon and posterior edge of the mantle shelf (siphon) (Figure 1–A1).

Animals were restrained in a small aquarium. The purple gland was pinned to a wax-covered lucite stage (Pinsker *et al.*, 1970) and the siphon was hooked and moored in place. Two Water Pics (Aquatec Products, Colo.) were positioned to deliver independently quantifiable jets of sea water to either part of the receptive field. A photocell was placed under the gill to monitor gill movement. Stimulus intensity was adjusted so that the first stimulus to either area produced a full gill-withdrawal reflex (Figure 1, A2). Interstimulus intervals (i.s.i.'s) ranged from 30 seconds to 2 minutes. Shorter inter-stimulus intervals could not be used in behavioral experiments because the duration of the unhabituated responses often exceeded 20 seconds.

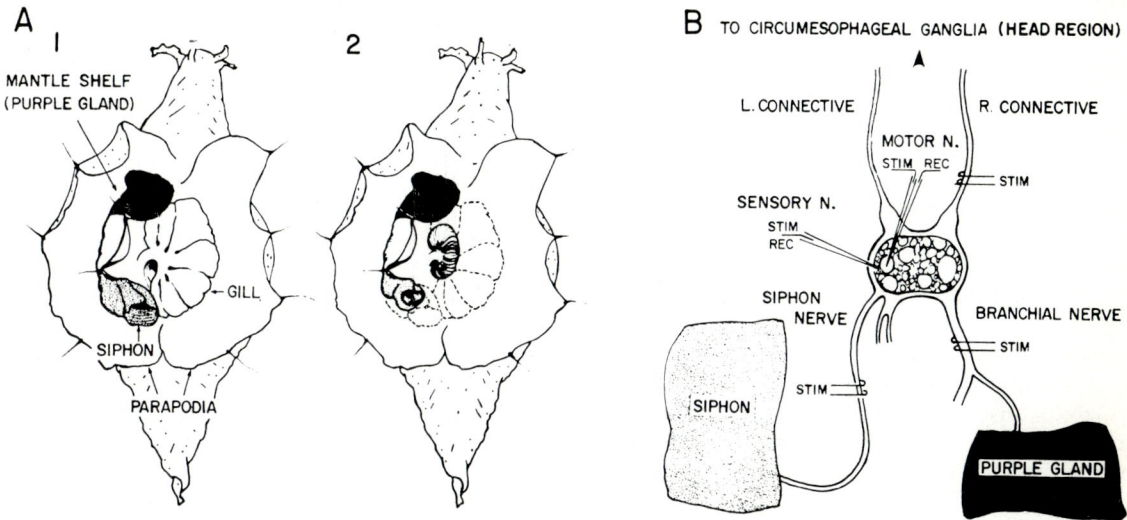

FIGURE 1   *A*. Dorsal view of an intact *Aplysia*. The parapodia and mantle shelf have been retracted to reveal the gill. *Part* 1: Position of organs in unstimulated condition. The anterior part of the mantle shelf and its edge, the purple gland are indicated in *solid black*. This area of the receptive field for the gill withdrawal reflex was used in the behavioral experiments and called *purple gland*. The siphon and the posterior part of the mantle shelf and of the purple gland are *stippled*. In the behavioral experiments they are called *siphon*. *Part* 2: Position of organs during withdrawal reflex following tactile stimulation within the receptive field. The mantle shelf, purple gland and the siphon contract along with the gill. Dotted lines indicate the initial position of the gill and siphon.
*B*. Composite diagram of the isolated ganglion preparation. In the neurophysiological studies, the siphon nerve, which innervates the siphon, and the branchial nerve, which innervates the anterior part of the mantle shelf and purple gland, were electrically stimulated. Stimulation of the right or left connective provided stimuli for hetero-synaptic facilitation. In some experiments the ganglion was removed from the animal with a piece of the siphon skin still attached through the siphon nerve. Intracellular microelectrodes were inserted into sensory neurons as well as into the motor neuron L7.

## II. *Neurophysiological Studies*

The abdominal ganglion (Figure 1B) and, in some experiments, a piece of the siphon skin connected to the ganglion via the siphon nerve, was removed and pinned to the paraffin floor of a lucite chamber as previously described (Frazier *et al.*, 1967; Castellucci *et al.*, 1970). The chamber was perfused with artificial sea water and kept at room temperature (22°C). In some cases a sea water solution with 90 mM $Ca^{++}$ (10 × normal concentration) was used to block polysynaptic input to the motor neuron (L7), or a solution with 220 mM $Mg^{++}$ (4.5 × normal concentration) was used to block chemical transmission in the ganglion and in peripheral structures. The solutions were prepared by adding isotonic solutions of $CaCl^2$ or $MgCl^2$ to artificial sea water.

Intracellular recordings from the sensory neurons were obtained with glass microelectrodes filled with 2M $K^+$ citrate having a resistance of 20–30 megohms. The electrode was connected to a Wheatstone bridge for current injection. The motor neuron L7 was impaled with either a double barrel (Mendelson, 1967) or with two independent microelectrodes, one for recording and the other for passing current. Nerves and connectives were stimulated with brief electrical pulses using silver-silver chloride electrodes.

As in the behavioral experiments, an inter-stimulus interval of 30 seconds was commonly used. However, since the duration of the EPSP was much less than the behavioral response, shorter interstimulus intervals of 10 and 15 seconds were also explored. EPSP decrement was obtained with all intervals; the shorter intervals, however, produced more rapid decrement.

## RESULTS

### I. *Behavioral Studies*

The reflex we studied involves the withdrawal of the gill to stimulation of a sensory receptive field consisting of the siphon and the edge of the mantle shelf which contains the purple gland (Figure 1). The anterior portion of the mantle shelf and the purple gland are innervated by a branch of the branchial nerve. The siphon and the posterior third of the purple gland and the mantle shelf are innervated by the siphon nerve (see Hoffmann, 1939, and Figure 1B). Therefore the siphon and the anterior third of the purple gland provide two anatomically distinct afferent pathways for a common motor response. We will simply refer to these two pathways as 'siphon' and 'purple gland' respectively.

To examine the relationship of dishabituation to sensitization we compared the effects of a 'dishabituatory' stimulus on both habituated and non-habituated responses by using these two different reflex pathways. If the dishabituation is due to a removal of habituation, the 'dishabituatory' stimulus should facilitate only the habituated pathway and bring it up to its original value. If, however, dishabituation is a special manifestation of sensitization, then: (1) a 'dishabituatory' stimulus might facilitate a non-habituated response as well as a habituated one, and (2) the habituated response might be capable of being facilitated *beyond* its control value.

It was first necessary, however, to determine whether the non-habituated pathway was affected by habituation of the other pathway.

## (a) *Lack of Generalization of Reflex Habituation*

To examine the consequence of habituating the gill-withdrawal reflex to stimulation of one afferent pathway (from the purple gland) on the response produced via another non-habituated pathway (from the siphon), we first delivered a single test stimulus to the siphon (T1). We then habituated the reflex by repeated stimulation of the purple gland at intervals ranging from 30–120 seconds. These rates of stimulation produced habituation ranging from 5–50% of the initial control value. Habituation of the purple gland pathway was followed by another stimulus to the siphon (T2) to test for generalization. After a period of rest, the experimental procedure was reversed: the previously habituated part of the receptive field (purple gland) was made the target of the test stimuli and the previously non-habituated part of the field (siphon) was made the target for the repetitive stimuli (Figure 2).

The data from 23 runs in 12 experiments with ınter-stimulus intervals of 2 min (N = 4) and 30 ₅ec (N = 8) were pooled. Each animal contributed

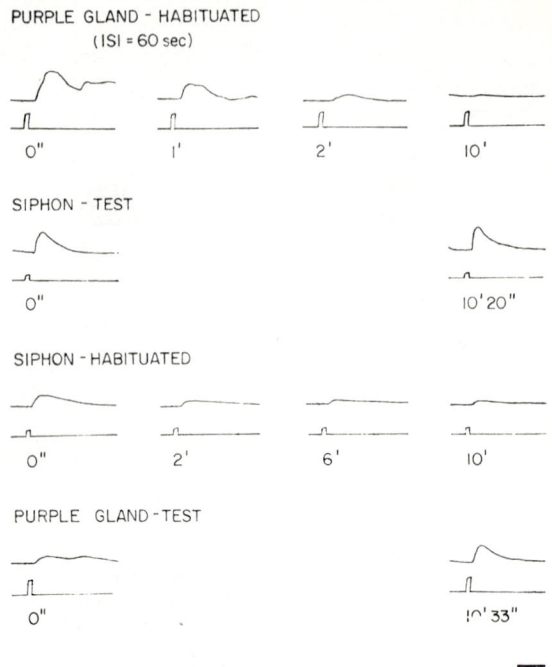

FIGURE 2  Lack of generalization of habituation from the purple gland to siphon. Stimuli were applied to the siphon and to the purple gland to produce a gill-withdrawal reflex (inter-stimulus interval = i.s.i. in this and all other figures). A single (test) stimulus was first presented to the siphon (0″). Repetitive stimuli were then applied to the purple gland which produced habituation of the reflex response (0″ to 10′). A second test stimulus now applied to the siphon (10′ 20″) produced an undecremented response, essentially the same reflex withdrawal as did the first test stimulus (0″), indicating that no generalization of habituation occurred from the purple gland to the siphon. After 40 minutes rest, the experimental procedure was reversed. The first test stimulus to the purple gland (0″) produced a response which is still partially habituated indicating incomplete recovery from the previous repetitive stimulation (compare purple gland habituated 0″ to purple gland test 0″). Even though the reflex was then habituated by repetitive stimulation to the siphon (0″ to 10′), a second test stimulus to the purple gland (10′ 33″) produced a response that was *further* recovered, indicating that no generalization of habituation occurred from the siphon to the purple gland. (Time calibration = 5 sec.)

a single score for generalization, the mean of T1–T2 for all runs within that preparation, and a single score for habituation (the first response in the habituation series minus the last response). A sign test revealed that although there was significant habituation (p < .001) there was no significant generalization of habituation. The results of 8 runs in 4 preparations at a single inter-stimulus interval of 2 min are summarized graphically in Figure 3.

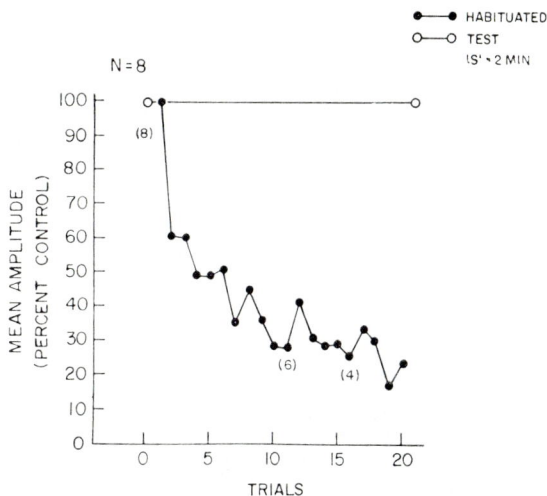

FIGURE 3 Summary of eight runs in four experiments designed to examine generalization of habituation between the habituated and test pathways. As indicated in parentheses, all runs had at least 10 trials; 6 had 15 trials, and 4 had 20 trials. The mean percent change from control for the second test stimulus applied to the non-habituated pathway was −.125% (S.D. = 12.75%). A sign test revealed that there was no significant difference between the second and first test stimulus even though significant habituation ($p < .001$) was obtained.

These experiments do not indicate that habituation of this reflex cannot generalize to stimuli that are more closely spaced within the receptive field. However, since no generalization of habituation occurred from the siphon to the purple gland, we could now compare the effects of a single 'dishabituatory' stimulus on both habituated and non-habituated responses.

(b) *Relation of Behavioral Dishabituation to Sensitization*

To examine the relationship of dishabituation to sensitization we presented a single test stimulus (T1) which produced a gill response, to either siphon or purple gland. We then habituated the reflex to 5 to 50% of control by repeated stimulation of the other pathway. Following a second test stimulus (T2) to the non-habituated pathway, a strong tactile stimulus (a strong jet of sea water or several vigorous brush strokes) was presented to the head or neck region.

In 19 runs (i.s.i. = 30 secs) in 8 preparations, dishabituation was seen in 63% of the runs and was observed at least once in every preparation. The

mean percent dishabituation was 65%. This was obtained by measuring the amplitude of the dishabituated responses (expressed as percentage of initial control) minus the amplitude of the last habituated response (expressed as percentage of initial control). Occasionally (16% of the runs) dishabituation was *larger* than the initial control value (see Figures 4A and 5).

In 6 of 15 runs (33%) in 8 preparations sensitization of non-habituated responses was observed. Even though there was no generalization of habituation from the siphon to the purple gland, the second test stimulus (T2) would, by chance alone, produce a response which was slightly less than the first test response (T1), approximately 50% of the time. Facilitation of such responses could conceivably be considered as a case of dishabituation. Since sensitization, by its most rigorous definition, represents the facilitation of *completely non-decremented* response, we have further examined only those cases (N = 8) in which there was no change or slight increment between test stimuli. Out of the 8 runs, sensitization by these criteria was observed in 3 runs (3 different preparations), i.e., in 37.5% of the total cases (Figures 4B, 5). The percent sensitization in this more restricted sample (37.5%) is comparable to that found in the larger sample (33%). In 5 of 6 cases where the neck stimulus produced sensitization (Figure 5) it also produced dishabituation. In the sixth case dishabituation could not be examined.

Several factors might account for the relatively low incidence of sensitization in these experiments. One, there may be a 'ceiling' effect whereby large (non-habituated) responses show less facilitation than do smaller responses. Two, the restraining of the animals may itself provide a sensitizing influence that makes further facilitation difficult to produce. Three, the sensitizing stimulus very often produced a large and long-lasting contraction during which it was impossible to measure facilitation of either the habituated response (dishabituation) or the non-habituated response (sensitization).

To summarize, the gill-withdrawal reflex in *Aplysia* shows, on occasion, both dishabituation beyond control and sensitization. Although both phenomena occur infrequently, every time a stimulus sensitized a non-habituated reflex response it dishabituated a habituated reflex. These data are therefore consistent with the notion that dishabituation and sensitization are different manifestations of a common facilitatory process. Why sensitization occurs only in a third of the cases is not clear, however.

**A**

HABITUATION (SIPHON)
(ISI = 30 sec)

0"　　　2' 30"　　　5'

NECK STIMULATION

0"

DISHABITUATION

2' 30"　　　3'

**B**

NO HABITUATION (SIPHON)
(ISI = 30 sec)

0"　　　5' 6"

NECK STIMULATION

0"

SENSITIZATION

30"　　　4'

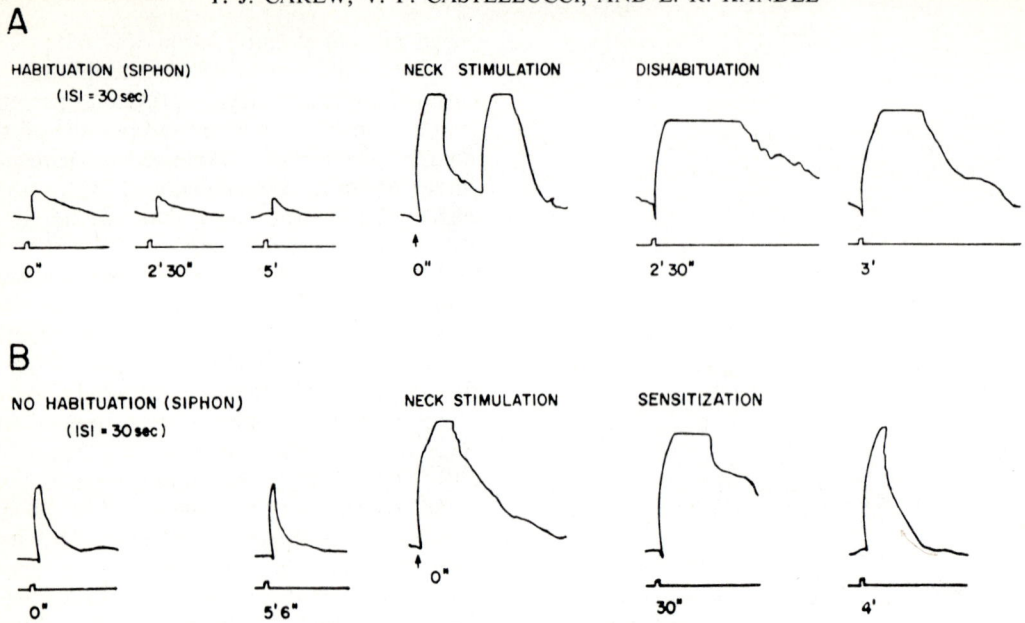

FIGURE 4　*Dishabituation above control and sensitization. Part A*: After 5 minutes of repeated stimulation to the siphon, the dorsal neck region was stimulated with 4 brush strokes (0") which produced several full contractions of the gill. When the habituating stimulus was again presented, the response was dishabituated above control (the traces are clipped due to blocking of the amplifiers). *Part B*: Two test stimuli were presented to the siphon at 5' 6" intervals to establish a control level. Four brush strokes were then applied to the neck region. Several test stimuli after neck stimulation showed sensitization. Parts A and B are from different experiments. (Time calibration = 5 secs.)

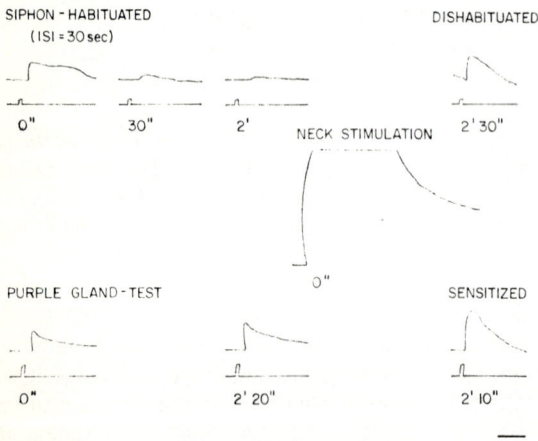

SIPHON - HABITUATED
(ISI = 30 sec)

0"　　　30"　　　2'

DISHABITUATED

2' 30"

NECK STIMULATION

0"

PURPLE GLAND - TEST

0"　　　2' 20"

SENSITIZED

2' 10"

FIGURE 5　Facilitation of both habituated and non-habituated responses by presentation of a single 'dishabituatory' stimulus. After a single test stimulus to the purple gland (0") repetitive stimuli were delivered to the siphon, producing habituation of gill-withdrawal. After a second test to the purple gland (2' 20") (indicating that no generalization of habituation had occurred) a series of vigorous brush strokes were presented to the neck region, producing a gill contraction. Subsequent presentation of a stimulus first to the purple gland and then to the siphon revealed facilitation of both non-habituated responses (sensitization) and habituated responses (dishabituation).

## II. *A Neuronal Analysis of the Relationship of Dishabituation to Sensitization*

We next examined how the neuronal correlates of dishabituation relate to those of sensitization. These neuronal correlates can be studied in the isolated ganglion by stimulating the appropriate afferent nerves or sensory neurons and recording intracellularly from gill-motor neurons (Kupfermann *et al.*, 1970; Castellucci *et al.*, 1970; Castellucci, Kandel and Schwartz, 1971). We recorded from L7, one of the two major motor neurons of the gill-withdrawal reflex, and stimulated the siphon and branchial nerves, containing the main afferent nerves from the siphon and anterior portion of the purple gland (see Figure 1). Brief electrical pulses (1.5 msec) applied to these nerves produced complex EPSPs in L7 that are quite similar to those produced by natural stimulation in a semi-intact preparation (see Figure 6). As was the case with natural stimuli in the semi-intact preparation, repeated electrical stimulation of a peripheral nerve in the isolated ganglion produced a decrement of the EPSP which paralleled the behavioral habituation (Figure 7).

A

GILL RESPONSE

L₇

TACTILE STIM.

40 mV

1 sec

B

L₇

SIPHON NERVE STIM.

20 mV

1 sec

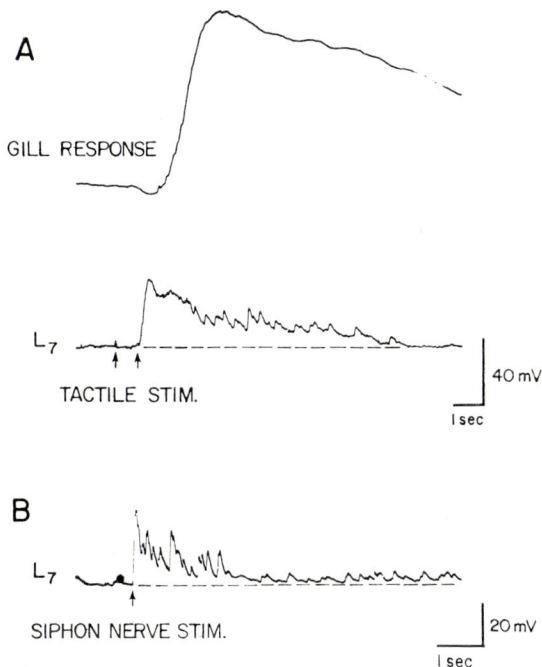

FIGURE 6   Comparison of EPSP produced in L7 in a semi-intact preparation in response to a tactile stimulus to the siphon and the EPSP produced in L7 in the isolated ganglion in response to a brief electrical stimulus to the siphon nerve.

*Part A*: Simultaneous record of gill contraction (obtained with a photocell) and of an EPSP in L7 in response to an 800 msec jet of sea water applied to the siphon. Cell L7 was hyperpolarized about 50 mV to prevent spike generation. The gill contraction is due to the activity of the other motor neurons in the reflex. When L7 fires, the reflex gill response is about twice as large. (From Kupfermann, Pinsker, Castellucci and Kandel, in preparation.)

*Part B*: EPSP produced in L7 by a 1.5 msec electrical shock to the siphon nerve. The cell was hyperpolarized about 30 mV to prevent firing. The general configuration and time course of the PSPs produced with natural and artificial stimulation was roughly comparable. (Notice the difference in time calibration.)

### (a)   *Lack of Generalization of EPSP Decrement*

We first examined the neuronal correlates for generalization of habituation following stimulation of the siphon and branchial nerves. The motor neuron L7 was hyperpolarized 20–30 mV to prevent it from firing and the stimulus intensity to both nerves was first adjusted so as to produce roughly equal EPSPs in L7. After 20 minutes' rest, a single test stimulus (T1) was delivered to one of the nerves. A second test stimulus (T2) was then delivered after an interval (usually 10–15 minutes) equal to the total duration of the subsequent

repetitive stimulation to the other nerve. Following the second test stimulus, the other nerve was stimulated 10–20 times, and the EPSP decremented to 20–60% of the initial control. The third test stimulus (T3) was then delivered and the effects of decrementing the EPSP of the other pathway on the test pathway were examined. After 30 minutes' rest, the 'test' and the repeatedly stimulated nerves were reversed and the experiment was repeated (Figure 7).

The results of 9 runs in 4 preparations are summarized in Figure 8. The mean percent decrement observed between the first and second test stimuli was 8.3% (Figure 8). The mean percent decrement observed between the second and third test stimuli (between which was interpolated EPSP decrement produced by repeated stimulation of the 'decremented' nerve) was −1.0%. If any generalization of response decrement had occurred between the two afferent nerves, one would have predicted that the decrement seen between the second and third test EPSPs would have been *greater* than that observed between the first and second. The lack of a difference indicates that no generalization of the response decrement occurred.

The finding that the EPSP decrement produced by repetitive stimulation of the siphon nerve does not generalize to the branchial nerve is compatible with the conclusion derived from the behavioral experiments, that habituation elicited via the siphon does not generalize to the purple gland. The lack of generalization of reflex habituation permitted us to examine, on a behavioral level, the relationship of habituation to dishabituation and of dishabituation to sensitization. The finding of a non-generalization of EPSP decrement permitted us to carry out a parallel analysis on a neurophysiological level.

### (b)   *Relation of Facilitation of a Decremented EPSP (Dishabituation) to Facilitation (Sensitization) of a Non-decremented EPSP*

Castellucci *et al.* (1970, 1971) have previously described that a decremented complex EPSP can be facilitated following a strong stimulus to either of the connectives and that at time the facilitated EPSP was even larger than control. The connectives are fiber tracts that interconnect the abdominal ganglion with the head (circumesophageal) ganglia that innervate the head and neck region.

We extended that analysis by comparing the facilitation produced by connective stimulation on decremented and non-decremented EPSPs. As in

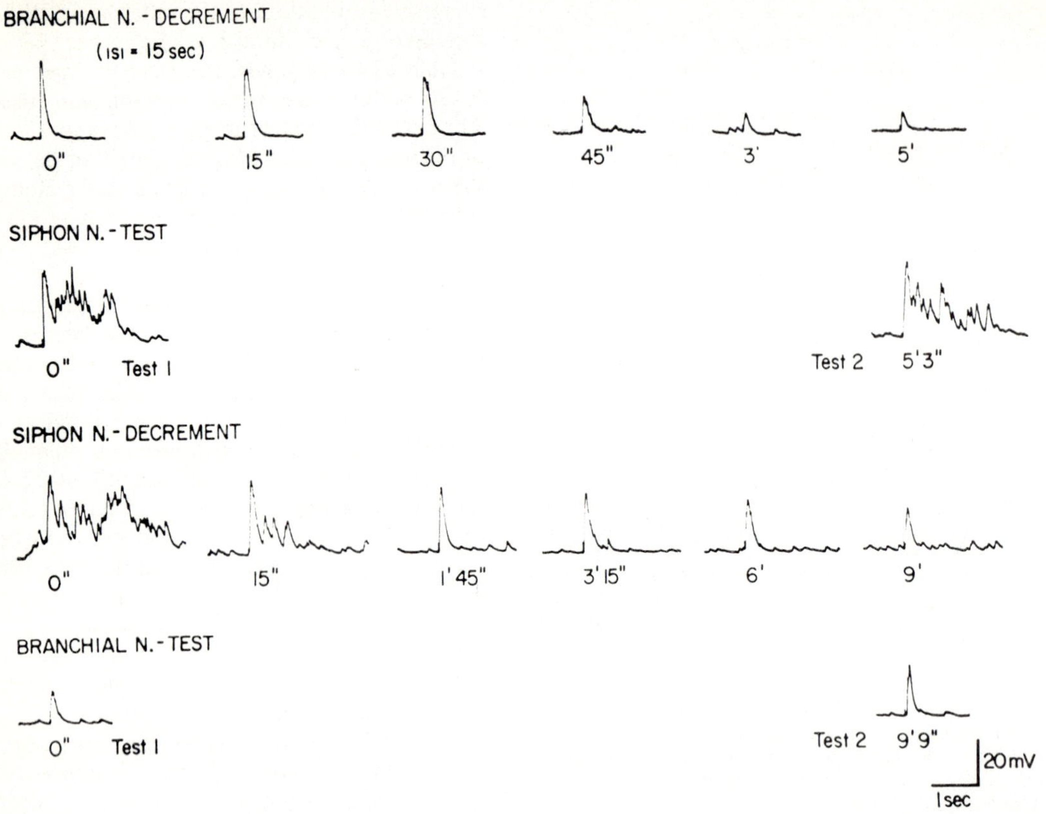

FIGURE 7    Lack of generalization of EPSP decrement. Stimuli were applied to siphon and branchial nerves that produced complex EPSPs in L7. A single test stimulus (Test 1, line 2) was first presented to the siphon nerve. Repetitive stimulation was then applied to the branchial nerve and produced EPSP decrement (top line). A second test stimulus then applied to the siphon nerve (Test 2, line 2) produced an EPSP in L7 which was comparable to control (Test 1). This indicates that no generalization of decrement occurred from the branchial nerve pathway to that of the siphon nerve. After a rest of 15 minutes, the experimental procedure was then reversed. The first test stimulus to the branchial nerve (Test 1, line 4) produced an EPSP which was still partially decremented (compare 0″ in lines 1 and 4), due to incomplete recovery from the prior repetitive stimulation (line 1). However, even though the EPSP produced by stimulation of the siphon nerve was then decremented by repetitive stimulation, a second test stimulus to the branchial nerve (Test 2, line 4) produced an EPSP that was *further* recovered. This indicates that no generalization of EPSP decrement had occurred from the siphon nerve pathway to the branchial nerve pathway. In this and all subsequent experiments in L7, the membrane potential was hyperpolarized about 30 mV, using a second microelectrode, to prevent spontaneous firing.

the behavioral experiments the presentation of the test stimuli alone sometimes produced slight EPSP decrement. The argument could therefore be made that the facilitation of these EPSPs should be considered as analogous to behavioral dishabituation rather than to behavioral sensitization. To avoid this confusion, we also restricted our analysis of the neural correlates of sensitization to thos test EPSPs which showed no change or increment between the first and last test presentations. In such cases the facilitation observed following a strong stimulus to the connective could be considered only as a correlate of sensitization.

In *all* cases (19 runs in 5 experiments) the decremented EPSP was facilitated, and in 70% of cases the facilitation was larger than the control (average 118%). The same stimulus to the connective also facilitated the non-decremented EPSP in 80% of the cases (average 119%) sometimes by as much as 200% (see Figure 9).

Both decremented and non-decremented PSPs showed similar increases (about 120%) compared to their *initial* controls. The decremented EPSPs obviously showed proportionately more facilitation (224%) when the last observed (decremented) response was used as control. These data suggest

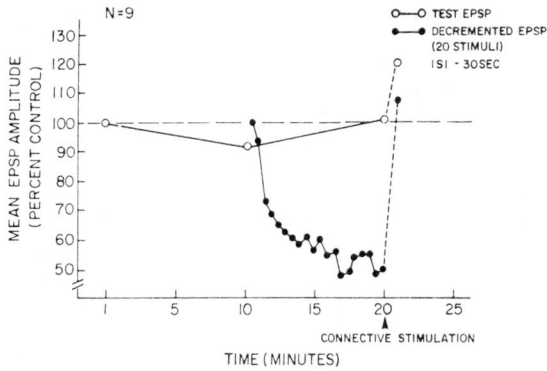

FIGURE 8 Summary graph of 9 runs from 4 different preparations indicating that there was no generalization of EPSP decrement between decremented and test inputs to the motor neuron (L7). Following a single train of stimuli (6/sec for 6 sec) to the connective, both decremented and non-decremented responses were facilitated (see text for further details).

hat the facilitating stimulus augments all responses toward a common ceiling and that decremented EPSPs were proportionately more facilitated because they were smaller (see also Kandel and Tauc, 1965a—Figure 15).

In summary, in the behavioral experiments, the same strong stimulus that produced dishabituation of a habituated pathway was also capable of producing sensitization of a non-habituated pathway, at least some of the time. In the neurophysiological experiments the same strong stimulus which produced facilitation of a decremented EPSP usually also produced facilitation of a non-decremented EPSP. Taken together, these two sets of data support the notion that dishabituation of the gill-withdrawal reflex is not simply the removal of habituation, but is an independent process which is actually a special case of sensitization. Dishabituation differs from sensitization only in involving a previously habituated response.

On the cellular level, facilitation analogous to dishabituation and sensitization were more readily

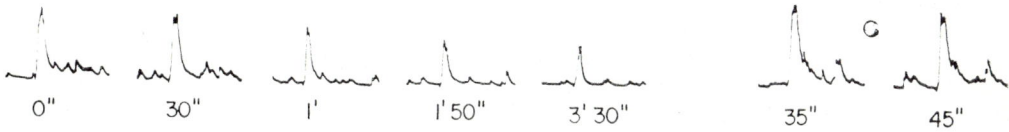

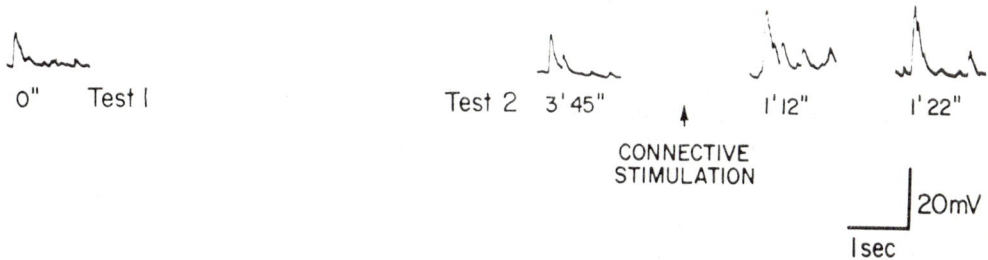

FIGURE 9 Facilitation of both decremented and non-decremented EPSPs by presentation of a single train of stimuli to the connective. A single test stimulus was delivered to the branchial nerve (Test 1, line 2), and produced an EPSP in L7. Repetitive stimulation was then applied to the siphon nerve to produce EPSP decrement (line 1). After a second test to the branchial nerve (Test 2, line 2) revealed that no generalization of EPSP decrement had occurred from the siphon nerve to the branchial nerve, a single train of stimuli (6/sec for 6 sec) was delivered to the left connective. Subsequent presentation of a stimulus, first to the siphon nerve (line 1) and then to the branchial nerve (line 2) revealed facilitation of both decremented and non-decremented EPSPs.

demonstrable than in the behavior studied in the restrained animal. Facilitation of *decremented EPSPs* (analogous to dishabituation) occurred invariably (100%) and as might be predicted for a common process, the percent occurrence of facilitation *beyond* initial control (70%) and the maximum amplitude of the facilitation (118%) were similar to values obtained for the facilitation of *non-decremented EPSPs* (80% and 119%, respectively). These findings indicate that the neurophysiological correlates may provide a more sensitive index than the behavioral ones. Alternatively, restraining animals may mitigate against optimal demonstration of dishabituation and sensitization.

## III. *Cellular Analysis of Sensitization:*
### *The Mechanism of Heterosynaptic Facilitation in L7*

Our behavioral and cellular neurophysiological experiments indicate that dishabituation is a special case of sensitization. At the cellular level this *common* process is correlated with an increase in the amplitude of the EPSP produced by stimulation of one pathway, following a train of strong stimuli to another pathway (heterosynaptic facilitation). An increase in EPSP amplitude can be caused by a number of different neuronal mechanisms. For example, facilitation of the EPSP in the motor neuron L7 could be due to a postsynaptic contribution such as an increased input resistance or an increased transmitter release produced in the presynaptic neurons by spike generation in the motor neuron. Alternatively, heterosynaptic facilitation in L7 could be due to a change in synaptic impingement resulting either from a removal of a synaptic inhibitory input that converges on the motor neuron or to an increase in the effectiveness of an excitatory input. We next examined these alternative mechanisms in an attempt to analyze the mechanism of heterosynaptic facilitation.

### (a) *The Passive Electric Properties of the Postsynaptic Cell During Facilitation*

The increase in EPSP amplitude associated with sensitization could be produced by an increase in the input resistance of the extrasynaptic membrane of the motor neurons. To test this possibility we measured the resistance of the motor neuron (L7) with intracellularly applied hyperpolarizing pulses and found that the facilitation of the EPSP did not result from a change in the input resistance of the neuron (Figure 10).

In most instances, the facilitatory stimulus produced a depolarization in the postsynaptic cell which lasted for 20–60 seconds. Since the resistance of the membrane in this range of membrane potential is non-linear, due to anomalous rectification (see Kandel and Tauc, 1966), the depolarization brought the cell's membrane into a region of increased resistance (Figure 11). This increase in input resistance could contribute slightly to the increment of the first several responses immediately following the heterosynaptic stimulus (Figure 10B). Facilitation still persisted, however, after the depolarization had ceased or when the change in membrane potential (and its secondary effect on the input resistance of the cell) was compensated for (Figure 10A).

These measurements of input resistance were made in the cell body, yet the synaptic impingement of molluscan neurons are on the axon and its processes. The question remained to what degree these electrotonic potential measurements provided an index of resistance changes in the synaptic region of L7. We therefore examined the effect of spontaneous synaptic activity on the electrotonic potential. We found that the characteristic burst of elementary IPSPs that occurred in L7, and has been attributed to Interneuron II (Frazier *et al.*, 1967), invariably produced large changes in the amplitude of the electrotonic potential (Figure 12). Thus electrotonic potential measurements are sensitive to resistance changes in at least some part of the synaptic region. However, the possibility still remains that changes in input resistance remote from the cell body may not be detected.

### (b) *Action Potential Generation of the Post-synaptic Cell During Facilitation*

The firing of the motor neurons following the heterosynaptic stimulus could conceivably produce a change in the synaptic efficacy of its own input (see, however, Kandel and Tauc, 1965b; Wurtz, Castellucci and Nusrala, 1967). To test this possibility we fired the motor neuron directly with intracellular pulses and found that, even at high frequencies, the action potentials in the motor neuron were incapable of facilitating the EPSP (Figure 13). By contrast, a train of stimuli presented to either connective regularly produced large facilitation.

These data indicate that neither changes in the passive properties of the motor neuron nor the generation of action potentials in the motor cell were likely to contribute to the facilitation. The

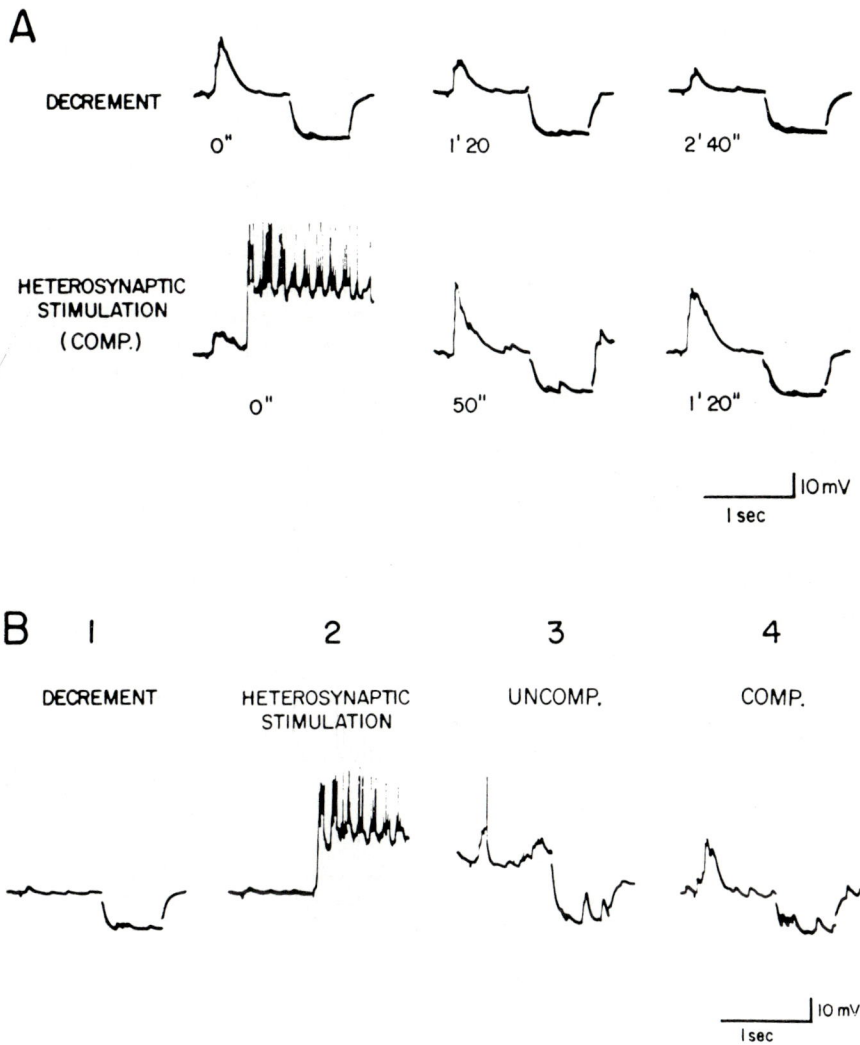

FIGURE 10   Absence of input resistance change in the soma of motor neuron L7 during EPSP decrement and heterosynaptic facilitation.

*Part A*: Decrement of the complex EPSP in L7. The siphon nerve was stimulated every 10 seconds and the EPSP recorded. A constant current hyperpolarizing electrical pulse was applied through another intracellular electrode and the resulting hyperpolarizing electrotonic potential and the synaptic activity were also recorded. There was no significant change in soma input resistance, as indicated by the relative constancy of the electrotonic potential, in L7 during EPSP decrement and facilitation. The heterosynaptic stimulation (left connective 6/sec for 6 sec) produced a depolarization in L7 but this was immediately compensated for and the membrane potential was returned to the initial reference level by passing current intracellularly through the second microelectrode.

*Part B*: Examples of a decremented EPSP and of its facilitation with and without compensation for membrane depolarization. The depolarization produced by the extrastimulus was not immediately compensated (B3) resulting in an increase in input resistance due to anomalous rectification. When the depolarization was compensated for (B4) by returning the cell to its original resting potential, the input resistance was similar to control (B1) yet the EPSP was still facilitated.

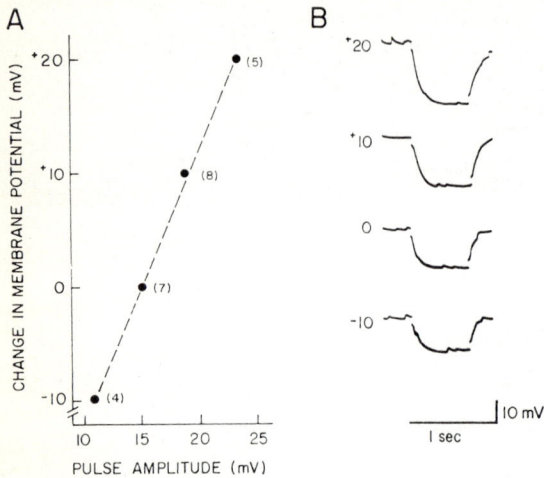

FIGURE 11   Anomalous rectification in gill motor neuron L7. *Part A*: Graph of amplitude of a hyperpolarizing electrotonic potential, used to monitor input resistance of the cell body, as a function of membrane potential. The cell's membrane potential was changed systematically and the electrotonic potential produced by constant hyperpolarizing current pulses were examined. The numbers of determinations for each point of the graph is indicated in parentheses. Reference zero is 30 mV below the firing level of the cell. *Part B*: Samples of recordings illustrated in the graph in *Part A*. The electrotonic potential was increased in amplitude as the membrane was depolarized. In this and the subsequent figures double-barrel electrodes were used to pass current and measure potential changes.

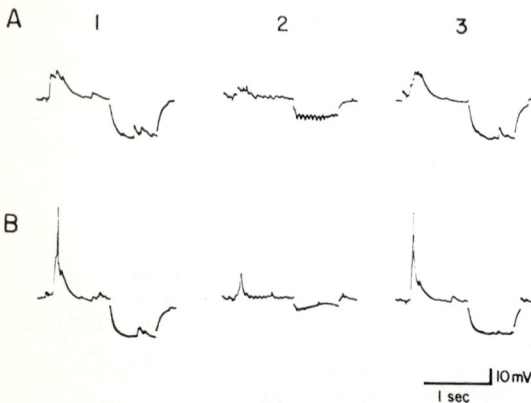

FIGURE 12   Resistance changes (measured by an electrotonic potential) in the cell body L7 as the result of synaptic activity attributed to Interneuron II.
*Part A*: Three successive traces (10 secs apart). A spontaneously occurring burst of IPSPs attributed to Interneuron II produced a large decrease in the input resistance as measured in the cell body. (Compare $A_1$ and $A_3$ to $A_2$.)
*Part B*: Three successive traces (10 secs apart) from another experiment. The IPSP burst occurred during frame $B_2$.

facilitation appeared to involve a change in the synaptic input to the motor neuron. This could result from one of two processes: (1) a decrease in inhibitory synaptic drive (disinhibition), or (2) an increase in excitatory synaptic drive.

### (c) The Role of Disinhibition or Increase in Excitatory Synaptic Drive

To examine the role of disinhibition we compared heterosynaptic facilitation near the resting level of the membrane potential to that obtained when the membrane potential was hyperpolarized considerably beyond the reversal level of the IPSPs (Waziri and Kandel, 1969). If the EPSP contained an occult IPSP, a decrease in the IPSPs as a result of the heterosynaptic facilitation might lead to a *decrease* in the amplitude of the complex PSP or at least a less obvious facilitation. Heterosynaptic stimulation at hyperpolarized levels of membrane potential nonetheless produced an increase in the EPSP amplitude quite comparable to that observed at the resting level. This test is not very direct, however, since disinhibition could increase the EPSP primarily by removing the conductance increase produced by the IPSP. Alternatively the disinhibition might involve a removal of a presynaptic inhibitory influence.

A more direct test of the role of disinhibition required examining the effects of a heterosynaptic stimulus on an elementary monosynaptic EPSP. An elementary EPSP would also permit one to distinguish recruitment of new excitatory elements from an increase in the synaptic effectiveness of previously active elements.

The opportunity to examine elementary monosynaptic EPSPs was provided by the finding that, as in the leech (Nicholls and Baylor, 1968), the cell bodies of primary mechanoreceptor neurons in *Aplysia* were located within the abdominal ganglion. These cells form a distinct cluster near L7; they have a large and stable resting potential of 50–70 mV and have no spontaneous background synaptic activity. These cells are normally silent but they can be discharged by mechanical stimulation of small areas (roughly 4–40 mm square) of the skin of the siphon or the purple gland. We have primarily studied the sensory cells from the siphon skin. These cells send their axons out the siphon nerve. Electrical stimulation of the siphon nerve or tactile (Figure 14A) or electrical stimulations (Figure 14B) of the siphon skin produced spikes, without prepotentials, in these cells (Figure 14). Even when the spike was blocked by hyperpolarizing the cell

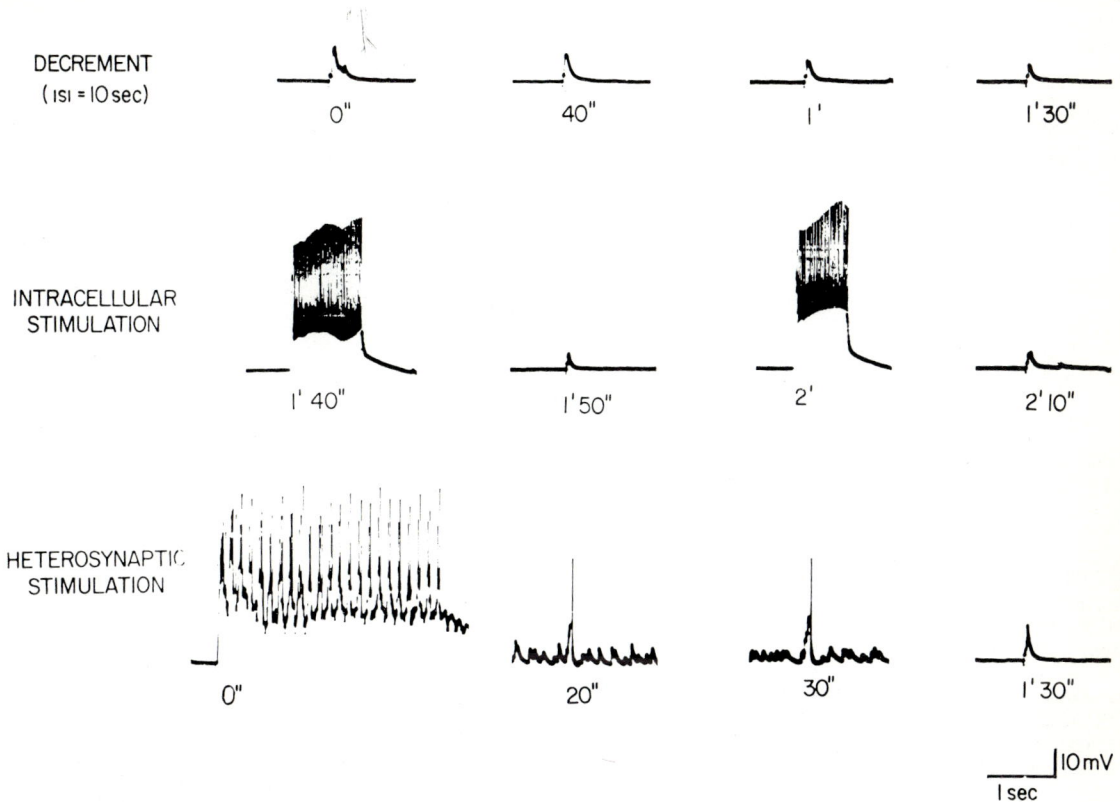

FIGURE 13    Failure of directly initiated spikes in the gill motor neuron L7 to produce facilitation of a decremented EPSP. Following decrement of the EPSP produced by stimulation of the siphon nerve (i.s.i. = 10 sec) the cell was stimulated intracellularly. A high frequency burst of spikes did not facilitate the decremented EPSP. Heterosynaptic stimulation of the cell by a train of stimuli to the left connective (7/sec, for 3.3 sec) facilitated the EPSP for several minutes. The facilitating stimuli were delivered immediately following the last EPSP (2′ 10″).

body, no underlying EPSPs were found (Figure 15). The cells followed high frequency stimulation (20/sec) of the siphon nerve without failure and tactile stimuli applied to the skin continued to elicit bursts of spikes in these neurons even when the ganglion and the skin were bathed in a high $Mg^{++}$ solution that blocked chemical synaptic transmission (del Castillo and Katz, 1954a) (Figure 15), including the synapses between these cells and the motor neuron L7. These results suggest that these cells are the primary mechano-receptor neurons of the gill-withdrawal reflex. The possibility still remains, however, that an electrical synapse exists at the periphery and that the cells are second-order sensory neurons (see Nicholls and Baylor, 1968) that function as first-order elements.

Each spike in the sensory neuron produced a large elementary EPSP in the gill motor neuron L7, often large enough to discharge a spike. The EPSP produced had a short and constant latency which was not affected by solutions of high $Ca^{++}$ content that increased the firing threshold of neurons by a factor of 2 to 4 and thereby *reduced* interneuron activity (Figure 16). These findings suggest that the mechano-receptor neurons make monosynaptic connections with the motor neuron L7. These EPSPs were clearly chemical in nature and were increased in amplitude by high $Ca^{++}$ solutions and were blocked by increasing the $Mg^{++}$ concentration of the artificial sea water solution (del Castillo and Katz, 1954a,b).

As was the case for the complex EPSP on gill motor neurons, the monosynaptic EPSP showed marked decrement with repeated stimulation (Figure 16B) (at frequencies ranging from 1 per minute to 1 per 10 seconds) and recovery with rest. The EPSP also exhibited heterosynaptic facilitation following a strong stimulus to the left or right connectives (Figure 17, line 3). This facilitation occurred without firing the sensory neuron (Figure 17, line 2) thereby excluding post-tetanic potentiation as a mechanism for the facilitation. In fact,

150 T. J. CAREW, V. F. CASTELLUCCI, AND E. R. KANDEL

a

50 mV
2 sec

b

50 mV
30 msec

0 mV

-7.5

-17.5

-30.0

50 mV
2 sec

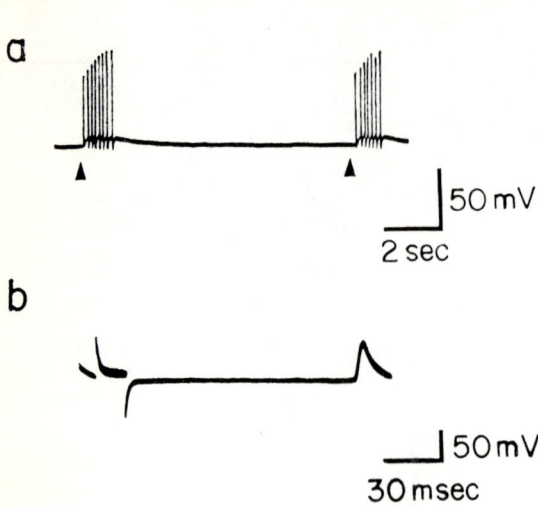

FIGURE 14 Properties of the mechanoreceptor neurons of the gill-withdrawal reflex.
*Part A*: The skin of the siphon area and the abdominal ganglion were isolated and bathed in artificial sea water. All nerves were cut except the siphon nerve (see Figure 1B). The arrows indicate the application of moderate pressure to the skin using a small glass probe (diameter 0.5 mm). Comparable bursts of spikes were elicited every time the probe touched a small receptive area (about 2 mm square).
*Part B*: Several superimposed traces from another sensory cell. Stimulation at 6/sec to show constant latency. A small bipolar platinum electrode was used to deliver an electrical stimulus over the receptive field of the mechanoreceptor. This cell followed frequencies up to 20/sec and did not change its latency. Size of the field: 3 mm square. The unit responded to a moderate pressure.

FIGURE 15 Blocking of the sensory neuron spike by intracellular hyperpolarization of the mechanoreceptor cell body. The skin and the abdominal ganglion were bathed in a solution of artificial sea water with high $Mg^{++}$ content (220 mM) for 35 minutes. A small glass probe was used to deliver moderate pressure over the receptive field (size: 2 mm square) of the cell. The cell body was gradually hyperpolarized and the spike produced by the tactile stimulus was gradually blocked without revealing underlying EPSPs.

the facilitating stimulus usually produced a small but long-lasting hyperpolarization in the sensory neuron that increased its threshold (Figure 17, line 2—arrows). Experiments are now in progress to examine the relationship of this hyperpolarization to the facilitation and to analyze further the mechanism of this heterosynaptic facilitation.

In summary, the excitatory postsynaptic potential produced at the synapse between the sensory and motor neuron decreased with repeated stimulation and was increased by the facilitating stimulus. The facilitation of this elementary EPSP suggests that removal of neither presynaptic nor postsynaptic inhibition are involved in the heterosynaptic facilitation. Neither does the facilitation require recruitment of new excitatory interneurons, although this may occur as an additional mechanism. The primary mechanism of the facilitation is an increase in the effectiveness of the excitatory synapses of the previously active sensory neurons.

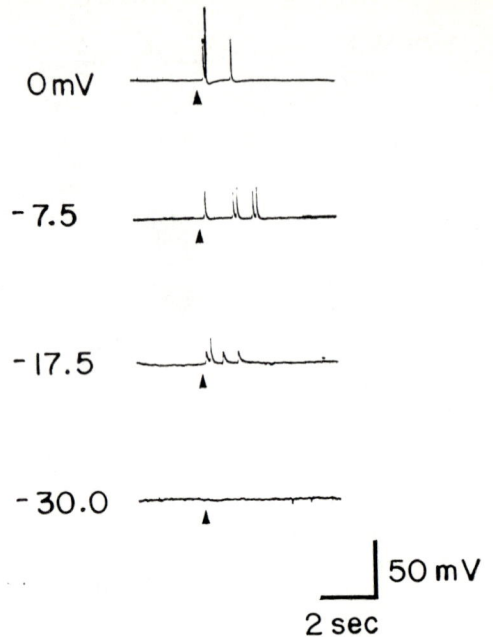

## DISCUSSION

A. *Dishabituation, Sensitization and Arousal*

Dishabituation is possibly the most interesting of the several parametric features of short-term habituation. It involves the effect of one type of stimulus on another and therefore provides a natural bridge between the habituation paradigm and higher associative learning such as classical conditioning. We have used the gill-withdrawal reflex as a model system for analyzing dishabituation and its relationship to sensitization. Our finding, that dishabituation is a special case of sensitization, confirms the results of Spencer, Thompson and Nielsen (1966) in the cat. Similar behavioral findings have now been made in several other higher forms including man (for review see Groves and Thompson, 1970). In none of these other cases has it yet proven possible to analyze the detailed neural mechanisms of dishabituation. The finding that dishabituation is a special case of sensitization in

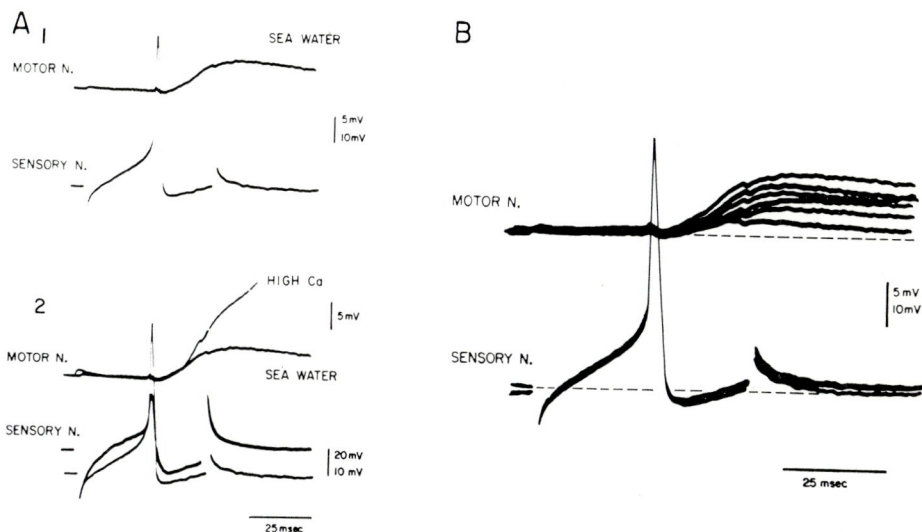

FIGURE 16   EPSP in gill motorneuron produced by the sensory neuron. *A*: *part* 1: Fast sweep recording to illustrate the short latency of the EPSP. *A*: *part* 2: Superimposed traces of EPSP elicited in artificial sea water with normal $Ca^{++}$ content (9.1 mM) and in a solution of high $Ca^{++}$ content (91 mM). The record was taken after the ganglion had been bathed in high $Ca^{++}$ sea water for 27 minutes. Because of the increased firing threshold of the cell in a high $Ca^{++}$ solution, the intracellular stimulation had to be increased considerably. The resulting large current artifact could not be completely balanced with the Wheatstone bridge and the gain was therefore reduced (see calibration). The latency for the EPSP produced by the sensory neuron in L7 was the same in normal and high $Ca^{++}$ sea water. *Part C*: Superimposed traces of EPSP decrement produced by firing the sensory neuron every two seconds. The first EPSP of the series is the same as in Part *A*. In this and the subsequent figure, intracellular stimulation of the sensory neuron was carried out by injecting current through the recording electrode using a Wheatstone bridge.

*Aplysia*, as it is in other forms, suggests, however, that the neural mechanisms of dishabituation in *Aplysia* may prove to be equally general. The recent finding of long-term habituation of a related reflex (Carew, Pinsker and Kandel, 1971) is therefore of particular interest because it suggests that studies of the gill-withdrawal reflex may also provide general insights into the neural mechanisms of long-term behavioral modifications.

Although dishabituation and sensitization involve a common cellular mechanism of which dishabituation is a special case, the phenomenological (behavioral) distinction between them may not be trivial. The sensitizing stimulus may tend to bring all responses toward a common ceiling, perhaps determined by the equilibrium potential of the EPSP, by the maximum number of spikes an EPSP can elicit, or by the physical properties of the responding organ (in this case, the gill). Small, habituated responses might therefore tend to undergo greater facilitation than would large, non-habituated responses. As a result the dishabituating effects of a given sensitizing stimulus

may often appear to be either more stable or more powerful than its sensitizing effect. Indeed, we have found behaviorally that dishabituation is more readily demonstrable than is sensitization, although the incidence for the two processes is about the same on the cellular level, where the resolution is greater.

Groves and Thompson (1970) have recently proposed that as dishabituation is a special instance of sensitization, so might sensitization be a specific instance of an even more general state, behavioral 'arousal'. It may be useful to divide behavioral 'arousal' into at least two distinct components, each of which can then be examined experimentally in simple preparations. These components are: (1) orientation, and (2) general, heightened responsiveness to stimuli. Sensitization is analogous to this second component of arousal. Both sensitization and arousal lead to increased responsiveness, and in each case the heightened responsiveness is generalized.

That sensitization increases responsiveness is implied in its definition. Less obvious, is the fact

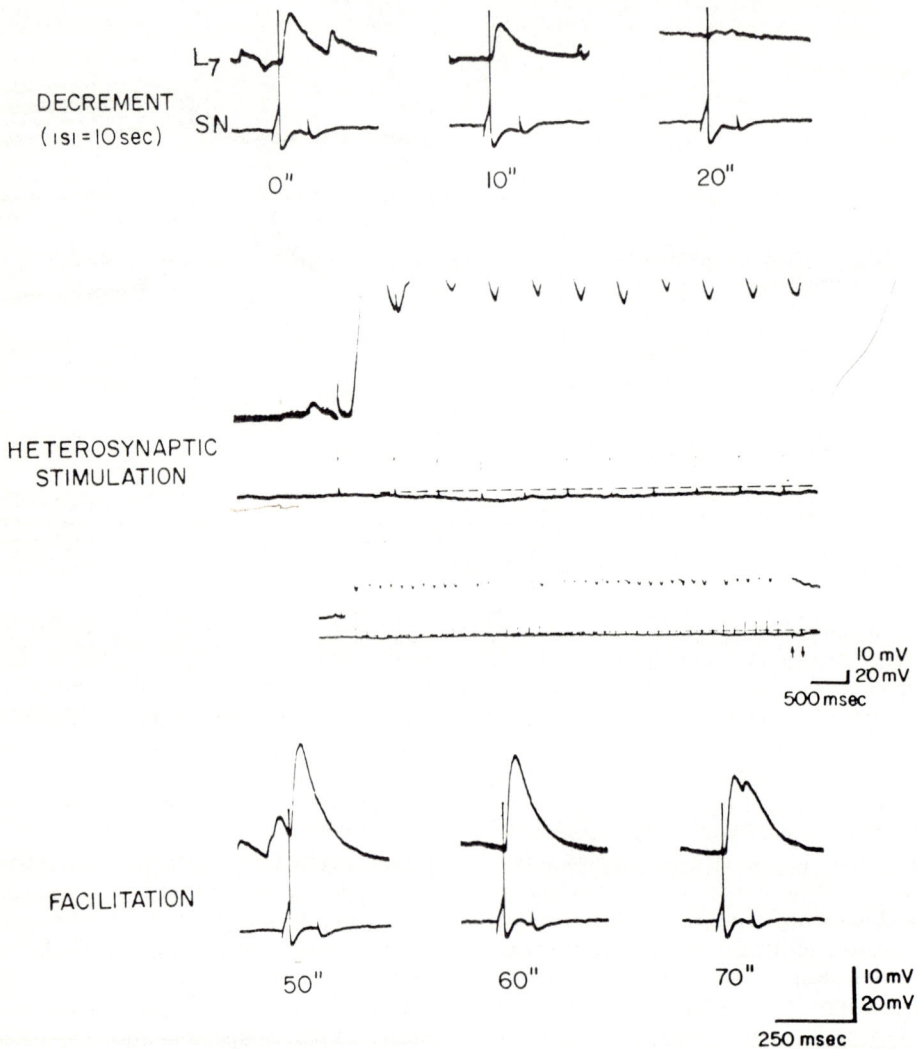

FIGURE 17 Decrement and facilitation of an elementary, presumably monosynaptic EPSP in the gill motor-neuron L7. The intracellular stimulation of the sensory cell (SN—lower trace of each pair) was repeated every 10 sec (top set of traces = decrement in L7). After 3 stimuli, heterosynaptic stimulation was applied to the left connective (7/sec, 5.5 sec). The second set of traces (HETEROSYNAPTIC STIMULATION) is shown first at normal gain and below at lower gain to show that the heterosynaptic stimulus, rather than firing the sensory neuron (note the stimulus artifact in SN trace) caused a sustained hyperpolarization that was particularly evident in the high gain record. The arrows in the low gain trace indicates an ineffective attempt to fire the sensory neuron. The hyperpolarization elevated the threshold to elicit a spike so that in successive tests, the strength of the intracellular stimulation had to be increased to produce spikes. After the heterosynaptic stimulation, the EPSP produced by the sensory neuron in L7 was facilitated for several minutes. All indicated times are referred to the first stimulus (0″).

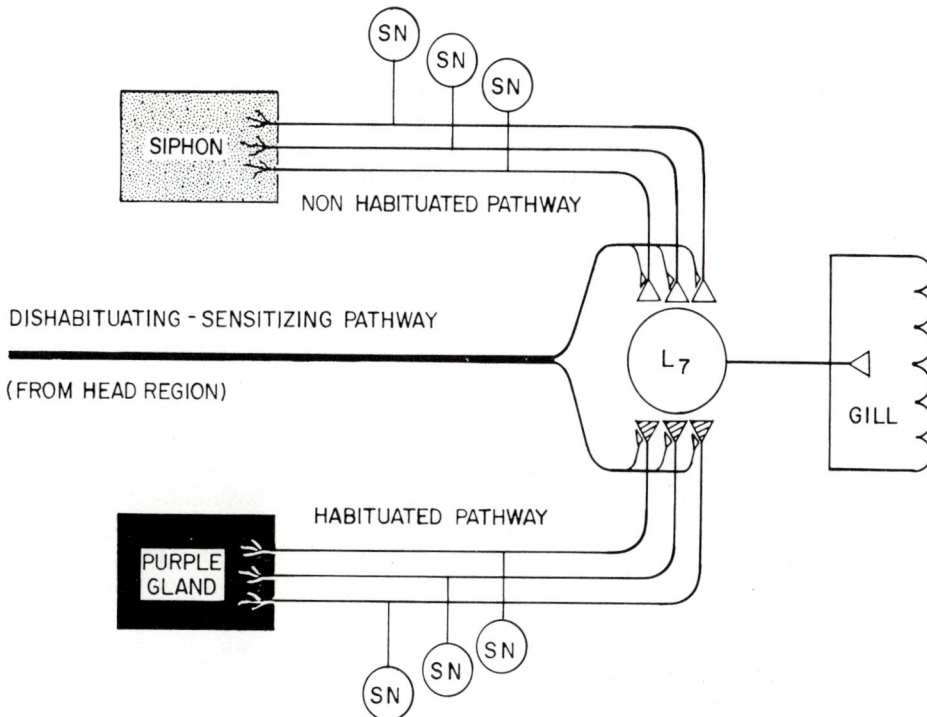

FIGURE 18   Schematic diagram illustrating converging input on the gill motor neuron L7. The difference in spatial extent on the pathways mediating habituation (shaded synapses) and sensitization—dishabituation explains the lack of generalization of habituation as well as the evident generalization of sensitization (dishabituation). The model postulates that the sensitizing pathway (which is not necessarily monosynaptic) mediates presynaptic facilitation on the synapse between the primary sensory neurons (SN) and the motor neuron L7. The model deals only with the monosynaptic components of the reflex. The polysynaptic components have not been analyzed but apparently do not contribute significantly to generalization.

that the effects of sensitization are usually widespread especially when contrasted to habituation. As was evident in our experiments, habituation showed limited generalization and was primarily restricted to the stimulated pathway. By contrast sensitization altered the responsiveness of several different reflex pathways.

Insofar as arousal may include sensitization, it provides an opportunity to substitute a somewhat vague, albeit useful behavioral concept (arousal) for one that is more restricted and more testable (sensitization). For example, 'arousal' produced by a high level of background stimulation is known to impair reflex habituation (Groves and Thompson, 1970). It would be of interest to see whether background sensory stimulation can affect habituation in *Aplysia* and whether this results from changes in the tonic level of heterosynaptic facilitation.

B. *Neural Analysis of Sensitization*
(1) *Heterosynaptic Facilitation as the Neural Process Underlying Sensitization and Dishabituation*
The data obtained from our cellular analysis indicate

that sensitization and dishabituation involve a common heterosynaptic facilitatory process leading to an increased effectiveness of the synapses between sensory and motor neurons of the gill-withdrawal reflex. We find that the neuronal correlates of habituation and dishabituation (sensitization) appear to involve two separate regulatory mechanisms acting on a common set of synapses. Habituation leads to a prolonged decrease in synaptic efficacy which is limited to the stimulated pathway; sensitization leads to a briefer, but more widespread heterosynaptic facilitation of synaptic efficacy involving both stimulated and unstimulated pathways.

The difference in the spatial extent of the pathways mediating habituation and sensitization (dishabituation) explains the *lack* of generalization of habituation as well as the evident generalization of sensitization (dishabituation) to non-habituated pathways. Our results suggest that the major plastic change in a habituated pathway of the gill-withdrawal response involves the synapses between the primary sensory neurons and the motor neurons.

Since the receptive field of the sensory neurons is relatively small, one would predict that as two stimuli within the receptive field of the reflex are separated by 10 mm or more they would show progressively less generalization of habituation. Consistent with this idea we have found that habituation of the gill-withdrawal reflex from stimulation of a single site such as the siphon does not generalize to the purple gland, some 50 mm away. However, the distribution of heterosynaptic facilitation produced from a single site such as the head is much more extensive and affects a number of closely related reflexes. As a result any pathway, habituated or not, can be facilitated (sensitized) by a stimulus that produces heterosynaptic facilitation. A model of how the spatial relationships of the pathways mediating habituation and sensitization and their synaptic properties can explain our behavioral results is illustrated schematically in Figure 18.

We cannot, as yet, precisely evaluate to what degree heterosynaptic facilitation can quantitatively account for all behavioral dishabituation or sensitization of this reflex. The gill-withdrawal reflex involves at least 4 motor neurons: two major ones (L7 and LD-G) account for about 45% each of the motor output and two minor ones (L9–1 and L9–2) that contribute to the remaining 10% (Kupfermann and Kandel, 1969; Kupfermann et al., 1971). All of our experiments on the sensory neurons were done only on L7. We have used the monosynaptic pathway between the mechanoreceptor neurons and motor neuron L7 as a model for studying the total reflex. Comparable experiments need to be done on LD–G, the other major motor neuron. However, the facilitation of the sensory neuron EPSP in L7 was large and could be greater than initial control, triggering spikes in the motor cell. Moreover, at least ten to fifteen sensory neurons appear to synapse on L7. It therefore seems likely that a substantial part of the facilitation of the complex EPSP in L7 was due to facilitation of the elementary EPSP produced by the sensory neurons. The firing of L7 contributes 40–45% of the total reflex. Since the increase in the EPSP in L7 during heterosynaptic facilitation in turn leads to increased firing of L7, a substantial part of at least the early component of the enhanced gill response can be attributed to heterosynaptic facilitation of the excitatory synapses between L7 and the sensory neurons.

## (2) Mechanism of Heterosynaptic Facilitation

When heterosynaptic facilitation was first described (Kandel and Tauc, 1965a and b) attention was drawn to its parametric similarity to behavioral sensitization (quasi-conditioning). Kandel and Tauc also provided evidence that heterosynaptic facilitation was a process distinct from post-tetanic potentiation and suggested that it involved a presynaptic facilitation of one pathway by the other. Recent experiments provide additional evidence that distinguishes heterosynaptic facilitation from post-tetanic potentiation (PTP). Epstein and Tauc (1970) showed that in R2, the cell studied by Kandel and Tauc, experimental manipulations that abolish PTP such as cooling or replacing extracellular $Na^+$ with $Li^+$ do not interfere with heterosynaptic facilitation. Our experiments in a different cell, L7, show that the stimulus that produces heterosynaptic facilitation of the synaptic potential produced by the sensory neuron on the motor neuron, does not fire the sensory neuron, thus excluding PTP as a mechanism. Even when the sensory neuron was fired repetitively with direct current pulses, we only found decrement of the synaptic potential. Thus, the reflex pathway of a known habituating response contains a critical set of synapses that does not undergo homosynaptic (post-tetanic) facilitation but only heterosynaptic facilitation. Indeed, the tendency for homosynaptic decrement in this system seems remarkable even by comparison to other decrementing synapses in Aplysia (Kandel and Tauc, 1965; Bruner and Kehoe, 1970).

While there now exists good evidence distinguishing heterosynaptic from post-tetanic facilitation, there has been relatively little progress in understanding the mechanism of heterosynaptic facilitation. Some additional information comes from experiments designed to interfere with active $Na^+$–$K^+$ transport (Epstein and Tauc, 1970) and with new protein synthesis (Schwartz, Castellucci and Kandel, 1971) and show that neither process is necessary for heterosynaptic facilitation to occur. Whereas these studies set broad limits on possible mechanisms of these short-term plastic changes, the specific model proposed by Kandel and Tauc, whereby one pathway controls the amount of transmitter released by the synaptic terminals of another pathway, has not yet been directly tested. As a result it is still not possible to distinguish with any degree of certainty between presynaptic facilitation and increased sensitization of the receptor in the postsynaptic cell. However, Coggeshall (1967) and Lewis et al. (1969) have now provided morphological evidence for many apparent presynaptic terminals ending on other terminals in the neuropile of the abdominal ganglion. In addition, there are experiments in dually innervated muscles of crayfish that provide independent evidence for presynaptic facilitation of the action of one axon by another. Atwood (1967) has found that if an impulse in one axon arrived just before an impulse in the second, the synaptic potential and the extracellularly recorded synaptic current produced by the second axon were enhanced. Experiments of this sort are needed in *Aplysia*. Our finding of a hyperpolarization of the presynaptic neuron as a result of heterosynaptic stimulation, is therefore of interest because it provides an opportunity for relating a potential change in the presynaptic neuron to a change in transmitter release at its terminals.

## ACKNOWLEDGMENTS

We thank Drs. Irving Kupfermann, Earl Mayeri, Harold Pinsker and Wm. Alden Spencer for their helpful comments on an earlier draft of this paper, and Miss Kathrin Hilten for her continued and excellent assistance with the illustrations. Dr. Kupfermann also kindly permitted us to use Figure 6A from an unpublished collaborative experiment. This research was supported by NIMH Grant MH 15980 and NS 09361–01. T. Carew was supported by a postdoctoral fellowship (MH 08638–06) awarded by the Department of Psychiatry, N.Y.U. Medical Center. V. Castellucci was supported by a Canadian Medical Council fellowship (100–2C–88). E. Kandel was supported by a Career Scientist Award No. 5K5-MH 18,558–03 from NIMH.

## REFERENCES

Atwood, H. L., 1967, Crustacean neuromuscular mechanisms, *American Zoologist* **7**: 527–551.

Brinley, F. J., Kandel, E. R., and Marshall, W. H., 1960, Potassium outflux from rabbit cortex during spreading depression, *Journal of Neurophysiology*, **23**: 246–256.

Bruner, J., and Kehoe, J., 1970, Long term decrements in the efficacy of synaptic transmission of mollucs and crustaceans. In: *Short Term Changes in Neural Activity and Behavior*, Horn, G., Hinde, R., eds., Great Britain: Cambridge University Press, pp. 323–359.

Carew, T. J., Crow, T., and Petrinovich, L. F., 1970, Lack of coincidence between neural and behavioral manifestations of cortical spreading depression, *Science*, N.Y. **169**: 1339–1342.

Carew, T. J., Pinsker, H., and Kandel, E. R., 1971, Long-term habituation of the siphon withdrawal reflex in *Aplysia*, Munich, XXV International Congress of Physiological Sciences.

Castellucci, V. F., and Goldring, S., 1970, Contribution to steady potential shifts of slow depolarization in cells presumed to be glia, *Electroencephalography and Clinical Neurophysiology* **28**: 109–118.

Castellucci, V. F., Kandel, E. R., and Schwartz, J. H., 1971, Macromolecular synthesis and the functioning of neurons and synapses. In: *Structure and Function of Synapses*, eds. Pappas, G., and Purpura, D., Appleton-Crofts, pub. (in press).

Castellucci, V. F., Kupfermann, I., Pinsker, H., and Kandel, E. R., 1970, Neuronal mechanisms of habituation and dishabituation of the gill withdrawal reflex in *Aplysia*, *Science*, N.Y. **167**: 1445–1448.

Coggeshall, R. E., 1967, A light and electron microscope study of the abdominal ganglion of *Aplysia californica*, *Journal of Neurophysiology* **30**: 1263–1278.

del Castillo, J., and Katz, B., 1954a, The effects of magnesium on the activity of motor nerve endings, *Journal of Physiology*, London, **124**: 553–559.

del Castillo, J., and Katz, B., 1954b, Quantal components of the end-plate potential, *Journal of Physiology*, London, **124**: 560–573.

Dodge, R., 1923, Habituation to rotation, *Journal of Experimental Psychology* **6**: 1–35.

Epstein, R., and Tauc, L., 1970, Heterosynaptic facilitation and post-tetanic potentiation in *Aplysia* nervous system, *Journal of Physiology*, London, **209**: 1–23.

Frazier, W. T., Kandel, E. R., Kupfermann, I., Waziri, R., and Coggeshall, R. E., 1967, Morphological and functional properties of identified cells in the abdominal ganglion of *Aplysia californica*, *Journal of Neurophysiology* **30**: 1288–1351.

Groves, P. M., and Thompson, R. F., 1970, Habituation: a dual process theory, *Psychological Review* **77**: No. 5, 419–450.

Harris, J. D., 1943, Habituatory response decrement in the intact organism, *Psychological Bulletin* **40**: 385–422.

Hoffman, H., 1939, In: *Klassen und Ordnungen des Tierreichs*, 3. Band: Mollusca, II. Abteilung: Gastropoda, 3. Buch: Opisthobranchia, Teil 1, 696–728.

Humphrey, G., 1930a, Le Chatelier's rule and the problem of habituation and dishabituation in *Helix albolabris*, *Psychologisch Forschung* **13**: 113–127.

Humphrey, G., 1930b, Extention and negative adaptation, *Psychological Review* **37**: 361–363.

Humphrey, G., 1933, *The Nature of Learning*. New York: Harcourt, Brace.

Kandel, E. R., and Tauc, L., 1965a, Heterosynaptic facilitation in neurones of the abdominal ganglion of *Aplysia depilans*, *Journal of Physiology*, London, **181**: 1–27.

Kandel, E. R., and Tauc, L., 1965b, Mechanism of heterosynaptic facilitation in the giant cell of the abdominal ganglion of *Aplysia depilans*, *Journal of Physiology*, London, **181**: 28–47.

Kandel, E. R., and Tauc, L., 1966, Anomalous rectification in the metacerebral giant cells and its consequences for synaptic transmission, *Journal of Physiology*, London, **183**: 287–304.

Konorski, J., 1948, *Conditioned Reflexes and Neuronal Organization*. London: Cambridge University Press.

Konorski, J., 1967, *Integrative Activity of the Brain: an inter-disciplinary approach*. Chicago: University of Chicago Press.

Kupfermann, I., and Kandel, E. R., 1969, Neural controls of a behavioral response mediated by the abdominal ganglion of *Aplysia*, *Science*, N.Y. **164**: 847–850.

Kupfermann, I., Pinsker, H., Castellucci, V. F., and Kandel, E. R., 1970, Neuronal correlates of habituation and dishabituation of the gill-withdrawal reflex in *Aplysia*, *Science*, N.Y. **167**: 1743–1745.

Kupfermann, I., Pinsker, H., Castellucci, V. F., and Kandel, E. R., 1971, Central and peripheral control of gill movements in *Aplysia* (in preparation).

Lewis, E. R., Everhart, T. E., and Zeevi, Y. Y., 1969,

Studying neural organization in *Aplysia* with the scanning electron microscope, *Science*, N.Y. **165**: 1140–1143.

Mendelson, M., 1967, A simple method of fabricating double-barreled micropipette electrodes, *Journal of Scientific Instrumentation* **44**: 549–550.

Nicholls, J. G., and Baylor, D. A., 1968, Specific modalities and receptive fields of sensory neurons in the central nervous system of the leech, *Journal of Neurophysiology* **31**: 740–756.

Pavlov, I. P., 1927, Conditioned Reflexes. (Trans. by G. V. Anrep) London: Oxford.

Pinsker, H., Castellucci, V. F., Kupfermann, I., and Kandel, E. R., 1970, Habituation and dishabituation of the gill-withdrawal reflex in *Aplysia*, *Science*, N.Y. **167**: 1740–1742.

Schwartz, J. H., Castellucci, V. F., and Kandel, E. R., 1971, The functioning of identified neurons and synapses. in the abdominal ganglion of *Aplysia* in the absence of protein synthesis, *Journal of Neurophysiology* (in press).

Sharpless, S. K., and Jasper, H., 1956, Habituation of the arousal reaction, *Brain* **79**: 655–680.

Sokolov, Y. N., 1963, *Perception and the Conditioned Reflex.* (Trans. by S. W. Waydenfeld) Oxford: Pergamon Press.

Spencer, W. A., Thompson, R. F., and Neilson, D. R., Jr., 1966a, Response decrement of flexion reflex in acute spinal cat and transient restoration by strong stimuli, *Journal of Neurophysiology* **29**: 221–239.

Spencer, W. A., Thompson, R. F., and Neilson, D. R., Jr., 1966b, Alterations in responsiveness of ascending and reflex pathways activated by iterated cutaneous afferent volleys, *Journal of Neurophysiology* **29**: 240–252.

Spencer, W. A., Thompson, R. F., and Neilson, D. R., Jr., 1966c, Decrement of ventral root electrotonus and intracellularly recorded postsynaptic potentials produced by iterated cutaneous afferent volleys, *Journal of Neurophysiology* **29**: 253–274.

Thompson, R. F., and Spencer, W. A., 1966, Habituation: A model phenomenon for the study of neuronal substrates of behavior, *Psychological Review* **173**: 16–43.

Wall, P. D., 1970, Habituation and post-tetanic potentiation in the spinal cord. In: *Short-Term Changes in Neural Activity and Behavior*, Jorn, G., and Hinde, R. A., eds., Great Britain, Cambridge University Press, pp. 181–210.

Waziri, R., and Kandel, E. R., 1969, Organization of inhibition in the abdominal ganglion of *Aplysia*. III. Interneurons mediating inhibition, *Journal of Neurophysiology* **32**: 520–539.

Wendt, E. R., 1936, An interpretation of inhibition of conditioned reflexes as competition between reaction systems, *Psychological Review* **43**: 258–281.

Wickelgren, B. G., 1967a, Habituation of spinal motoneurons, *Journal of Neurophysiology* **30**: 1404–1423.

Wickelgren, B. G., 1967b, Habituation of spinal interneurons, *Journal of Neurophysiology* **30**: 1424–1438.

Wurtz, R., Castellucci, V. F., and Nusrala, J., 1967, Synaptic plasticity: the effect of the action potential in the postsynaptic neuron, *Experimental Neurology* **18**: 350–368.

# THE SELECTIVE REGIONAL STIMULATION BY HEMICHOLINIUM-3 OF THE FORMATION OF CEREBRAL CYTIDINE DIPHOSPHOCHOLINE *IN VIVO*†

M. V. GOMEZ,‡ O. Z. SELLINGER, JOSEPHINE C. SANTIAGO and E. F. DOMINO

*Michigan Neuropsychopharmacology Research Program,*
*Department of Pharmacology and the Mental Health Research Institute,*
*University of Michigan Medical Center, Ann Arbor, Michigan 48104*

Hemicholinium-3 was administered into the left ventricle of dogs and 4 hours later the left and right caudate nuclei, thalami, hippocampi and amygdala were dissected out. The specific activity of the enzyme CTP-phosphocholine cytidyltransferase (E.C.2.7.7.15) determined in homogenates of these brain structures was higher by about 40% in the left (ipsilateral) as compared to the right (contralateral) caudate nucleus, thalamus and hippocampus, while, conversely, it was about 70% lower in the left than in the right amygdala.

The selective regional effect of hemicholinium-3 on the synthesis of cerebral cytidine disphosphocholine *in vivo* was confirmed at the level of the microsomes, for its specific activity was markedly higher in the fraction derived from the left than in that isolated from the right caudate nucleus.

Long and Schueler (1954) designed the hemicholiniums in an attempt to produce effective antagonists of acetylcholine. Among the compounds they synthesized, hemicholinium-3, the bishemiacetal form of α, α′-dimethylethanolamino-4, 4′-bisacetophenone, proved to be the most potent and has hence been used extensively in studies of the biosynthesis of acetylcholine (Gardiner, 1961; MacIntosh, 1963; Hebb *et al.*, 1964) and of the uptake of choline by membrane-enclosed systems. Diaugustine and Haarstad (1970) recently showed the hemiacetal function of the hemicholinium-3 molecule to be requisite for effective inhibition of the biosynthesis of acetylocholine.

Current studies employing HC-3 have focused on its interference with the uptake of choline by sympathetic ganglia (Collier and MacIntosh, 1969), cortical synaptosomes (Marchbanks, 1968, 1969; Potter, 1968; Diamond and Kennedy, 1969) and by the phrenic nerve-diaphragm preparation (Chang and Lee, 1970). However, this drug also has non-cholinergic effects. For example, it has been shown that it blocks the action potential of a squid axon;

an effect which could not be antagonized by either choline or acetylcholine (Frazier, Narahashi and Moore, 1969).

Recent work done in our laboratories has demonstrated that, 4 hours after the intraventricular administration of [$^{14}$C]-HC-3 to dogs, the drug is tightly bound to the subcellular membranes of the caudate nucleus and the hippocampus (Sellinger, Domino, Haarstad and Mohrman, 1969). Subsequently, (Gomez, Domino and Sellinger, 1970 a and b) we reported that the administration of HC-3 causes an acceleration of the conversion *in vivo* of [$^{14}$C] choline to phosphorylated metabolites in the caudate nucleus of the dog, resulting in [$^{14}$C]-phosphatidylcholine of higher specific radioactivity than that isolated from animals not receiving HC-3. On the other hand, Rodriguez de Lores Arnaiz, Zieher and de Robertis (1970) have noted no effect of HC-3 on the incorporation of [$^{32}$P] into the phospholipids of the cerebral cortex of the rat. Pursuing our analysis of the stimulatory effect of HC-3 on the formation *in vivo* of cerebral phosphatidylcholine (Gomez *et al.*, 1970b), we now report a selective, regional stimulation *in vivo*, but not *in vitro*, of the enzyme activity catalyzing the formation of CDP-choline from CTP and phosphorylcholine, namely CTP-phosphocholine cytidyltransferase (EC. 2.7.7.15).

† This research was supported by the United States Public Health Service Grants MH 11846 (to EFD) and NB 06294 (to OZS).

‡ Present address: Department of Biochemistry, Faculdade de Medicina, Universidade Federal de Minas Gerais, Belo Horizonte, M.G., Brazil.

## EXPERIMENTAL PROCEDURES

*Materials.* Hemicholinium bromide was purchased from the Aldrich Chemical Co., Milwaukee, Wis. Its purity was checked by paper chromatography as previously described (Gomez *et al.*, 1970a). CTP and CDP-choline were products of PL-Bio-chemicals, St. Louis, Mo. Phosphorylcholine, 1, 2-[$^{14}$C] chloride (9.44 mCi/mmole) was purchased from Tracerlab, Waltham, Mass. and [$^{12}$C]-phosphorylcholine, calcium salt, from Sigma Co., St. Louis, Mo. The calcium was removed by passage through a column of Chelex-100, 200–400 mesh, Na form (Bio-Rad Laboratories, Richmond, California). 3-phosphoglyceric acid, tricyclohexyl-ammonium salt, was a product of Calbiochem., Los Angeles, Calif.

*Animals* Beagle-like mongrel dogs of either sex (10–15 kg) were used. They were anesthetized with $N_2O$-oxygen, paralyzed with 1 mg/kg of decame-thonium, intubated and artificially respired as described by Dren and Domino (1968). The animals were sacrificed by air injection 4 hours after HC-3 (5 mg) was given into the *left* ventricle. The brains were rapidly removed and the following neuroanatomical areas were hemi-dissected from the left (experimental) and the right (control) side of the brain: caudate nucleus, hippocampus, thalamus and amygdala.

*Preparation of Homogenates and of Microsomal Fraction* The brain areas were weighed and were homogenized in ice-cold 0.25 M sucrose as previously described (Gomez *et al.*, 1970a). The microsomal fraction was also obtained as per Gomez *et al.* (1970a).

*Enzyme Assay* The standard incubation mixture contained (in μmoles): Tris-succinate buffer, pH 7.4: 5.0; $MgCl_2$: 4.0; CTP: 0.65; ATP: 0.6; 3-phos-phoglyceric acid: 1.5; NaF: 8.5; phosphoryl-choline (containing about 79,000 dpm of $C^{14}$-phosphorylcholine): 0.9 and tissue in a final volume of 0.25 ml. The tubes were covered during incubation at 37°, while shaking at about 120 oscillations/min. The incubation (60 min unless otherwise stated) was stopped by placing the tubes in boiling water for 5 min. After cooling to room temperature, 2 ml of an aqueous suspension of Norit charcoal (15 mg/ml) were added to each tube. After 10 min, during which time the tubes were buzzed 3 times on a Vortex-Junior mixer, the charcoal was washed with 3 successive 5 ml portions of distilled water. The charcoal was sedimented by brief centrifugation after the last water wash and the supernatant was discarded. CDP-choline was quantitatively eluted from the charcoal with 3 successive portions of a mixture of ethanol-conc. ammonia and water (60:37:3). The volume of the pooled eluate was 8 ml. Two ml of this eluate were pipetted into scintillation vials and the volume brought to about 0.5 ml under a stream of nitrogen. Twelve ml of 10% (v/v) Biosolv-3 (Beckman Co., Fullerton, Calif.) in toluene (containing 5 g of PPO and 0.1 g of dimethyl POPOP per liter of toluene) were added to each vial for counting. The activity of CTP-phosphocholine cytidyltransferase is expressed in mμmoles of CDP-choline synthesized/hr/g of wet weight and the specific activity is obtained by dividing by the protein content in milligrams. Additional details and properties of the enzyme are described elsewhere (Gomez, *et al.*, 1971).

*Radioactivity* This was measured in a Nuclear-Chicago Unilux II spectrometer. Quenching corrections were applied using the channels ratio method and the cpm were converted to dpm by means of a PDP-8 computer. The counting efficiency ranged between 80 and 90%. When aliquots of the charcoal eluate containing the [$^{14}$C]-CDP-choline generated during incubation were chromatographed as previously described, all of its radioactivity migrated with authentic [$^{12}$C]-CDP-choline. Standards of [$^{14}$C]-CDP-choline carried through the assay procedure were routinely included and were recovered in excess of 90%.

*Analytical Procedures* Protein was determined according to Lowry, Rosebrough, Farr and Randall (1951) with bovine serum albumin as standard.

## RESULTS

Following its administration into the left ventricle, [$^{14}$C]-HC-3 is swiftly taken up by the cerebral structures close to and ipsilateral with the site of injection and can be localized there by autoradiography (Domino and Cassano, unpublished observations). Consequently, the left caudate nucleus, hippocampus, thalamus and amygdala become rapidly labelled, while the contralateral, right caudate nucleus and hippocampus are labelled to a significantly lesser extent (Sellinger *et al.*, 1969). This predominant localization on the ipsilateral side of HC-3 *in vivo* provided a unique opportunity to follow up the recently observed effect of HC-3 on the formation of cerebral phosphatidylcholine (Gomez *et al.*, 1970b), since it made the target structures on the left side the experimental and those on the right side, the "built-in" control test systems. Moreover, since a previous study (Gomez *et al.*, 1970a) has clearly indicated an effect of HC-3 on the overall conversion of choline to CDP-choline *in vivo* and since more recent experiments ruled out effects of HC-3 on the conversion of CDP-choline to phosphatidylcholine, the intermediate enzymatic step leading to the formation of CDP-choline from CTP and phosphorylcholine (Porcellati, 1969; Porcellati and Arienti, 1970) was selected for study.

Table I illustrates the selective, stimulatory effect of HC-3 injected *in vivo* on the CTP-phosphocholine cytidyltransferase of the left, but not of the right, caudate nucleus, hippocampus and thalamus. Although the left-right differences varied from experiment to experiment (Table I), the specific activity of the enzyme on the left exceeded that of the enzyme on the right side in every experiment. Interestingly, however, HC-3 appeared to exert an opposite effect, namely inhibit the CTP-phosphocholine cytidyltransferase in the amygdala for, in 3 successive experiments (Table II), the specific activity of the enzyme was markedly higher in the right than in the left half of this structure.

Recently, we established the microsomal localization of CTP-phosphocholine cytidyltransferase in the canine caudate nucleus (Gomez *et al.*, 1971), as well as in two other regions of the rat brain. In order to confirm the selective nature of the stimulatory effect of HC-3 *in vivo* on the formation of CDP-choline at the subcellular level, microsomes were isolated from the left and the right caudate nucleus of four of the dogs used to obtain the data shown in Table I and the activity of CTP-phosphocholine cytidyltransferase was determined (Table II). In every animal, the specific activity of the enzyme was clearly higher on the left than on the right side,

TABLE I

The effect of HC-3 *in vivo* on CTP-phosphocholine cytidyltransferase in four canine brain regions

| Brain Region | Experiment No. | Tissue Side | | |
|---|---|---|---|---|
| | | Left (L) | Right (R) | L/R |
| | | s.a† | s.a. | % |
| Caudate Nucleus | 1 | 28.0 | 22.1 | +27 |
| | 2 | 28.3 | 21.8 | + 7 |
| | 3 | 21.2 | 15.8 | +34 |
| | 4 | 22.1 | 13.4 | +65 |
| | 5 | 34.0 | 17.1 | +99 |
| | 6 | 30.4 | 25.6 | +18 |
| | | 26.5‡ = mean | 19.3‡ | +41.7 |
| Thalamus | 2 | 8.7 | 8.0 | + 8 |
| | 4 | 13.9 | 9.1 | +53 |
| | 5 | 20.9 | 18.6 | +12 |
| | 6 | 27.0 | 15.6 | +72 |
| | | 17.6 = mean | 12.8 | +36.2 |
| Hippocampus | 2 | 15.7 | 11.2 | +40 |
| | 4 | 17.4 | 16.1 | + 8 |
| | 5 | 30.4 | 18.7 | +62 |
| | | 21.2 = mean | 15.3 | +36.7 |
| Amygdala | 4 | 14.2 | 23.1 | −61.5 |
| | 5 | 19.6 | 23.2 | −84.5 |
| | 6 | 14.7 | 20.6 | −71.5 |
| | | 16.2 = mean | 22.3 | −72.5 |

† M$\mu$moles of CDP-choline/hr/mg of protein.

‡ The right caudate and left caudate of an uninjected dog had specific activity values of 14.3 and 16.3, respectively.

### TABLE II

The stimulation by HC-3 of the caudate microsomal CTP-phosphocholine cytidyltransferase

| Eperiment No.† | Microsomes Isolated From: | | |
|---|---|---|---|
| | Left CN (L)<br>s.a‡ | Right CN (R)<br>s.a | L/R<br>% |
| 2 | 50.9 | 28.4 | +78 |
| 4 | 88.4 | 60.0 | +47 |
| 5 | 68.9 | 36.0 | +90 |
| 6 | 67.2 | 40.7 | +65 |
| | Mean: 68.8 | 41.3 | +70.0 |

+ CN: Caudate nucleus.
† The numbers of the experiments refer to those listed in Table I.
‡ M$\mu$moles of CDP-choline/hr/mg of protein.
The specific activity of the microsomal fraction of the left caudate of an uninjected dog was 44.5.

even though the microsomal values were markedly higher than those shown in Table I in both halves of the caudate.

Separate experiments showed that the addition of HC-3 ($5 \times 10^{-5}$M and $5 \times 10^{-7}$M) to homogenates and microsomal fractions of the left caudate nucleus of dogs injected with the drug *in vivo* caused no further increase of the activity of the enzyme. Under the conditions of these experiments, the enzyme was similarly unaffected in the right caudate as well.

An additional test of the capacity of HC-3 to affect the formation of CDP-choline *in vitro* consisted in incubating homogenates and microsomal fractions of caudate nuclei of the cerebral cortex of intact dogs and rats with HC-3 ($5 \times 10^{-7}$ and $5 \times 10^{-5}$M) and measuring the activity of CTP-phosphocholine cytidyltransferase. The effects of the drug were erratic and inconsistent, and the effect observed *in vivo* could not be confirmed.

## DISCUSSION

The observed effects of intraventricularly administered HC-3 point to the complexity of action of this drug. A major factor in the interpretation of our data is the assumption that the ipsilateral side of the brain is exposed to considerably larger concentrations of the drug than the contralateral side, so that the latter may be used as the control side and the former as the experimental side. Although autoradiographic evidence as well as our previous findings (Sellinger *et al.*, 1969) support

this notion, admittedly the contralateral side of the brain does contain measurable amounts of HC-3, particularly in the regions proximal to the site of injection (Domino and Cassano, unpublished observations). Therefore, our results presumably compare effects of large *vs* small, and not nil, concentrations of HC-3 on the regional formation of cytidine diphosphocholine. Direct evidence for this contention is provided by the specific activity values of CTP-phosphocholine cytidyltransferase of the left and right caudate of uninjected dogs (see legend, Table I) which were somewhat lower (mean of 15 m$\mu$ moles of cytidine diphosphocholine/hr/mg of protein), than the mean value of the right caudate of injected dogs (Table I).

The failure of Rodriguez de Lores Arnaiz *et al.* (1970) to observe stimulatory effects of HC-3 on the incorporation of [$^{32}$P] into the phospholipids of the cerebral cortex of the rat may be ascribed to the relatively low dose (0.25–0.35 mg/kg) tolerated by the rat which, consequently, these investigators were obliged to use, to the subarachnoid route of injection which certainly does not favor penetration of the injected material into the deep structures and, lastly, to the fact that measurements of radioactivity were made exclusively on the total extract of cortical phospholipids rather than on the water-soluble fraction which contains cytidine diphosphocholine (Gomez *et al.*, 1970a). Moreover, only the cerebral cortex was examined by Rodriguez de Lores Arnaiz *et al.*, (1970), a region of the brain notably unaffected by HC-3 (Sellinger *et al.*, 1969).

It is well-known (Shellenberger and Domino,

1967) that HC-3 causes markedly different effects on the EEG of the dog depending on the dose. Large intraventricular doses (5 mg) such as the one used in the present study cause marked EEG seizures, especially subcortically. Although the mechanism of these seizures and of their genesis by HC-3 are totally unknown, it may well be that the stimulatory effect of HC-3 on the formation of cytidine diphosphocholine is partly mediated by or represents a consequence of, such locally elicited hyperactive states.

The present findings further illustrate the complexity of the action spectrum of HC-3 in the central nervous system for they add to the long list of its known effects. Some of these are: blockage of choline transport across membranes (Schuberth, Sundwall, Sorbo and Lindell, 1966; Browning and Schulman, 1968; Diamond and Kennedy, 1969); reduction of ACh levels (Dren and Domino, 1968; Wilson, Mohrman and Domino, 1969); action as a false neurotransmitter (Rodriguez de Lores Arnaiz et al., 1970); inhibition of acetylcholinesterase (Shellenberger and Domino, 1967) and action as a non-cholinergic blocking agent (Frazier, Narahashi and Moore, 1969).

The paradoxical regional effect of HC-3 whereby the formation of cytidine diphosphocholine was stimulated in three regions of the brain and was inhibited in one, namely the amygdala (Table I), is at present unexplained. It is reminiscent, however, of similar regional effects of other drugs such as, for example, reserpine, phenylisopropylhydrazine and d-amphetamine sulfate which affect the regional metabolism of cerebral catecholamines (Glowinski, Axelrod and Iversen, 1966). It is possible that the regional differences in the effects of these and other drugs may actually be the result of subtle, regional differences in the uptake and the consequent disposition of those brain metabolites which are primarily affected by the drugs in question (Snyder and Coyle, 1969).

## REFERENCES

Browning, E. T., and Schulman, M. P., 1968, *J. Neurochem.* **15**: 1391.

Chang, C. C., and Lee, C., 1970, *Neuropharmacol.* **9**: 223.

Collier, B. and MacIntosh, F. C., 1969, *Can. J. Physiol. Pharmac* **47**: 127.

DiAugustine, R. P. and Haarstad, V. B., 1970, *Biochem. Pharmac.* **19**: 559.

Diamond, I., and Kennedy, E. P., 1969, *J. biol. Chem.* **244**: 3258.

Dren, A. T., and Domino, E. F., 1968, *J. Pharmac. exp. Ther.* **161**: 141.

Frazier, D. T., Narahashi, T., and Moore, J. W., 1969, *Science, Wash.* **163**: 820.

Gardiner, J. E., 1961, *Biochem. J.* **81**: 297.

Glowinski, J., Axelrod, J., and Iversen, L. L., 1966, *J. Pharmac. exp. Ther.* **153**: 30.

Gomez, M. V., Domino, E. F., and Sellinger, O. Z., 1970, *Biochim. biophys. Acta* **202**: 153.

Gomez, M. V., Domino, E. F., and Sellinger, O. Z., 1970, *Biochem. Pharmacol.* **19**: 1753.

Gomez, M. V., Domino, E. F., Santiago, J. S., and Sellinger, O. Z., *Neurobiology*, (*Copenhagen*) in press, 1971.

Hebb, C. O., Ling, G. M., McGeer, E. G., McGeer, P. L., and Perkins, D., 1964, *Nature*, Lond. **204**: 1309.

Long, J. P., and Schueler, F. W., 1954, *J. Amer. pharm. Assoc.* **43**: 79.

Lowry, O. H., Rosebrough, N. J., Farr, A. L., and Randall, R. J., 1951, *J. biol. Chem.* **193**: 265.

MacIntosh, F. C., 1963, *Can. J. Biochem. Physiol.* **41**: 2555.

Marchbanks, R. M., 1968, *Biochem. J.* **110**: 533.

Marchbanks, R. M., 1969, *Biochem. Pharmacol.* **18**: 1763.

Porcellati, G., 1969, *Acta Neurologica* (*Napoli*) **24**: 119.

Porcellati, G., and Arienti, G., 1970, *Brain Res.* **19**: 451.

Potter, L. T., 1968, In *The Interaction of Drugs and Subcellular Components in Animal Cells* (Edited by Campbell, P. N.) p. 293. Little, Brown, Boston.

Rodriguez de Lores Arnaiz, G., Zieher, L. M., and De Robertis, E., 1970, *J. Neurochem.* **17**: 221.

Schuberth, J., Sundwall, A., Sorbo, B., and Lindell, J. O., 1966, *J. Neurochem.* **13**: 347.

Sellinger, O. Z., Domino, E. F., Haarstad, V. B., and Mohrman, M. E., 1969, *J. Pharmac. exp. Ther.* **167**: 63.

Shellenberger, M. K., and Domino, E. F., 1967, *Int. J. Neuropharmac.* **6**: 283.

Snyder, S. H., and Coyle, J. T., 1969, *J. Pharmac. exp. Ther.* **165**: 78.

Wilson, A. E., Mohrman, M. E., and Domino, E. F., 1969, *The Pharmacologist* **11**: 291.

# BLOOD FLOW IN RED AND WHITE MUSCLE
# IN EARLY DEVELOPMENT

G. F. WOOTEN and DONALD J. REIS

*Department of Neurology, Cornell University Medical College, 1300 York Avenue,
New York, N.Y. 10021*

The fractional distribution of muscle blood flow was measured in the adult and during postnatal development in chronically prepared unanesthetized rabbits by the isotope dilution method. In the adult rabbit blood flow ranged from 5.86 to 56.12 ml/min/100 gm of muscle. Flow was approximately four times greater in grouped red than white limb muscles. Myoglobin concentration ranged from 0.44 to 3.55 mg/gm muscle. The myoglobin concentration of grouped red exceeded that of white muscle and varied in each muscle directly and linearly with blood flow with a high correlation (P < .001). At birth there was no significant difference in fractional blood between pairs of red and white muscles. However, by the third to fourth week of life significant flow differences appeared which were sustained until adult values were attained. The development of blood flow differences between red and white muscles occurred in a cranio-caudal direction. The changes underlying the postnatal differentiation of blood flow between red and white muscles took place predominantly in white muscle.

Blood flow; red and white muscle; postnatal development; myoglobin.

In the adult cat nutrient blood measured in quiet wakefulness or under anesthesia is approximately three times as great in red as in white skeletal muscle (9, 14, 22) and is directly related both to the degree to which a muscle depends upon aerobic metabolism for energy and its twitch duration (23). At birth mammalian skeletal muscles are histochemically and mechanically homogeneous with all having slow contraction times (1, 2, 6, 12, 20, 21). Differentiation into red and white muscles occurs during early development with histochemical and physiological maturity being achieved in rat and cat by the fourth to sixth week of life and somewhat earlier in the rabbit (1, 2, 6, 12, 17, 21).

It has recently been observed that in the kitten there are no differences in blood flow between the red soleus and white gastrocnemius until after the development of differences in the speed of contraction and enzymatic activities (13). The time course and rostrocaudal pattern of the changes in blood flow distribution in muscle, and whether such changes occur selectively within red or white muscles or both, remains to be determined. In this study we have investigated the development of changes in the distribution of blood flow to selected red and white muscles in the fore- and hindlimb of the chronically prepared unanesthetized rabbit during development. It will be demonstrated that there are no differences in blood flow between red and white muscles at birth, that differences develop with a characteristic time course in early development in a rostrocaudal direction, and that the principal changes of flow appear to occur in white muscle.

## METHODS

### Measurement of Blood Flow by $^{86}Rb$ Method

Blood flow distribution to skeletal muscles of unanesthetized, chronically prepared rabbits was measured by the isotope dilution method of Sapirstein (27) using $^{86}Rb$ as the indicator. The principle of this method is based on the fact that Rb, like K, will rapidly distribute itself in the intracellular compartment of tissue (with the exception of the central nervous system) and remain at a constant concentration for up to two minutes. Since most of the injected $^{86}Rb$ is extracted in one circulation in most organs, the radioactivity of any organ or part of an organ is proportional to the percentage of the cardiac output reaching that

163

tissue. The ratio of organ activity to injected activity is termed the fractional flow (*FF*) and is calculated:

$$FF = \frac{\text{organ activity (cpm/gm)} \times 100}{\text{total injected activity (cpm)}}$$

If the cardiac output is known it is possible to estimate absolute blood flow by the calculation:

Absolute flow (ml/min/100 gms) =

$$= \frac{\text{organ activity (cpm/gm)}}{\text{total injected activity (cpm)}} \times \frac{\text{cardiac}}{\text{output}} \times 100$$

The use of $^{86}$Rb for measurement of nutrient blood flow to muscle has been validated in anesthetized and unanesthetized animals (10, 22). Values for red and white muscle flow obtained in this manner are comparable with those measured by the technique of $^{85}$Kr washout (22) and measurement of venous effluent (9, 14). When flow measurements are made on several muscles simultaneously in the same animal, the cardiac output can be considered a constant. Thus comparison of fractional flow values alone can serve to measure differences in blood flow distribution to various muscles in the same animal and has been so used in this study.

The distribution of blood flow was measured in developing Swiss albino rabbits ranging from 6 to 42 days of age and in adult rabbits of both sexes. Body weight ranged from 98 g to 1178 g in the developing rabbits and from 2.4 to 2.9 kg in the adult. In a preliminary operation the rabbits were anesthetized with 3% halothane anesthesia and a polyvinyl cannula was threaded down the right external jugular vein until the tip was estimated to be just above the right atrium. In the smaller rabbits a dissecting microscope was necessary for this procedure. The cannula was fixed in deep tissue, brought out posteriorly through a stab wound in the neck, and fixed by a collar. The cannula was filled with normal saline, plugged with a trochar and flushed periodically to avoid clotting. After three or four days, when fully recovered from the operation, the unrestrained quiet, alert animals were injected with $^{86}$Rb in a dosage of 1 $\mu$C per 125 g of body weight in a minimum volume of isotonic saline, usually from 0.2 to 3 cc. One minute later the animal was killed instantaneously by injection of 2 cc saturated KCl into heart through the jugular cannula. Samples from various muscles were removed, blotted dry, weighed, and then counted in a Nuclear-Chicago automated gamma well-type scintillation counter on the integral mode with a base setting of 200 KEV. *FF* was then calculated.

## Measurement of Myoglobin Concentration

Myoglobin was measured spectrophotometrically by a modification of the method of Reynafarje (24). Freshly excised tissue samples weighing approximately 1 g were cleaned free of fat and connective tissue, blotted to remove excess blood, weighed, and placed in a test tube to which was added 5 cc of 0.04 M phosphate buffer at pH = 6.6. The tissue was then homogenized in a high speed homogenizer with the tube immersed in an ice water bath. Additional phosphate buffer was then added so that the homogenate contained 19.25 ml/gram of muscle tissue. The muscle homogenate was then centrifuged at 25,000 x g at $-4°$C for 2 hr after which 3–4 cc of the clear supernatant was transferred to a 10 cc test tube. Carbon monoxide gas was bubbled through the solution for 8 min. A pinch of dry sodium dithionate was added to the solution to insure complete reduction of myoglobin and carbon monoxide was passed through the solution for two more min. The solution was then quickly transferred to a cuvette and the optical density was read at 568 m in a spectrophotometer. Myoglobin concentration was calculated by the following equation (24):

$$C_{m/l}^{Mb} = \frac{O.D._{538} - O.D._{568}}{E^{Mb}_{538} - E^{Mb}_{568}}$$

where $C_{m/l}^{Mb}$ is concentration of myoglobin in moles/liter $O.D.$ is the optical density, and $E$ is the molar extinction coefficient. Since there is a constant ratio of buffer to fresh muscle (19.25 ml/gm) and by substituting the extinction coefficients of carboxymyoglobin at 538 and 568 m (14.8 $\times 10^3$ and 11.8 $\times 10^3$ respectively, the equation becomes:

$$C_{mg/gm}^{Mb} = (O.D._{538} - O.D._{538}) \times 117.3$$

This technique insures that the result for myoglobin concentration is essentially free of contamination by hemoglobin (24).

## RESULTS

*Fractional and absolute blood flow and myoglobin concentration in skeletal muscles and myocardium in quiet alert adult rabbit* The fractional distribution of blood flow (*FF*) and myoglobin concentration measured in selective muscles of four chronically prepared unanesthetized adult rabbits are seen in Table I. These muscles were selected to include samples of red, white and intermediate color from limb and trunk as well as left and right ventricular myocardium. The absolute blood flow was estimated by assuming an average cardiac output for the adult rabbit of 488 ml/min (7, 18). As in the cat (23), the blood flow of limb and trunk muscles falls onto a continuum ranging over almost a 10-fold range. Myoglobin concentrations also fall onto a continuum with more than a 3-fold difference between extremes. When the muscles are grouped according to color (Table II) there are significant differences in the blood flow and myoglobin con-

centrations between each group. As in cat and rat, myocardial blood flow greatly exceeds that of all skeletal muscles.

The relationship between *FF* and myoglobin concentration in individual muscles is plotted in Figure 1. Regression analysis (30) indicates that in rabbit as in cat (23) there is a highly correlated direct linear relationship between these two variables in all skeletal muscles. The slopes of these relationships are quite similar for cat and rabbit. The relationship of myocardial flow to myoglobin concentration falls off the curve having myoglobin

## TABLE I

Fractional flow (*FF*), absolute blood flow, and myoglobin concentration in skeletal muscles and myocardium of alert, adult rabbit. Abbreviations: R, red; I, intermediate; W, white; S.M., short medial head; D, deep portion; S, superficial portion; E.C.R., extensor carpi radialis; L.I., long internal head; C.O., cardiac output

| Muscle | Color | *FF* ± S.E. (n = 4) (% c.o./gm muscle) | Absolute blood flow ± S.E. (ml/min/100 gm muscle) | Myoglobin concentration ± S.E. (n = 4) (mg/gram muscle) |
|---|---|---|---|---|
| *Skeletal muscles of limbs and trunks* | | | | |
| Triceps (S.M.) | R | .115 ± .008 | 56.12 ± 3.90 | 2.52 ± 0.27 |
| Diaphragm | R | .108 ± .012 | 52.70 ± 5.86 | 2.76 ± 0.08 |
| Crureus | R | .090 ± .008 | 43.92 ± 3.90 | 3.34 ± 0.11 |
| Soleus | R | .088 ± .005 | 42.94 ± 2.44 | 3.55 ± 0.27 |
| Intercostal | I | .055 ± .010 | 26.84 ± 4.88 | 1.47 ± 0.23 |
| Supraspinatus (D) | I | .053 ± .013 | 25.86 ± 6.34 | 2.02 ± 0.27 |
| Sacrospinalis | I | .050 ± .011 | 24.40 ± 5.37 | 1.64 ± 0.23 |
| Gastrocnemius | W | .040 ± .005 | 19.52 ± 2.44 | 1.55 ± 0.25 |
| E.C.R. | W | .040 ± .012 | 19.52 ± 5.86 | 1.35 ± 0.18 |
| Supraspinatus (S) | W | .039 ± .012 | 19.03 ± 5.86 | 1.14 ± 0.18 |
| Triceps (L.I.) | W | .037 ± .008 | 18.06 ± 3.90 | 1.06 ± 0.15 |
| Tibialis anterior | W | .033 ± .005 | 16.10 ± 2.44 | 1.32 ± 0.09 |
| Semitendonosus | W | .018 ± .002 | 8.78 ± 0.98 | 1.09 ± 0.15 |
| Adductor femoris | W | .017 ± .002 | 8.30 ± 0.98 | 1.03 ± 0.06 |
| Semimembranosus | W | .012 ± .002 | 5.86 ± 0.98 | 0.44 ± 0.11 |
| Longissimus dorsi | W | .012 ± .002 | 5.86 ± 0.98 | 0.94 ± 0.17 |
| *Myocardium* | | | | |
| Right ventricle | | .170 ± .009 | 82.96 ± 4.39 | 2.52 ± 0.10 |
| Left ventricle | | .315 ± .034 | 153.72 ± 16.59 | 2.61 ± 0.23 |

## TABLE II

Mean values for fractional flow (*FF*), absolute blood flow, and myoglobin concentration in grouped red, intermediate, and white muscles in rabbit. Abbreviations as in Table I

| Muscle color | *FF* ± S.E. (n) (% c.o.) (gm muscle) | Absolute blood flow ± S.E. (ml/min/100 gm) | Myoglobin concentration ± S.E. (n) (mg/gm) |
|---|---|---|---|
| Red | .097 ± .004 (16) | 47.3 ± 2.0 | 3.04 ± .14 (16) |
| Intermediate | .053 ± .006† (12) | 25.9 ± 3.0† | 1.71 ± .14† (12) |
| White | .023 ± .003‡ (36) | 11.2 ± 1.5‡ | 1.07 ± .09‡ (36) |

† Difference from red muscle significant (P < .001) by student *t* test.
‡ Difference from both red and intermediate muscles significant (P < .001) by student *t* test (29).

values less than would be predicted by the high blood flow as in cat (23).

*Change in FF during development*    During the first six weeks of life there is a progressive diminution in *FF* in all skeletal muscles and myocardium (see Table III). When pairs of closely apposed muscles, one destined to become red and the other white,

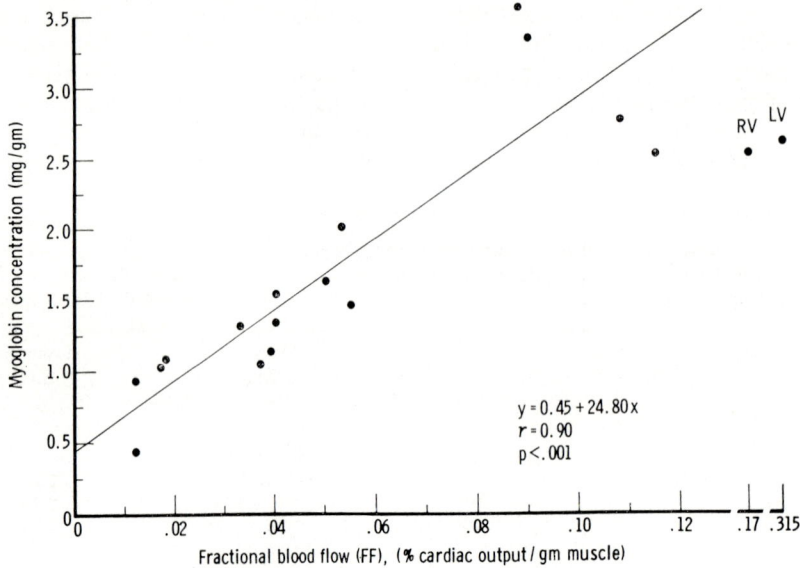

FIGURE 1   Relationship between fractional blood flow (*FF*) and myoglobin concentration in different muscles of quiet, alert rabbit. The slope and *Y*-intercept of the curve were calculated by the method of least squares (29). Abbreviations: RV, right ventricle; LV, left ventricle.

TABLE III

Fractional flow (*FF*) in several red and white skeletal muscles and myocardium during postnatal development in rabbit. Abbreviations as in Table I

| Muscle | (Color) | Fraction of cardiac output/100 grams muscle ± S.E. (*n*) at different days of age | | | | | | |
|---|---|---|---|---|---|---|---|---|
| | | 6 days | 12 days | 21 days | 28 days | 35 days | 42 days | Adult |
| Triceps, S.M. | (R) | 1.062±.224 (5) | .618±.074 (5) | .352±.037† (3) | .295±.053† (4) | .215±.022† (4) | .110±.011† (3) | .115±.008† (4) |
| Triceps, L.I. | (W) | .797±.167 (5) | .523±.072 (5) | .199±.014† (3) | .166±.023† (4) | .124±.015§ (4) | .050±.008† (3) | .037±.008† (4) |
| Crureus | (R) | .690±.116 (5) | .479±.138 (5) | .299±.045 (3) | .265±.031‡ (4) | .207±.016‡ (4) | .093±.021‡ (3) | .090±.008‡ (4) |
| Semimembranosus | (W) | .545±.065 (5) | .413±.067 (5) | .144±.037 (3) | .091±.017‡ (4) | .057±.017‡ (4) | .016±.001‡ (3) | .012±.001‡ (4) |
| Soleus | (R) | .623±.026 (5) | .483±.065 (5) | .328±.093† (3) | .251±.011§ (4) | .189±.007§ (4) | .127±.023§ (3) | .088±.005§ (4) |
| Gastrocnemius | (W) | .576±.024 (5) | .466±.058 (5) | .290±.050 (3) | .182±.022§ (4) | .124±.015§ (4) | .058±.012§ (3) | .034±.005§ (4) |
| Left ventricle | | 2.534±.150 (5) | 1.925±.109 (5) | 1.337±.039 (3) | .991±.248 (4) | .721±.078 (4) | .503±.315 (3) | .315±.034 (4) |

S.M., short medial head of triceps. L.I., long, internal head of triceps. Adult muscle color: R = red; W = white.
† Difference between triceps, S.M., and triceps, L.I., significant (P < .05).
‡ Difference between crureus and semimembranosus significant (P < .05).
§ Difference between soleus and gastrocnemius significant (P ± .05).

are compared, differences in the rate of change of *FF* are observed. As illustrated in Figure 2 there is no significant difference in *FF* between the red short medial head and white long internal head of the triceps of the forelimb up to 12 days of age. Between 12 to 21 days, however, a significant difference in *FF* appears which is maintained throughout the remainder of development until adult values are approximated by the 42nd day. Similar changes also occur in red and white hindlimb muscles (Figure 3). However, the differences in *FF* in hindlimb muscles occur about a week later than in the forelimb primarily during the period from 21 to 28 days. Thus the differences in *FF* between red and white muscle characteristic of the adult are not present at birth but appear to evolve in a rostrocaudal direction.

*Change in skeletal muscle FF during development relative to FF in left ventricular myocardium* The development of the difference in blood flow between red and white muscles could result from (a) a disproportionate increase in flow to red muscle, (b) a disproportionate decrease in flow to white muscle, or (c) a combination of both (a) and (b). To determine which mechanism may account for the changes in blood flow distribution between red and white muscle we have related the changes in flow in limb muscles to that of the left ventricular myocardium. The left ventricular myocardium was selected as a reference point on the basis of the following evidence:

During early development the *FF* in left ventricular myocardium decreases at a rate describing a hyperbolic curve (Figure 4). The reciprocal of the *FF* plotted against body weight (Figure 5A) describes a highly correlated straight line, indicating that left ventricle is growing at the same rate as the whole body. The percentage of the cardiac output perfusing the whole left ventricular myocardium plotted against body weight during development results in a flat curve (Figure 5B) indicating that the percentage of the cardiac output perfusing the whole left ventricle does not change as body weight increases. Since left ventricular weight varies directly with body weight it may be assumed that the decrease in *FF* in the left ventricle during development varies directly with its increase in weight. When *FF* values of red and white skeletal muscles are expressed as a percentage of ventricular *FF* during development, as in Figure 6A for the red soleus and white gastrocnemius, the ratio of *FF* of red muscle to ventricle does not change while

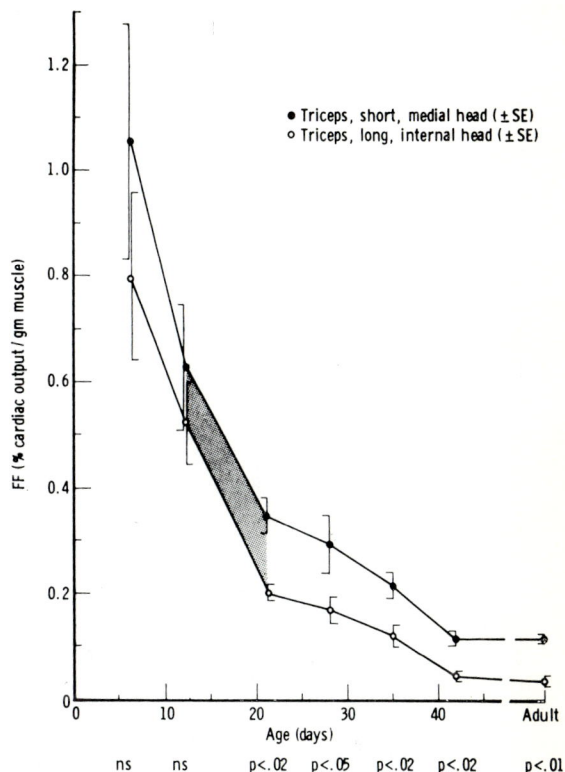

FIGURE 2   Changes in fractional blood flow (*FF*) during postnatal development in the red, short, medial head (closed circle) and white, long, internal head (open circle) of triceps in rabbit. Significance (P < .05) calculated by student *t* test of the difference between red and white muscle at different times is indicated by P value below abscissa. Shaded area represents time through which significant differences in blood flow occur. Abbreviation: ns, not significant. Brackets represent ± S.E.M.

the ratio of the *FF* in the white muscle progressively decreases. That this is the general case for red and white limb muscles is demonstrated in Figure 6B, in which the ratio of *FF* of grouped red and white muscles to left ventricular myocardium is displayed. Since the ventricular *FF* is changing at the same rate as organ growth the results suggest that the *FF* in red muscles diminishes during development at a rate equal to general body growth while white muscle *FF* decreases at a faster rate than would be predicted by growth alone. As a result of the different rates of change of red and white muscle the adult picture of a 2- to 3-fold difference in nutrient flow in red and white muscle is ultimately achieved.

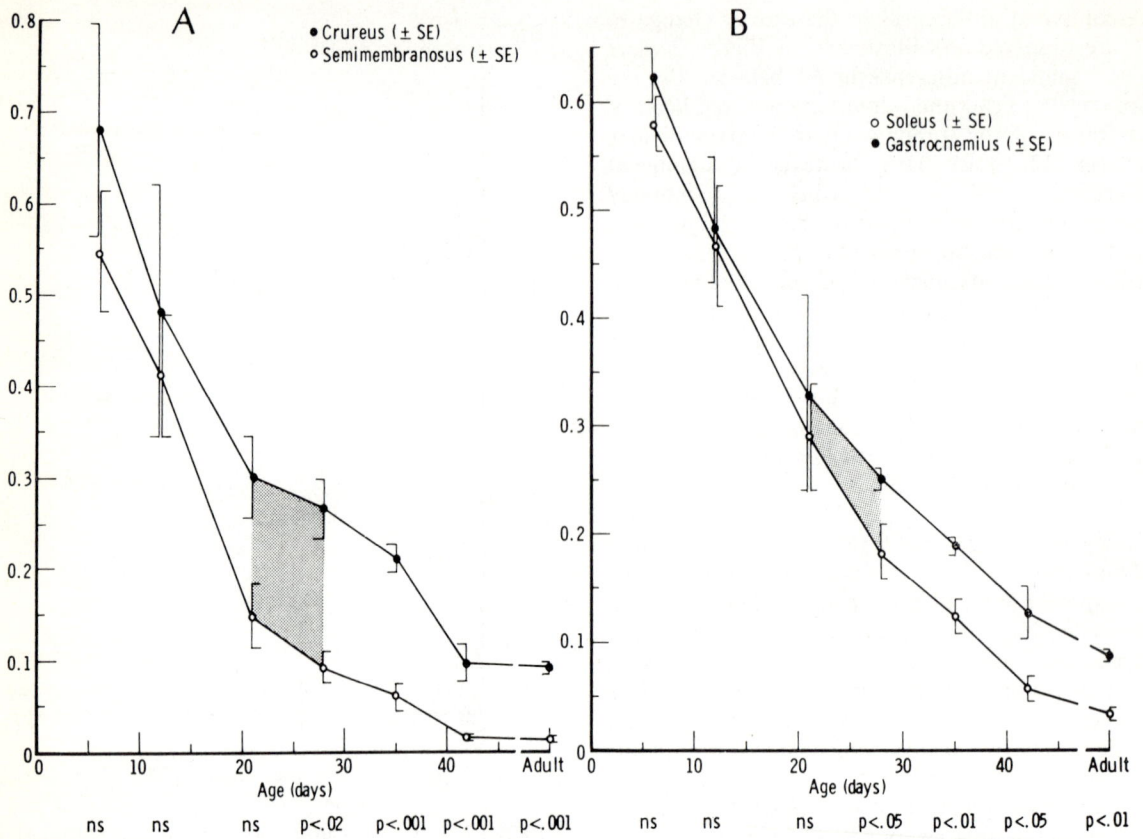

FIGURE 3    Changes in fractional blood flow (*FF*) during postnatal development in several paired red and white hindlimb muscles of rabbit.
A. Crureus (red; closed circles) and semimembranosus (white; open circles).
B. Soleus (red; closed circles) and gastrocnemius (white; open circles).
Significance ($P < .05$) of the difference between muscles at different times is indicated by P value below abscissa. Shaded area represents time during which significant differences in blood flow occur. Abbreviations: ns, not significant. Brackets represent $\pm$ S.E.M.

## DISCUSSION

This study indicates that in the adult rabbit, like the cat (22, 23), nutrient blood flow in red is generally three times that in white skeletal muscle and is directly related to the concentration of myoglobin. Since myoglobin concentration is a reliable indicator of the degree of oxidative metabolism in any muscle (19) the linear relationship between *FF* and myoglobin concentration suggests that nutrient muscle blood flow in the rabbit is related to the degree to which a muscle depends upon aerobic metabolism for its energy.

In all muscles examined in this study there was a progressive diminution in fractional blood flow (*FF*) during development. This decline may be attributed in large measure to an increase in organ weight. Thus while the fraction of the cardiac output perfusing a whole organ or anatomically distinct portion of an organ might remain fairly stable, as for example in the left ventricular myocardium, the fraction of the cardiac output perfusing a unit weight of the organ will decrease as the organ grows. It is likely that the absolute blood flow per unit weight of tissue is probably relatively constant during development because of a concomitant increase in cardiac output at a rate which, at least in the human (16), is directly proportional to the rate of growth. Since absolute blood flow is the product of *FF* and cardiac output (27), a relatively stable nutrient blood flow may be maintained in an organ during development by reciprocal changes of *FF* and cardiac output.

At birth there is no difference in the nutrient

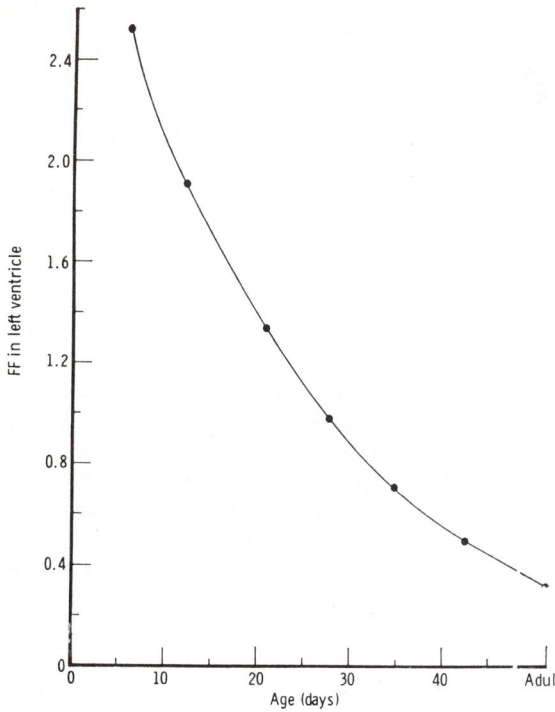

FIGURE 4   Changes in fractional blood flow (*FF*) during postnatal development in left ventricular myocardium of rabbit.

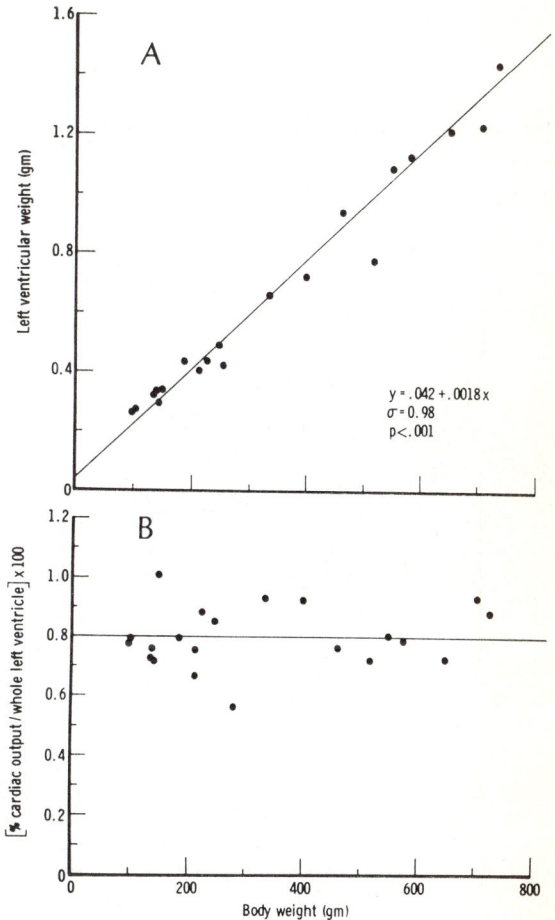

FIGURE 5   Relationship of whole left ventricular weight (A) and percentage of cardiac output perfusing whole left ventricular myocardium (B) to total body weight during first six weeks of life in rabbit. The slope and Y-intercept of the curve in part A were calculated by the method of least squares (29).

blood flow between red and white skeletal muscle. The difference in flow between muscles characteristic of the adult evolves, like the twitch and histochemical differences (1, 2, 5, 6, 12, 20, 21, 31) during the first few weeks of life, and like histochemical differentiation (21), proceeds rostrocaudally. The postnatal differentiation of the histochemical and twitch characteristics of muscle predominantly result from disproportionate change in these properties of white muscle relative to the neonatal pattern (1, 2, 6, 12, 20, 21). Similarly, most of the changes which result in the differences in blood flow between adult muscle types also appear to occur in white muscle. In development the *FF* in white muscle appears to gradually diminish relative to body growth in contrast to the *FF* of red muscle which increases in proportion to body growth.

There are three possible mechanisms to account for the disproportionate fall in the *FF* of white muscle: (a) There could be a greater net increase in fiber size of white muscle relative to red muscle in the presence of a capillary bed of constant size. The disproportionate growth would result in fewer capillaries surrounding white fibers as exists in

the adult (25), and would result in less blood flow per unit weight in white than in red muscle. This mechanism seems unlikely since the average fiber size in red is greater than in white muscle throughout development (21); (b) The greater capillary density of red muscle (23, 25, 29) could result from a decrease in capillary proliferation in white relative to red muscle during growth and development. The failure of capillary growth would occur concomitantly with the metabolic differentiation of fibers into oxidative and glycolytic types so that the density of the capillary network around every fiber is matched to the oxidative metabolism of that fiber (25). That the capillary network can

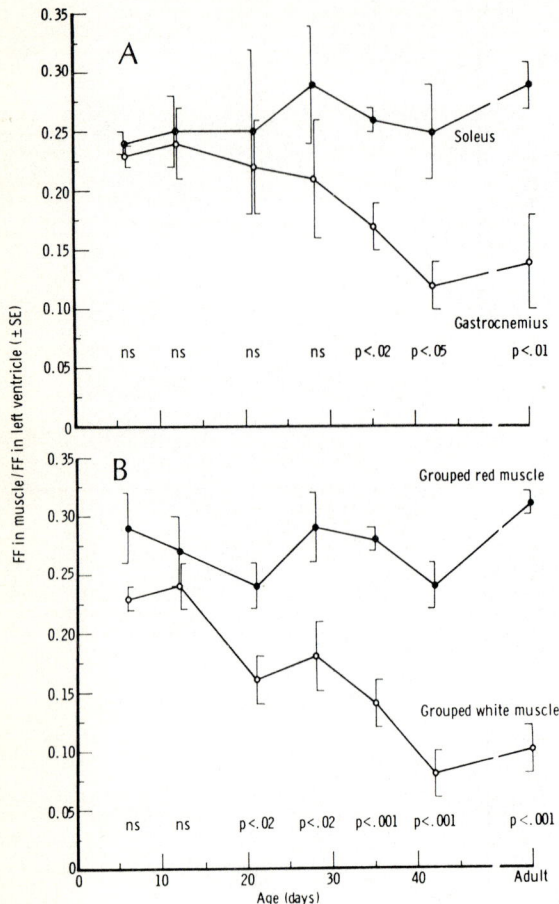

FIGURE 6 The ratio between *FF* in different skeletal muscles and left ventricular myocardial *FF* during postnatal development.
A. Ratio between soleus (red; closed circles) and gastrocnemius (white; open circles).
B. Ratio between grouped red muscle (closed circles) (see Table III) and grouped white muscle (open circles). Brackets represent ± S.E.M. Significance (P < .05) of the difference by students *t* test indicated above abscissa. Abbreviations: ns, not significant.

change with biochemical evolution in muscle has been demonstrated (26); (c) The development of greater vasoconstrictor tone in white muscle would account at least in part for the observed flow differences. In the adult cat a greater range of vasoconstrictor control is found in white muscle (9). This finding coupled with the fact that autonomic reflexes are not mature at birth (4, 8, 11) suggests that the decrease in blood flow in white relative to red muscle during development may in part be due to a disproportionate increase in basal vasoconstrictor tone of white muscle.

Since rabbit muscle shows some histochemical differentiation at birth (6), and blood flow differences between red and white muscles do not appear until the third to fourth week of life, it is likely that metabolic differentiation precedes that of blood flow in rabbits as in chicks and kittens (13, 15). This observation lends further credence to the view (25) that the metabolic requirements of individual muscle fibers may exercise a measure of control over the degree of capillarity and blood flow in their immediate vicinity.

The apparent rostrocaudal course of development of blood flow differences between red and white muscles found in this study are interesting in view of the fact that spontaneous and reflex skeletal movements also develop rostrocaudally with essentially the same time course (28). One would predict that this would be the case such that tonically active muscles would begin to maintain a higher basal blood flow than phasically active muscles. Whether the postnatal differentiation in blood flow between red and white muscles is neurally directed by maturation of dilator and constrictor mechanisms or secondary to changes in capillarity induced by metabolic differentiation of muscle fibers remains to be determined. Both mechanisms may well be operative.

## ACKNOWLEDGMENTS

This study was supported by NIH grants NS 04876, NS 03346, a predoctoral fellowship to G.F.W., and a Research Career Development Award (NS 31756) to Dr. Reis.

## REFERENCES

1. Buller, A. J., J. C. Eccles, and R. M. Eccles. Differentiation of fast and slow limb muscles in the cat hind limb. *J. Physiol.* **150:** 399–416, 1960.
2. Close, R. Dynamic properties of fast and slow skeletal muscles of the rat during development. *J. Physiol.* **173:** 79–95, 1964.
3. Denny-Brown, D. On the nature of postural reflexes. *Proc. Roy. Soc. London Ser. B.* **104:** 252–301, 1929.
4. Downing, S. E. Baroreceptor reflexes in newborn rabbits. *J. Physiol.* **150:** 201–213, 1960.
5. Dubowitz, V., and A. G. E. Pearse. A comparative histochemical study of oxidative enzyme and phosphorylase activity in skeletal muscle. *Histochem.* **2:** 105–117, 1960.
6. Dubowitz, V. Enzymatic maturation of skeletal muscle. *Nature* **197:** 1215, 1967.
7. Edwards, A. W. T., P. I. Korner, and G. D. Thorburn. The cardiac output of the unanesthetized rabbit and the

effects of preliminary anestnesia, environmental temperature and carotid occlusion. *Quart. J. Exptl. Physiol.* **44:** 309–321, 1959.

8. Ekholm, J. Postnatal changes in cutaneous reflexes and in the discharge patterns of cutaneous and articular sense organs. A morphological and physiological study in the cat. *Acta Physiol. Scand.* Suppl. 2, **97:** 1–130, 1967.

9. Folkow, B., and H. D. Halicka. A comparison between 'red' and 'white' muscle with respect to blood supply, capillary surface area, and oxygen uptake during rest and exercise. *Microvasc. Res.* **1:** 1–14, 1968.

10. Friedman, J. J. Muscle blood flow and ⁸⁶Rb extraction: ⁸⁶Rb as a capillary flow indicator. *Am. J. Physiol.* **214:** 488–493, 1968.

11. Friedman, W. F., P. E. Pool, D. Jacobowitz, C. Seagreen, and E. Braunwald. Sympathetic innervation of the developing rabbit heart. Biochemical and histochemical comparisons of fetal, neonatal, and adult myocardium. *Circulation Res.* **23:** 25–32, 1968.

12. Germino, N. I., H. D'Albora, and J. P. Wahrmann. Succinic dehydrogenase in the development of skeletal muscles of chicks. *Acta Anat.* **62:** 434–444, 1965.

13. Hajek, I., O. Hudlicka, and V. Vitek. The relation between blood flow and enzymatic activities in slow and fast muscles during development. *J. Physiol.* **204:** 86–87P, 1969.

14. Hilton, S. M., and G. Vrobova. Absence of functional hyperemia in the soleus muscle of the cat. *J. Physiol.* **194:** 86–87P, 1968.

15. Hudlicka, O. Resting and postcontraction blood flow in slow and fast muscles of the chick during development. *Microvasc. Res.* **1:** 390–402, 1969.

16. Jegier, W., P. Sekelj, P. A. M. Auld, R. Simpson, and M. McGreggor. The relation between cardiac output and body size. *Brit. Heart J.* **25:** 425–430, 1963.

17. Karpati, G., and W. K. Engel. Neuronal trophic function. *Arch. Neurol.* **17:** 542–545, 1967.

18. Korner, P. I. The effect of section of the carotid sinus and aortic nerves on the cardiac output of the rabbit. *J. Physiol.* **180:** 266–271, 1965.

19. Lawrie, R. A. The relation of energy rich phosphate in muscle to myoglobin and cytochrome oxidase activity. *Biochem. J.* **55:** 305–309, 1953.

20. Nystrom, B. Succinic dehydrogenase in developing cat leg muscles. *Nature* **212:** 954–955, 1966.

21. Nystrom, B. Histochemistry of developing cat muscles. *Acta Neurol. Scand.* **44:** 405–439, 1968.

22. Reis, D. J., G. F. Wooten, and M. Hollenberg. Differences in nutrient blood flow in red and white skeletal muscle in cat. *Am. J. Physiol.* **213:** 592–596, 1967.

23. Reis, D. J., and G. F. Wooten. The relationship of blood flow to myoglobin, capillary density, and twitch characteristics in red and white skeletal muscles in cat. *J. Physiol.* **210:** 121–135, 1970.

24. Reynafarje, B. Simplified method for the determination of myoglobin. *J. Lab. Clin. Med.* **61:** 138–145, 1963.

25. Romanul, F. C. A. Capillary supply and metabolism of muscle fibers. *Arch. Neurol.* **12:** 497–509, 1965.

26. Romanul, F. C. A., and J. P. van der Muelen. Slow and fast muscles after cross-innervation. *Arch. Neurol.* **17:** 387–402, 1967.

27. Sapirstein, L. A. Regional blood flow by fractional distribution of indicators. *Am. J. Physiol.* **193:** 161–168, 1958.

28. Skoglund, S. On the postnatal development of postural mechanisms as revealed by electromyography and myography in decerebrated kittens. *Acta Physiol. Scand.* **49:** 299–317, 1960.

29. Smith, D., and R. P. Giovacchini. The vascularity of some red and white muscles of the rabbit. *Acta Anat.* **28:** 342–358, 1956.

30. Snedecor, G. W., and W. G. Cochran. *Statistical Methods.* Ames, Iowa: Iowa University Press, 1967.

31. Stein, J. M., and H. A. Padylkula. Histochemical classification of individual skeletal muscle fibers in the rat. *Am. J. Anat.* **110:** 103–123, 1962.

32. Wirsin, C., and K. S. Larsson. Histochemical differentiation of skeletal muscle in the fetal and newborn mice. *J. Embryol. Exptl. Morph.* **12:** 759–767, 1964.

# DETERMINATION OF AMINO ACIDS IN SINGLE IDENTIFIABLE NERVE CELLS OF HELIX POMATIA

G. BRIEL and V. NEUHOFF

*Max-Planck-Institut für experimentelle Medizin,*
*Arbeitsgruppe Neurochemie, Göttingen (Germany)*
*and*
N. N. OSBORNE

*Wellcome Laboratories of Pharmacology,*
*Gatty Marine Laboratory, University of St. Andrews (Fife, Scotland)*

Microchromatography of dansylated compounds was used to study the distribution of free amino acids, GABA, serotonin and 5-hydroxyindole in the metacerebral serotonergic cell and one of the giant cells in the buccal ganglia of *Helix pomatia*. The distribution of some of the substances in each of the cell types varied depending upon the isolation procedure. Metacerebral cells dissected in the presence of nialamide contained large amounts of serotonin. The same neurons isolated in the absence of nialamide, but dissected with methylene blue contained no serotonin. The distribution of ornithine, glycine, alanine, arginine, ε-lysine, α-amino-histidine and cystine in each cell type varied depending upon the method of isolation.

Generally the distribution of dansylated substances is similar; GABA is present in each cell type but in low concentrations. However, there are some exceptions. The serotonergic cell contains less ornithine and more glycine than do the buccal cells. In addition the metacerebral cells have high amounts of an unknown substance, which may be a metabolite of serotonin.

Furthermore, 5-hydroxyindole is always found in the serotonergic cell, whilst it is present only in buccal neurons which have been dissected in the presence of nialamide.

## INTRODUCTION

The giant neurons of gastropod molluscs are at present attracting the attention of many neurologists. This interest is due both to the convenience of micro electrode recording techniques, and to the possibility of isolating single neurons and of conducting electro-physiological (Peterson and Kernell, 1970) and biochemical investigations (Cottrell and Osborne, 1970, McCamon and Dewhurst, 1970, Osborne, Ansorg and Neuhoff, 1971) in successive experiments on single defined neurons. Compared with the vast amount of information concerning the electrophysiology (see Tauc, 1967) and pharmacology (see Cottrell and Laverack, 1968, Sakharov, 1970) of individual gastropod neurons, only a small number of biochemical studies have been reported. This is due mainly to the absence of suitable micro methods which possess also a high degree of sensitivity.

Neuhoff and co-workers (Neuhoff, von der Haar, Schlimme and Weise, 1969, Casola, Weise and Neuhoff, 1969, Neuhoff, 1970, Neuhoff and Weise, 1970) have recently described a micro method for the detection of amino acids in quantities of tissue weighing less than 0.1 mg. Furthermore, the method is extremely sensitive when compared with other processes (e.g. ninhydrin), locating as little as $10^{-12}$M of amino acid. The technique involves the reaction of dansyl chloride with tissue free amino acids, and the subsequent separation of dansyl compounds on $3 \times 3$ cm polyamide layers. Using two or three solvent systems, individual amino acids can be separated and viewed under ultra violet light. This method was employed to study the amino acid composition in single identifiable neurons of the snail.

Our experiments were performed on a single and easily identifiable giant neuron in the metacerebral ganglion of *Helix pomatia*. The electrophysiological properties of this neuron are already known (Kandel and Tauc, 1966, Cottrell, 1970) and histochemical studies have shown that the cell contains serotonin (see Cottrell, 1970, Osborne and

Cottrell, 1971). The amino acid composition of this neuron is compared with that of a non-serotonergic cell in the buccal ganglion.

## METHODS

*Helix pomatia* were either collected locally in Göttingen, or purchased in Scotland from L. Haig & Co. Ltd., Newdigate, Surrey. On arrival at the laboratory the animals were used immediately, or kept in a glass tank at room temperature and fed on lettuce.

## DISSECTION OF NEURONS

The buccal and circumoesophageal ganglia were rapidly removed from an animal, immersed in a saline solution (Meng, 1960), and pinned by the edges of connective tissues, ventral side upwards, to the paraffin base of a small chamber. The remainder of the dissection was carried out under a stereomicroscope. Connective tissues around the neurons were carefully removed with fine forceps. The position of the metacerebral giant cells has been described by Kandel and Tauc (1966). The relevant giant buccal neurons are situated close to the entry of each cerebral-buccal connective. Since each buccal ganglion has three giant cells, care had to be taken in dissecting the correct one. Single giant nerve cells from either the metacerebral or buccal ganglia were carefully freed from the surrounding nervous tissue using thin tungsten needles. Cells were then individually lifted free by suction from the ultra thin tip (15 $\mu$ in diameter) of a glass pipette, attached to the mouth by rubber tubing.

In Göttingen, 10% methylene blue was used to assist in the removal of glia from the cells. The cells were then stored in a deep-freeze for two weeks before being analysed. The dissection of other neurons, which was carried out in Scotland, did not involve the use of methylene blue. Furthermore nialamide was added to the snail saline used for dissection (100 $\mu$g/ml of Meng's saline). The neurons were then freeze-dried in Scotland and sent to Göttingen to be analysed.

## DETERMINATION OF AMINO ACIDS

Isolated neurons (4–8) from either the buccal or metacerebral ganglia were transferred to a 5 $\mu$l

microcap (Drummond microcaps) which had been sealed at one end and which contained 0.05M sodium bicarbonate pH10. The neurons were then sedimented by centrifuging the microcap for 5 minutes at 15,000 rpm in a Hämatokritzentrifuge equipped with a special adaptor to centrifuge capillaries, and subsequently homogenized with a nerve canal drill (Beutelrockbohrer) which has a diameter of 0.25 mm (for details see Neuhoff, 1970). The microcap was centrifuged again for 15 minutes. Proteins were precipitated by an equal volume of cold acetone and the precipitation completed by placing the sample in a freezer ($-20°$ C) for 30 minutes. Once more the microcap and its contents were centrifuged for 15 minutes, and the supernatant transferred to a clean 10 $\mu$l microcap to be analyzed. Equivalent volumes of solution and $^{14}$C dansyl chloride (Centre d'Etudes Nucleaires de Saely, France: 3 mg/ml in acetone, specific activity 49 mc/mM) in acetone (1:3 by volume) where mixed in a 10 $\mu$l capillary and subsequently incubated for 30 minutes at 37° C. Aliquots of dansylated products were then carefully applied to the corner of a 3×3 cm polyamide layer (Carl Schleicher und Schüll, DC-Fertigfolie F 1700 Mikropolyamid) and developed in an ascending fashion. Water/formic acid (100:3v/v) was used for the first dimension, and benzene/acetic acid (9:1v/v) for the second (see Figures 1a, 2a, 3a and 4a). A third chromatography in the second dimension, using the solvent ethyl acetate/methanol/acetic acid (20:1:1v/v/v) was necessary for the separation of some amino acids (see Figures 1b, 2b, 3b and 4b). Standard dansylated amino acids were added to the extracts or chromatographed separately to assist in the identification. Radioautograms of microchromatograms were obtained in the dark using highly sensitive film (Dupont Dosimeterfilm 046/545) exposure time being 4 days.

For quantitative measurements the fluorescent spots were marked with a soft pencil under U.V. light, scraped with a special razor-splinter knife and transferred (for details see Neuhoff, 1970) directly into counting vials. Each sample was suspended in a scintillation liquid (4 grm PPO and 0.1 grm POPOP per litre toluene) and counted in a Packard Model 3380.

## RESULTS

Radioautograms showing the occurrence of $^{14}$C-dansylated substances are shown in Figures 1–4. Figure 1a is a picture of a radioautogram, in

## TABLE I

Composition of dansylated compounds separated by microchromatography

| Spot No. | Substance | A: 8 Buccal cells dissected in the presence of nialamide (in CPM) | B: 8 Metacerebral cells dissected in the presence of nialamide (in CPM) | C: 4 Buccal cells dissected without nialamide (in CPM) | D: 4 Metacerebral cells dissected without nialamide (in CPM) | Content of A in residues per 100 total residues | Content of B in residues per 100 total residues | Content of C in residues per 100 total residues | Content of D in residues per 100 total residues |
|---|---|---|---|---|---|---|---|---|---|
| 1 | Startpoint | | | | | | | | |
| 2 | Dansyl-OH | | | | | | | | |
| 3 | N-Tyrosine | 783 | 1629 | 200.2 | 343.5 | 2.75 | 3.43 | 2.33 | 1.86 |
| 4 | Tryptophane | 350.5 | 866.5 | 48.6 | 181.2 | 1.23 | 1.82 | 0.56 | 0.98 |
| 5 | Unknown comp. | 244.5 | 5689.9 | — | — | 0.86 | 11.98 | — | — |
| 6 | Bis-Ornithine | 6948 :2 | 5793 :2 | 523 :2 | 1653 :2 | 12.19 | 6.1 | 3.04 | 4.47 |
| 7 | Bis-Lysine | 1592.5 :2 | 1559.5 :2 | 707.2 :2 | 895 :2 | 2.80 | 1.64 | 4.11 | 2.42 |
| 8 | Methionine | 327 | 653 | 113.8 | 119 | 1.15 | 1.38 | 1.32 | 0.64 |
| 9 | Phenylalanine | 1436 | 3219,5 | 573.4 | 1079.6 | 5.04 | 6.78 | 6.66 | 5.83 |
| 10 | Leucine | 1507.5 | 2343 | 726.8 | 1023.8 | 5.29 | 4.93 | 8.45 | 5.53 |
| 11 | Isoleucine | 540 | 1373 | 436.8 | 738.2 | 1.9 | 2.89 | 5.08 | 3.99 |
| 12 | Unknown comp. | 2072.5 | 2980.5 | 103.6 | 1048 | 7.27 | 6.27 | 1.20 | 5.66 |
| 13 | Bis-Tyrosine | 491.5 :2 | 485 :2 | 56.2 :2 | 64.2 :2 | 0.86 | 0.51 | 0.33 | 0.17 |
| 14 | 5-OH-Indol | 626.5 | 577 | — | 118.8 | 2.2 | 1.22 | — | 0.64 |
| 15 | Valine | 754 | 1329.5 | 521.6 | 785.4 | 2.65 | 2.8 | 6.06 | 4.24 |
| 16 | Proline | 650 | 2638 | 484.2 | 664.4 | 2.28 | 5.55 | 5.63 | 3.59 |
| 17 | Unknown comp. | 910.5 | 754.5 | 480.8 | 485.8 | 3.2 | 1.59 | 5.59 | 2.62 |
| 18 | Alanine | 5347 | 4048.5 | 523.8 | 3551.6 | 18.77 | 8.52 | 6.09 | 19.19 |
| 19 | Glycine | 2497 | 5763 | 1268 | 2945.8 | 8.76 | 12.13 | 14.74 | 15.92 |
| 20 | Glutamic Acid | 1583 | 934 | 193.7 | 520.8 | 5.56 | 1.97 | 2.25 | 2.81 |
| 21 | Aspartic Acid | 565.5 | 346.5 | 83.9 | 166.4 | 1.99 | 0.73 | 0.98 | 0.9 |
| 22 | Glutamine, Serine, Threonine | | | | | | | | |
| 23 | Arginine, ε-Lysine, α-amino-Histidine, | 2613.5 | 3005.5 | 646 | 1263.2 | 9.17 | 6.33 | 7.51 | 6.83 |
| 24 | Cystine | 1065 | 5316.5 | 1554.6 | 2165.4 | 3.74 | 11.19 | 18.07 | 11.7 |
| | GABA | 102.3 | 119.8 | 84.4 | 125.2 | 0.36 | 0.25 | 0.97 | 0.67 |
| Total | | 28491.3 | 47505.95 | 8687.4 | 18632.3 | | | | |
| 25 | N-Serotonin | | 10674.6 | | | | 18.14 | | |
| 26 | Bis-Serotonin | | 1301.5 :2 | | | | 1.11 | | |

original size, of buccal cells (dissected in Göttingen) separated by two-dimensional chromatography, together with a map showing the distribution of substances, to be compared with Table 1. In order to separate spot No. 20–23 a third chromotography in the second dimension is required. Figure 1b is a picture of a chromatogram showing the separation of glutamine, threonine and serine, glutamic acid and aspartic acid, argenine, ε-lysine, α-amino-histidine and cystine as noted in Table I.

Chromatograms of cells from the metacerebral ganglia (also dissected in Göttingen) show a number of differences when compared with the cells of buccal ganglia (see Figures 2a and 2b and Table I). Of particular interest is the absence of 5-hydroxyindole in the buccal cells. However, of greater importance is the absence of serotonin in both cell types. Furthermore, the presence of small amounts of GABA in both the buccal and metacerebral

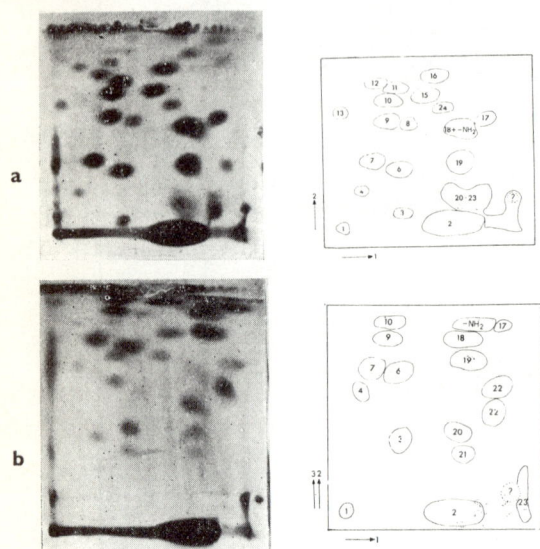

FIGURE 1 Autoradiograms in original size after microchromatography of substances in buccal cells (dissected without nialamide) after having reacted with $^{14}$C dansyl chloride. Exposure 5 days.
(a) after chromatography in two solvent systems;
(b) after chromatography in a third solvent system.
The direction of chromatography is indicated by arrows. 1st direction: water/formic acid (100:3), 2nd direction: benzene/acetic acid (9:1), 3rd direction: ethylacetate/methanol/acetic acid (20:1:1). In addition a map of each chromatogram is shown to assist in the identification of individual substances.
The numbers on each map correspond to the dansyl-compounds listed in Table I. Unmarked spots on chromatograms separated by two solvent systems belong to impurities of $^{14}$C dansyl chloride, to other unknown compounds or to a token in the x-ray film (broken line).

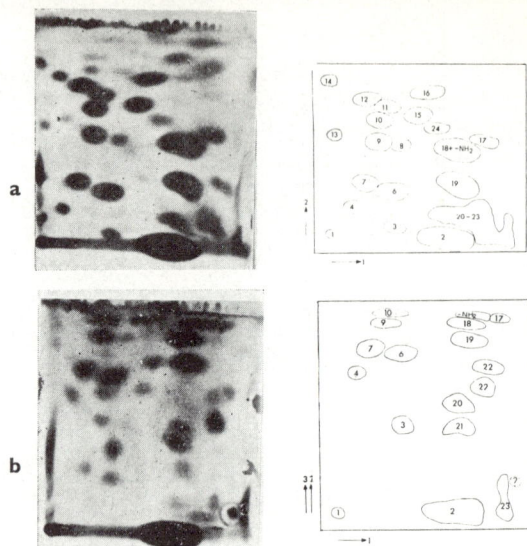

FIGURE 2 Autoradiograms and corresponding maps of $^{14}$C dansyl-compounds in metacerebral cells (dissected without nialamide). Exposure 5 days. For further explanation see Figure 1.

cells is to be noted. Quite high amounts of glycine are also present in each of the cell types. The high amount of alanine in the metacerebral cells, as compared with the buccal cells is difficult to explain.

Figures 3a and 3b are radioautograms of the same cell extract from the buccal ganglion, together with maps showing the distribution of substances when compared with Table I. In this case the cells were dissected in Scotland, without using methylene blue but nevertheless adding nialamide (a monoamine oxidase inhibitor) in the dissecting solution (see methods). Although a spot (No. 5) is present in the position of N-serotonin after two dimensional chromatography (see Figure 3a), it does not migrate to the position of N-serotonin after a third chromatography in the second dimension (see Figure 3b). N-serotonin is characterized by its migration in the third run (see spot No. 25). It is clear from the position of spot No. 5 (see Figure 3b) that the buccal cells do not contain serotonin. The buccal cells do however contain high amounts of glycine, ornithine and alanine, although the GABA content is low. Some 5-hydroxyindole is also present.

Cells from the metacerebral ganglia dissected in Scotland show clearly the occurrence of serotonin. Figure 4b indicates the presence of both bis-serotonin (spot No. 26) and N-serotonin (spot No. 25). The migration of N-serotonin (spot No. 25) after a third chromatography is shown in

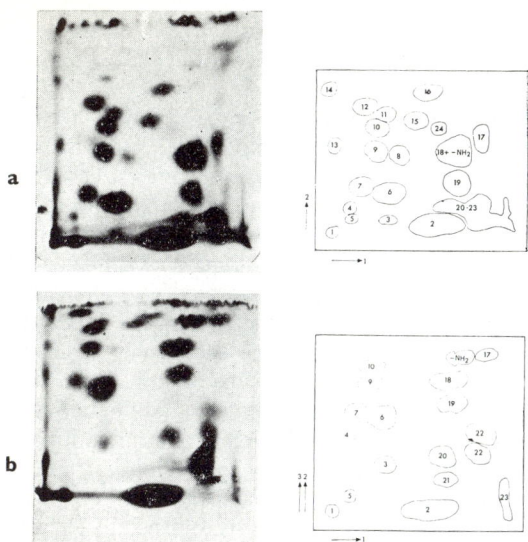

FIGURE 3   Autoradiograms and corresponding maps of $^{14}C$ dansyl-compounds in buccal cells (dissected in the presence of nialamide).

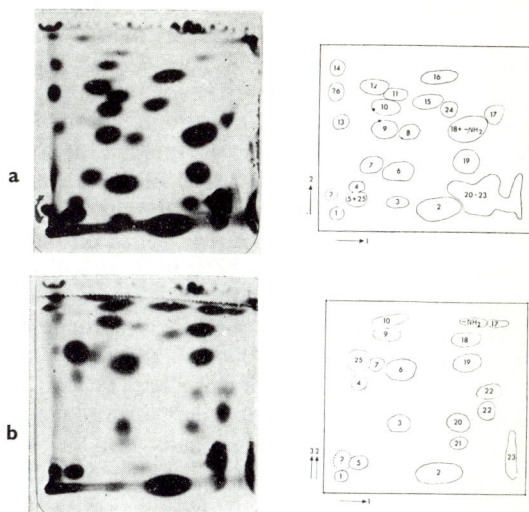

FIGURE 4   Autoradiograms and corresponding maps of $^{14}C$ dansyl-compounds in metacerebral cells (dissected in the presence of nialamide).

Figure 4b, which also demonstrates the separation of spot No. 20–23. Furthermore, the unknown substance (spot No. 5), which migrates to a position similar to that of N-serotonin after two dimensional chromatography and which is present in the buccal cells is also shown to occur in the metacerebral cells (see Figures 4a and 4b). In comparison with the buccal cells the glycine content in the metacerebral cells is also high, while the content of ornithine and alanine is lower. The GABA content of both the buccal and metacerebral cells is low.

## DISCUSSION

It has been stressed that in order to understand more about the nervous system, chemical analysis must be done on repeatedly identifiable isolated neurons rather than brain tissue, which often contains a heterogeneous population of neurons as well as glia and muscle tissue (see Rose, 1968, Giacobini, 1969). The one giant serotonin-containing neuron in each metacerebral ganglia of *Helix pomatia* is also present in a number of other gastropods (see Osborne and Cottrell, 1971). Each serotonergic neuron makes direct synaptic contacts with at least two of the three repeatedly identifiable giant neurons which lack biogenic amines in the buccal ganglia (Cottrell, 1970a). All the available data suggest that the serotonin within these meta-

cerebral neurons is used as a transmitter substance (Cottrell and Osborne, 1970, Cottrell, 1970a, b). The content of individual metacerebral cells from the slug (*Limax maximus*) has been estimated by bioassay to contain 1 ng of serotonin. A comparable amount of serotonin is also calculated to be in the giant cell of *Helix pomatia* (Osborne, 1971). The giant metacerebral neuron and the buccal neuron therefore represent two different types of nerve cells.

One of the most striking features of the results is the apparent difference in the content of some substances in the same cell sample which itself only varies in the way it has been isolated. The most obvious difference is the occurrence of vast amounts of serotonin in the metacerebral cells isolated in Scotland, as compared with those dissected in Göttingen. Another notable example is the presence of the unknown substance (spot No. 5) which occurs in both the buccal and metacerebral cells which were dissected in Scotland. Two other examples must be mentioned: the apparent abundance of ornithine in buccal cells dissected in Scotland, in contrast to the comparatively small amounts of this same amino acid found in cells obtained in Göttingen. Furthermore, the arginine, ε-lysine, α-amino-histidine and cystine content in both buccal and metacerebral dissected in Göttingen is far higher than that in cells dissected in Scotland.

Even to take into account the different ways in

which each cell sample was isolated can not fully explain these results. There is no evidence which would indicate that methylene blue has any effect on a cell's chemical contents. Use of nialamide in the samples dissected in Scotland could probably account to some extent for the preservation of serotonin in the metacerebral cells, although previous studies indicate that gastropod ganglia lack enzymes capable of inactivating serotonin (see Gerschenfeld and Stefani, 1968). It is also likely that keeping the metacerebral cells in a deep-freezer for a long period, and their subsequent thawing before analysis resulted in the enzymatic breakdown of some of the serotonin.

It is probable, however, that the main source of differences is elsewhere. The snails obtained in Göttingen were active, whilst those used in Scotland were hibernating. Another important factor involves the isolation of cells by free hand dissection. The possibility of contamination must be considered, especially in the samples where methylene blue was not used. Examination of isolated cells under phase contrast microscopy often revealed the occurrence of fibres from other cells and also glia elements sticking to the cell body. It could therefore be argued that use of methylene blue provides a purer cell sample. On the other hand, in opposition to this view is the length of time taken to remove from the cell body all contaminations which could be seen in methylene blue treated preparations. It is probable that during this procedure some membrane damage occurred, thus allowing leakage of chemicals from the cell and subsequently destroying the integrity of the neurons.

The low amounts of GABA in each of the cell types is difficult to explain, especially in relation to a possible transmitter role. The substance has a similar distribution in both the metacerebral and buccal cells. Enzymes capable of forming GABA from glutamic acid are known to exist in the snail brain (Osborne, Briel, and Neuhoff, 1971). However, the evidence for serotonin as being the transmitter substance in the metacerebral cell is very good (Cottrell, 1970a).

The occurrence of another dansylated substance should be mentioned. It is the unknown substance (Spot No. 5) present only in those neurons dissected in Scotland, which involved the use of a monoamine oxidase inhibitor (nialamide). The high amount of unknown substance in the serotonergic cell compared with the buccal cell seems to indicate that the substance is a metabolite in the metabolism of serotonin. Initial experiments have

shown that it is not 5-hydro xyindole acetic acid, a major metabolite of serotonin, in the vertebrate brain.

The results reported here reveal the heterogenity of neurons within the gastropod brain with respect to the amino acids and serotonin content. However further experiments are required to explain some of these results before deciding upon the exact chemical content of each cell type. It is difficult, for example, to determine 5-hydroxytryptophan, the immediate precursor of serotonin, because the bis-dansyl-compound of this substance, which is mainly formed at pH 10, runs only in the second dimension where it is mostly contaminated with dansyl-peptides. Therefore the determination of 5-hydroxytryptophan was impossible in these experiments, although this substance is known to exist in the snail's brain (Osborne, Briel and Neuhoff, 1971).

## REFERENCES

Casola, L., Weise, M., and Neuhoff, V., 1969, In vitro protein synthesis by optic nerves, Hoppe-Seyler's Z. Physiol. Chem. 350: 1175.
Cottrell, G. A., 1970a, Direct postsynaptic responses to stimulation of serotonin-containing neurones, Nature 225: 1060–1062.
Cottrell, G. A., 1070b, Actions of LSD–25 and reserpine on a serotonergic synapse, J. Physiol. 208: 28–29.
Cottrell, G. A., and Laverack, M. S., 1968, Invertebrate pharmacology, Ann. Rev. Pharmacol. 8: 273–298.
Cottrell, G. A., and Osborne, N. N., 1970, Serotonin: Subcellular localization in an identified serotonin-containing neuron, Nature 225: 470–472.
Gerschenfeld, H. M., and Stefani, E., 1968, Evidence for an excitatory transmitter role of serotonin in molluscan central synapse, Adv. in Pharmacol 6A: 369–392.
Giacobini, E., 1969, Chemistry of isolated invertebrate neurons, Handbook of Neurochem 2: 195–239.
Kandel, E. R., and Tauc, L., 1966, Input organization of two symmetrical giant cells in the snail brain, J. Physiol 183: 269–287.
McCaman, R. E., and Dewhurst, S. A., 1970, Choline acetyltransferase in individual neurons of Aplysia californica. J. Neurochem 17: 1421–1426.
Meng, K., 1960, Untersuchungen zur Störung der Herztätigkeit bei Helix pomatia, Zool. Jahr 68: 539–566.
Neuhoff, V., 1970, Determination of amino acids in picomole range. In: Manual of 1. EMBO-course on Micromethods in Molecular Biology, Göttingen, Max-Planck-Gesellschaft, Dokumentationsstelle, pp. 44–52.
Neuhoff, V., von der Haar, F., Schlimme, E., and Weise, M., 1969, Zweidimensionale Chromatographie von Dansyl-Aminosäuren im pico-Mol-Bereich, angewandt zur direkten Charakterisierung von Transfer-Ribonucleinsäuren, Hoppe-Seyler's Z. Physiol. Chem. 350: 121–123.
Neuhoff, V., and Weise, M., 1970, Determination of Picomole Quantities of γ-Amino-Butyric-Acid (GABA) and Serotonin, Arzneim-Forsch. (Drug Res.) 20: 368–372.

Osborne, N. N., 1971, A microchromatographic method for the detection of biogenic monoamines in isolated neurons, *Anal. Biochem.* In press.

Osborne, N. N., Ansorg, R., and Neuhoff, V., 1971, Micro-disc electrophoretic separation of soluble proteins from neurons and other tissues of *Helix* (pulmonate molluscs), *Intern. J. Neuroscience.* **1**: 259-264.

Osborne, N. N., Briel, G., and Neuhoff, V., 1971, Distribution of GABA and other amino acids in different tissues of the gastropod mollusc *Helix pomatia*, including *in vitro* experiments with ¹⁴C glucose and ¹⁴C glutamic acid, *Intern. J. Neuroscience.* **1**: 265-272

Osborne, N. N., and Cottrell, G. A., 1971, Distribution of biogenic amines in the slug *Limax maximus*, Z. Zell-forsch **112**: 15–30.

Peterson, P. R., and Kernell, D., 1970, Effect of nerve stimulation on the metabolism of ribonucleic acid in a molluscan giant neurone, *J. Neurochem* **17**: 1075–1085.

Rose, S. P. R., 1968, In: Applied Neurochemistry (A. Davison and J. Dobbing). Publishers: Blackwell Scientific, Oxford and Edinburgh, pp. 332–355.

Sakharow, D. A., 1970, Cellular aspects of invertebrate neuropharmacology, *Ann. Rev. Pharmacol* **10**: 335–352.

Tauc, L., 1967, Transmission in invertebrate and vertebrate ganglia, *Ann. Rev. Physiol* **47**: 521–593.

# BETA ADRENERGIC MECHANISMS INFLUENCING BRAIN STEADY POTENTIAL IN CATS AND RHESUS MONKEYS

JOHN H. HUBBARD,† W. STEPHEN CORRIE,‡ HARRY K. THOMPSON, and WADE H. MARSHALL

*Laboratory of Neurophysiology, National Institute of Mental Health,*
*and*
*National Institute of Neurological Diseases and Stroke, National Institutes of Health,*
*Bethesda, Maryland* 20014

Beta adrenergic catecholamines and $CO_2$ produce large steady potential (SP) shifts in the brains of both cats and monkeys. Shifts caused by $CO_2$ are blocked at the peak of an adrenergic response. The beta-aminergic shift is rapidly reversed by propranol.

These SP shifts are unrelated to changes in cerebral blood flow, brain impedance, or depth of anesthesia. Evidence points to effects on active transport across capillary endothelium or glia adjacent to the vascular lumen. Paired microelectrodes which straddle the blood brain barrier (BBB), even when placed within 1 millimeter of each other to record between sagittal sinus and subarachnoid space, give results identical with those obtained by less direct methods. These potential shifts across the barrier are sufficiently different from most previously described SP phenomena to be distinguished from them; the term 'intracranial transvascular potential (ITVP)' shift is proposed to describe them.

## INTRODUCTION

In cats, large negative shifts in brain steady potential develop coincident with carbon dioxide administration (Mottschall and Loeschcke, 1963). These results are the converse of observations by others in rats, rabbits, dogs, and goats (Tschirgi and Taylor, 1958; Held *et al.*, 1964). Woody and his colleagues (1970) confirmed the findings in cats and have extended their studies to include the Rhesus monkey with similar results. In both species the $CO_2$ negative potential response is relatively independent of electrode location, provided one of the nonpolarizable pair contacts the dura, arachnoid, or brain surface while the other lies external to the bony confines of the cranio-spinal axis.

Besson *et al.* (1970) have suggested that these steady potential transients are generated by changes in cerebral blood flow (CBF), since epinephrine injection causes negative DC shifts, and also competes with those caused by high tissue $CO_2$ concentrations. However, the authors do not rule out the possibility of a blood-brain barrier origin for the response.

Several of our experiments were designed to explore the latter possibility. Among other DC recording techniques, we have placed micropipettes across the barrier interface at a point between sagittal sinus and adjacent subarachnoid space with results identical to those of less direct methods. This report describes the DC responses observed with injection of certain biogenic amines. The pharmacodynamics of these electropotential effects on blood-brain barrier (BBB) will be discussed. Moreover, we shall demonstrate that there is no enduring correlation between CBF and brain steady potential (SP).

## METHODS

Thirty cats (1.8–4.5 kg) and eighteen Rhesus monkeys (3.5–6 kg) were studied. They were anesthetized either with pentobarbital (30–35 mg/kg I.P.) or with halothane-oxygen mixtures (0.8–3% halothane in $O_2$). Pentothal (20–40 mg/kg I.V.) was

† Present address: Division of Neurosurgery, University of Pennsylvania School of Medicine, and Veterans' Hospital, Philadelphia, Pa. 19104.

‡ Present address: Department of Neurology, Washington University, School of Medicine, St. Louis, Mo. 63110.

employed for anesthetic induction in those monkeys subsequently maintained on halothane. After tracheostomy and, in most animals, neuromuscular blockade with gallamine triethiodide (Flaxedil), breathing was controlled with a positive pressure, constant volume respirator. Concentrations of air, oxygen, and carbon dioxide in the gas mixture were determined by adjustment of venturi type flow-meters. Expiratory $CO_2$ was continuously monitored with an infra-red analyzer (Godart). Body temperature was maintained, and blood pressure recorded in all experiments.

1. In cats we injected drugs directly into the brachiocephalic trunk after passing a No. 50 polyethylene catheter through the right thyro-cervical artery to the intrathoracic point of origin of both carotid arteries (2.7–3.3 cm caudal to the first rib). Because the right vertebral artery is also a tributary of that trunk, a single injection delivers a test compound to all areas of cerebrum and brainstem simultaneously. The left vertebral artery is not infused by this technique; consequently, pontomedullary and cerebellar drug concentration reaches only about fifty percent of that in more rostral locations. This method provides for optimal central action of many compounds in doses causing little or no systemic effect.

In monkeys, lacking a comparable trunk, injections were made through catheters placed at the carotid bifurcations after ligation of external carotid branches. In recent experiments, however, stainless steel cannulae (27–30 ga.) were slipped into the internal carotid vessels without ligation of external branches. This modification preserves the normal hemodynamic balance between extra- and intracranial blood flow while routing the administered drugs only to the brain. In either species the volume injected through each catheter was limited to 0.6 cc over 3–8 seconds, with 30–40 minutes for recovery between injections.

2. Cerebral blood flow measurements were performed in monkeys with modification of the technique for torcular venous outflow in dogs (Rapela and Green, 1964)†. The junction of right lateral and sigmoid sinus was exposed by craniectomy with meticulous waxing of diploe to prevent air embolism. An extracorporeal circuit was employed; heparinized silastic tubing (10–12 inches long) was fitted with a 3 mm diameter electromagnetic flow-through probe (Statham 4001 flow system). The distal end of this shunt was first

† Dr. Rapela demonstrated this flow method to the authors and helped us work out the preliminary steps for its adaption to monkeys.

introduced via the rt. internal jugular vein to the junction of innominate vein and superior vena cava. The proximal end was then inserted into the lateral cerebral venous sinus coincident with opening of the by-pass circuit. The opposite sigmoid sinus was then occluded. Outflow values ranged from 20–35 cc/min at basal $pCO_2$ (35–40 mmHg). These values represent 50–70% of total CBF (Meyer et al., 1964).

3. For DC stability one of the following types of electrode was used, depending on the objective of a particular experiment:

A. Ag–AgCl wire placed in quartz glass tubing (300 $\mu$ tip) filled with physiologic saline or artificial CSF. The 'intracranial' electrode of each pair was counter-balanced on cerebral convexity or intact dura, while the other electrode, wrapped in saline soaked cotton lay between fascia and subcutaneous tissue in the scalp or neck.

B. A glass micropipette (1–10 $\mu$ tip) inserted into cisterna magna referred to an agar lead within the posterior facial vein. This pair was connected to the electrometer either through calomel half-cells or Ag–AgCl wires.

C. Beckman biopotential electrodes (sintered Ag–AgCl) in both acute and chronic experiments. 1–2 mm holes were drilled in the skull. The opening over suprasylvian gyrus was extended to dura; bone was heavily waxed. This opening remained free of contamination with blood. The other opening was deepened to enter the bony transverse vascular sinus (cats) or a diploic lake (monkeys). Both cavities were plugged with saline soaked cotton over which the electrodes were placed and bonded to the skull with cyanoacrylate glue.

D. Paired 3M KCl or 2.5M NaCl filled glass micropipettes (1 $\mu$ tip diameter) fitted with Ag–AgCl wires or connected to calomel half-cells were inserted across the sagittal sinus wall 1 cm behind the coronal suture. After a $3 \times 3$ mm craniectomy, each pipette was inserted at 45°, one within the lumen of the sinus and the other within subarachnoid space 1 mm lateral to the vascular lead.

Type A electrode pairs had a typical DC resistance of 15–50 kilohms; type B, 20–250 kilohms; type C, 10–50 kilohms; and type D, 1–3 megohms. Each pair was matched within one millivolt, and drift usually did not exceed 0.25 millivolts per hour. The results obtained with each are quite comparable, and no further distinction will be made between them in presenting our data. A high impedance ($10^{11}$ ohm) differential electrometer[1], consisting of

[1] The electrometer used in these studies was designed and fabricated by Anthony Bak of our electronics laboratory.

TABLE I

Magnitude and duration of steady potential shifts produced by three catecholamines with prominent beta adrenergic effects.

| Compound | Species | Dosage | DC Response | T½ Recovery | EXP./INJ. |
|---|---|---|---|---|---|
| Epinephrine | Cat | 0.5–1.5 μg/kg | −3.6 mV ± 0.8 | 1.5 minutes | N = 20/45 |
| Isoproterenol | Cat | 0.5–1.5 μg/kg | −5.6 mV ± 1.0 | 3.0 minutes | N = 24/50 |
| Nylidrin | Cat | 30–60 μg/kg | −5.7 mV ± 1.0 | 20–120 mins. | N = 18/22 |
| Epinephrine | Monkey | 1.0–3.0 μg/kg | −3.6 mV ± 1.0 | 3.0 minutes | N = 10/30 |
| Isoproterenol | Monkey | 1.0–3.0 μg/kg | −4.5 mV ± 0.7 | 4.0 minutes | N = 14/30 |
| Nylidrin | Monkey | 30–500 μg/kg | No effect | —— | N = 10/36 |

The entire dose of epinephrine and isoproterenol in cats was injected into the brachiocephalic trunk; half of that in monkeys was delivered through each internal carotid artery. Nylidrin was given IV. Values for DC response include ± 2 S.D. T½ indicates time required for return of potential shift half way to baseline. 'N' in the last column denotes total experiments (EXP.) with each drug and species, and cumulative injections (INJ.) of drug in each species.

dual field effect transistors, and an isolated-battery powered solid state operational amplifier served as the input to the oscillograph preamplifiers. R–C coupling of one penwriter channel provided an electrocorticogram (ECoG) from the DC recording system.

Placement of a nonpolarizable ground electrode on the neck musculature had virtually no effect on the results. It did, however, abolish occasional electrostatic artifacts which were especially troublesome at low humidity.

4. Brain pH was recorded using previously calibrated $10^9$ ohm Nims-Dole type electrodes with greater than 50 millivolt/pH unit sensitivity. Though absolute cortical pH measurement was not possible, changes were detected within .01 pH unit. Arterial pH, $pCO_2$, and $pO_2$ were determined intermittently on Instrumentation Laboratories apparatus, or by continuous monitoring with a Radiometer DS66014 flow-through assembly (1.5 mm I.D.). Impedance measurements were made in both cortex and internal capsule and compared with those during catecholamine related DC shifts; the theory and techniques of this method have been previously described (Li et al., 1968). All information was recorded by means of an 8-channel Beckman Type R oscillograph.

RESULTS

Our results are summarized in Table I. Epinephrine produced significantly smaller DC responses compared to those following an equivalent (wt/kg) dose of isoproterenol. 'T½ recovery,' the time elapsed when the potential returns half way to control level, is relatively brief with epinephrine. While the average DC reaction to isoproterenol is −5.6 mv, we have occasionally seen negative DC shifts exceeding −10 mv. Column 3 (Dose/Route) indicates the range of dose yielding maximal shifts in steady potential. Responses of fifty to ninety percent peak value can be elicited by as little as .10–.20 μg/kg of either epinephrine or isoproterenol. Since the brachiocephalic trunk in cats carries about one-third of the cardiac output, more than seventy cc dilute the drug bolus by the time peak DC response is attained. In the case of epinephrine, the resultant concentration may be similar to that associated with physiologic stress.

Note the prolonged 'T½ recovery' value for nylidrin (Table I). We have found, that for each drug, this value is linearly dose related. The higher doses of nylidrin and isoxsuprine required for shifts equivalent to those with isoproterenol is striking. However, the effects of the former compounds endure 10–40 times as long. These drugs are as effective at a given dose by either the intravenous or brachiocephalic route, in sharp contrast to epinephrine and isoproterenol.

*CBF and Steady Potential Transients*

In experiments conducted using Rhesus monkeys, CBF and cerebrovascular resistance (CVR) were unrelated to the large negative DC shifts seen with injection of epinephrine or isoproterenol. Figure 1 portrays the effects of pressor doses of epinephrine on CBF and DC response. In animals with an intact autoregulation, epinephrine hypertension should trigger the myogenic response (Bayliss effect) after a passive flow over-shoot which parallels

MONKEY #114 Pentobarb.
3-18-69

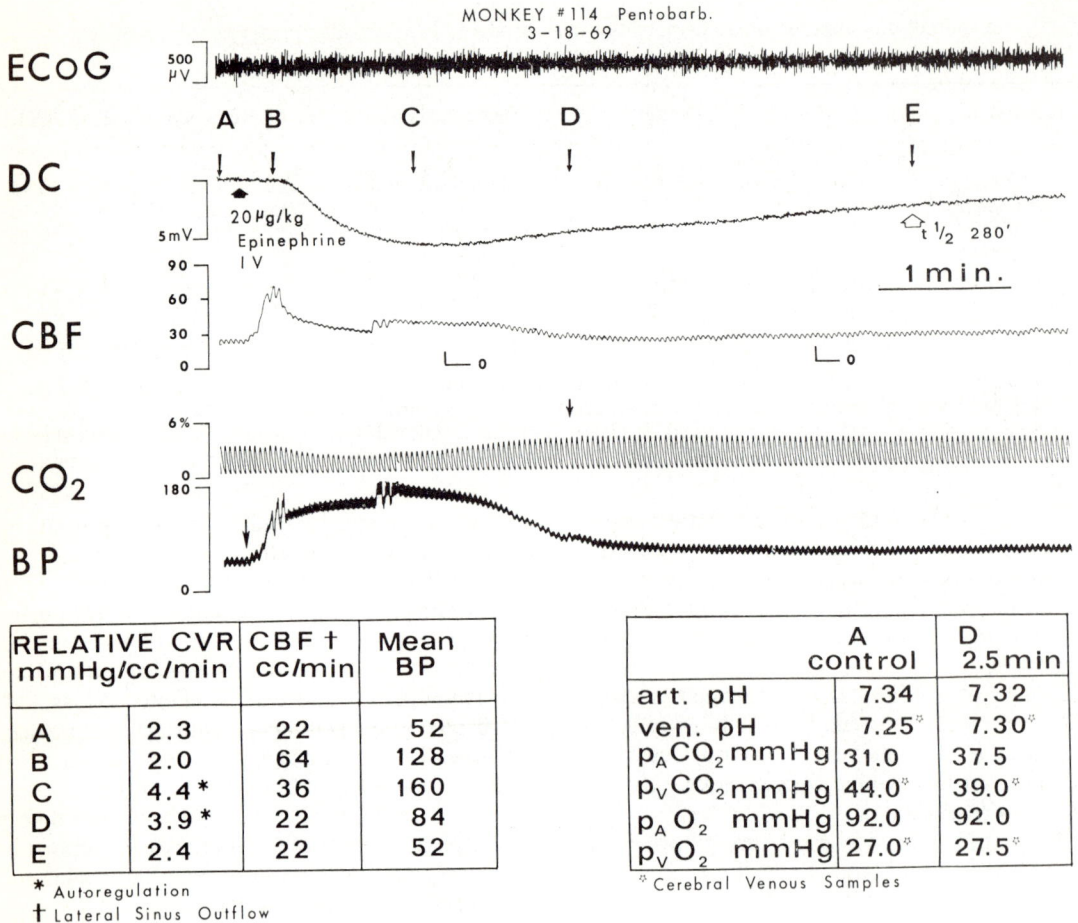

| RELATIVE CVR<br>mmHg/cc/min | CBF †<br>cc/min | Mean<br>BP |
|---|---|---|
| A    2.3 | 22 | 52 |
| B    2.0 | 64 | 128 |
| C    4.4* | 36 | 160 |
| D    3.9* | 22 | 84 |
| E    2.4 | 22 | 52 |

\* Autoregulation
† Lateral Sinus Outflow

| | A<br>control | D<br>2.5 min |
|---|---|---|
| art. pH | 7.34 | 7.32 |
| ven. pH | 7.25° | 7.30° |
| $p_A CO_2$ mmHg | 31.0 | 37.5 |
| $p_V CO_2$ mmHg | 44.0° | 39.0° |
| $p_A O_2$  mmHg | 92.0 | 92.0 |
| $p_V O_2$  mmHg | 27.0° | 27.5° |

° Cerebral Venous Samples

FIGURE 1   The independence of SP shift and CBF. The DC response exceeds $-5$ mV following a large dose of epinephrine. Prolonged hypertension triggers autoregulation of CBF (myogenic response). Two and one half minutes later (D) when CBF has returned to control level, DC response persist at $-4.5$ mV. Comparison at points A and D reveals a decrease in $p_{V-A} CO_2$ difference and drop in $p_V CO_2$ even though $p_A CO_2$ has risen slightly.

the jump in arterial pressure. The control of CBF by an increase in CVR (Figure 1) is as one might expect. Nevertheless, a large DC shift develops. Peak CVR and negative potential coincide (Figure 1–C). However, in other experiments when epinephrine was injected after alpha blockade with phenoxybenzamine (2–5 mg/kg I.V.), mild hypotension occurred without significant change in CVR; DC shifts ($-3.6$ mv $\pm 1$) were unaffected. We have noted that intracarotid injection of minute quantities of epinephrine (0.1–0.2 $\mu$g/kg), doses insufficient to cause hypertension, result in very transient CBF reduction (10–30%). Negative DC responses of 2–3 millivolts were not uncommon, even with such small doses.

Comparison of points A and D at which CBF is identical (Figure 1) reveals that $p_V CO_2$ has

actually decreased and that the A–V $pCO_2$ difference has narrowed. These results in several animals make unlikely the possibility that SP shifts with beta adrenergic amines are caused by metabolic $CO_2$ build-up in the brain.

Isoxsuprine has no effect on SP in monkeys, but produces a large transient increase in CBF (Figure 2C). The opium alkaloid, papaverine (0.5–1.0 mg/kg), is also a powerful dilator of cerebral vessels and produces a flow pattern indistinguishable from that illustrated for isoxsuprine. This compound, similarly, has no effect on steady potential.

Isoproterenol, which transiently dilates cerebral vessels, had little residual relaxant effect (in low doses) when DC shifts are maximal (Figure 2A and B). Figure 2D represents a series of bilateral

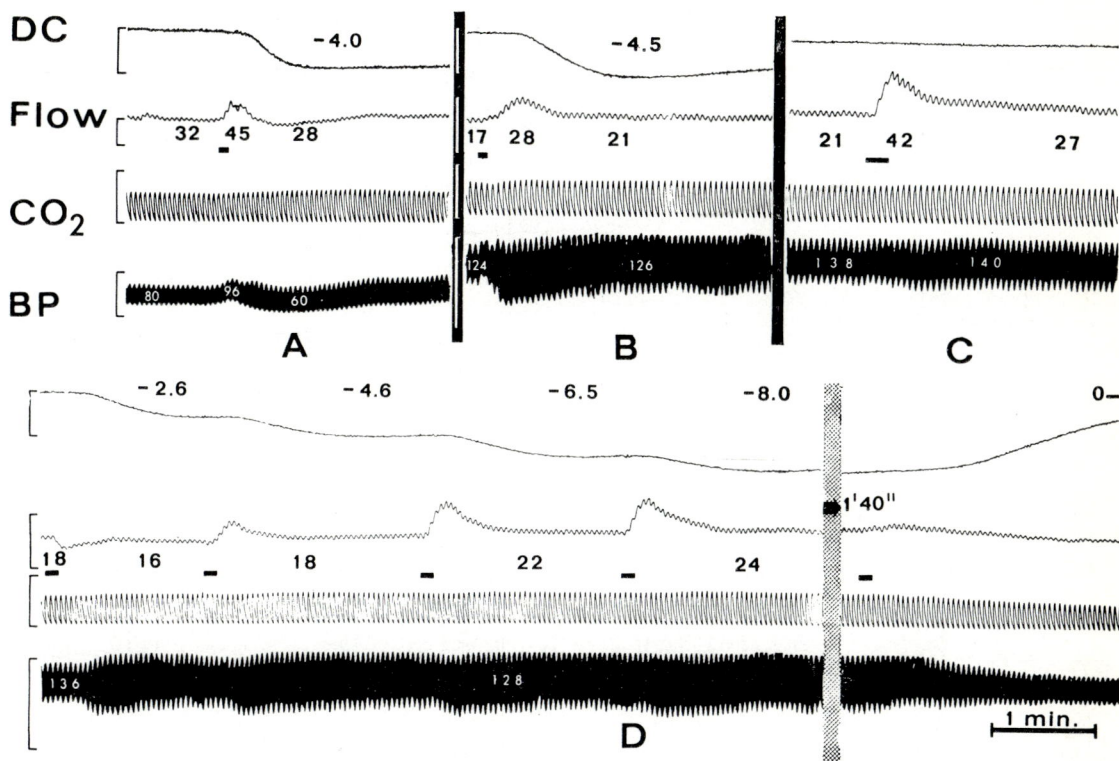

FIGURE 2   A: Isoproterenol injection (1.5 μg/kg); BP has dropped below limits of autoregulation; a sizeable SP shift develops unrelated to flow. B: In another monkey, isoproterenol (1.0 μg/kg) induces a DC shift with minimal change in mean pressure. C: Isoxsuprine (60 μg/kg) results in large CBF increase without DC shift. D: With sequential doses of isoproterenol (2.0 μg/kg), the SP shift grows to −4.6 mV without sustained flow increase. As blood concentration of this drug rises (16 μg delivered in 5′ 40″), the DC response levels off at 8 mV, while CBF has increased 30%. The decrease in flow with the initial injection cannot be explained. Propranolol reverses the DC shift (final injection–D) to control level within 3′ 5″ (not shown). A: Lateral sinus outflow (Monkey No. 121–4 kg). B. C. D: Rt. common carotid flow with simultaneous injection into both internal carotids (Monkey No. 311–6 kg). Calibrations: DC = 0–5 mV, Flow = 0–30 cc, $CO_2$ = 0–6%, and BP = 0–200 mmHg. Blocks superimposed on the flow baseline indicate injections.

internal carotid injections of isoproterenol. After each 4 μg of drug (< 1 μg/kg) there is growth of the DC shift. Even when the SP response has reached − 4.6 mV, there is no sustained increase in blood flow. Additional injections cause a persistent increase in flow, probably associated with accumulation of circulating catecholamine. When the total dose was delivered at the outset (not shown), the independence of CBF and SP shifts was obscured, with flow tracing appearing as the mirror image of the SP shift. The transient increases in flow associated with injection of the drug, must represent effects of momentary high concentration in cerebral arterioles. In monkeys, as in cats, propranolol reverses the catecholamine induced DC response (Figure 2D).

*Relationship Between Drug Structure and Bioelectrical Activity*

The beta-hydroxy phenylethylamine nucleus is depicted in Figure 3. Alongside the molecular skeleton are listed the compounds causing negative SP shifts, and the substituents which appear to confer this activity upon them. Nylidrin, the most potent of these substances, contains a 4–OH group and a large N–alkyl sidechain. The 3–OH catechol linkage alone is weak (e.g., phenylephrine). Sympathomimetic amines and precursors, which differ chemically from those illustrated (Figure 3), have no effect on SP. Such compounds include norepinephrine (0.5–2 μg/kg), metaraminol, mephentermine, ephedrine, L–dopa, dopamine, and tyramine.

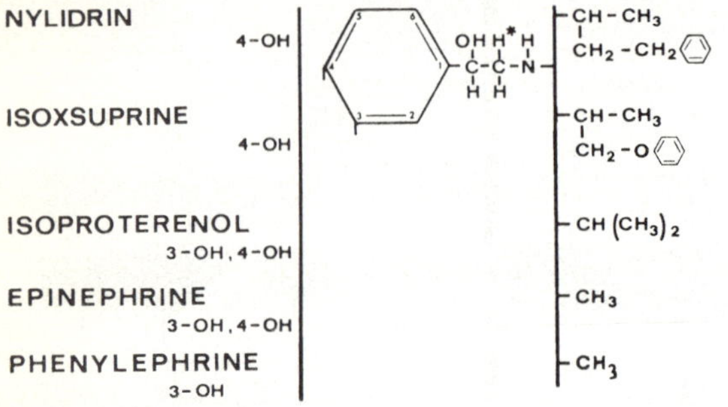

FIGURE 3   Compounds producing large steady potential shifts in order of decreasing potency. Both nylidrin and isoxsuprine have a methyl ($-CH_3$) substituent on the $\alpha$ carbon atom in place of $-H$ (*). Potency depends upon size of the N-alkyl sidechain, and on presence of a 3-OH or 4-OH group on the benzene ring.

Larger doses of norepinephrine ($>2$ $\mu g/kg$) frequently cause transient 1–2 mv positive shifts.

Nylidrin and isoxsuprine, curiously, have no effect on steady potential in monkeys. By contrast, their prolonged action in cats makes feasible the study of many factors upon which aminergic DC shifts depend. The large negative response following nylidrin injection can be quickly reversed by propranolol, a beta adenergic blocking agent (Figure 4). Propranolol administered alone has no effect on SP of the normal brain. Because this drug exhibits anesthetic properties in peripheral nerve (Vaughan Williams, 1967), we decided to compare the ability of procaine and lidocaine with that of propranolol to reverse the negative response caused by nylidrin. Neither compound was effective.

*Acid-base Balance and the Aminergic SP Shift*

A characteristic response to administration of 10%

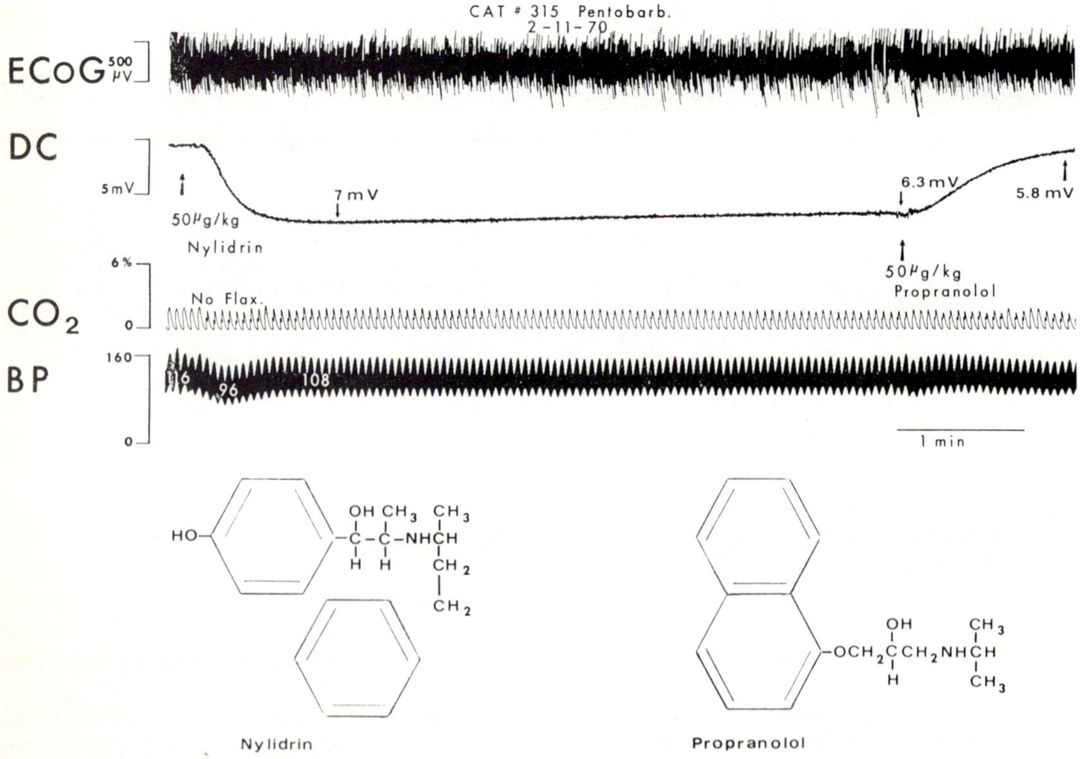

FIGURE 4   The DC shift produced by IV nylidrin in an unparalyzed cat, with reversal by propranolol. Its reappearance is prevented for over 2 hours, even with additional nylidrin injections. Gallamine triethiodide (Flexedil) has no effect on results. Abbreviations: ECoG = electrocorticogram, DC = brain steady potential, $CO_2$ = expiratory $CO_2$ level, BP = femoral arterial pressure. Figures within BP record denote mean arterial pressure. ECoG is condensed to better illustrate events with prolonged time course. Traces were obtained at 2.5 cm/sec in most experiments, confirming waveform and amplitude of ECoG seldom change with beta agents or with inspired $CO_2$, not exceeding ten percent.

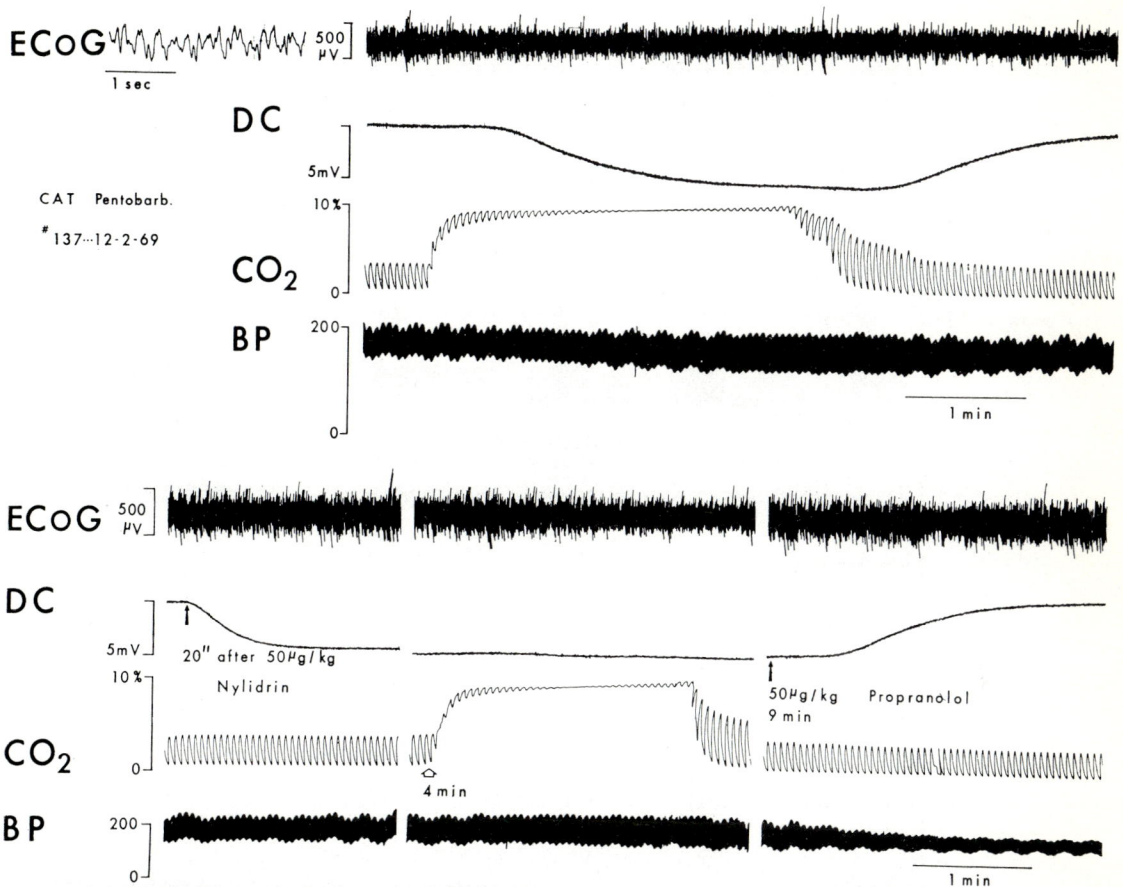

FIGURE 5    Occlusion of the carbon dioxide induced shift by a beta catecholamine. A typical SP shift is produced by administration of 10% $CO_2$ (above). Following recovery to baseline, nylidrin causes a similar shift; 4 minutes later $CO_2$ produces no further deflection (below). After propranolol has reversed beta adrenergic SP effects, the response to breathing $CO_2$ returns (not shown).

carbon dioxide is depicted in Figure 5. Soon after return to baseline, on cessation of $CO_2$ breathing, nylidrin was injected. Four minutes later, $CO_2$ inhalation caused no further shift. Reversal on the nylidrin negative potential (Figure 5, last panel) by propranolol permitted reappearance of a negative shift with carbon dioxide (not shown) identical to the control response. Conversely, at the nadir of a $CO_2$ curve there is no additional response to catecholamine injection. With depression of pH to the same extent ($7.05 \pm .05$) by .1N HCl or lactic acid infusion, the isoproterenol and nylidrin responses are but slightly diminished. The shift caused by epinephrine, on the other hand, is abolished along with most metabolic and cardiovascular effects of this substance in low doses.

Isoproterenol and other beta adrenergic amines usually cause slight cortical acidification (Figure 6, cat No. 311). The dip in diastolic pressure is characteristic of the potent vasodilator capacity of this drug. To dissociate hemodynamic changes from DC phenomena, we have employed brachiocephalic trunk injections of less than 600 nanograms which cause appreciable DC shifts while not widening pulse pressure. The administration of pilocarpine (0.2 mg/kg) about 15 minutes before giving isoproterenol was even more effective. It eliminates and often reverses the hypotensive effect of isoproterenol by stimulating sympathetic ganglionic transmission (Daniell and Bagwall, 1969). Beta amine injections, afterward, produced identical DC responses.

Paradoxical acidification of the cortex occurs on injection $NaHCO_3$ (1 cc, 0.9N). The barrier system provides an obstacle to bicarbonate influx, and unbalanced passage of the reaction product, $CO_2$, into the brain causes the pH drop. The phenomenon seems to be a reliable index of structural blood-drain

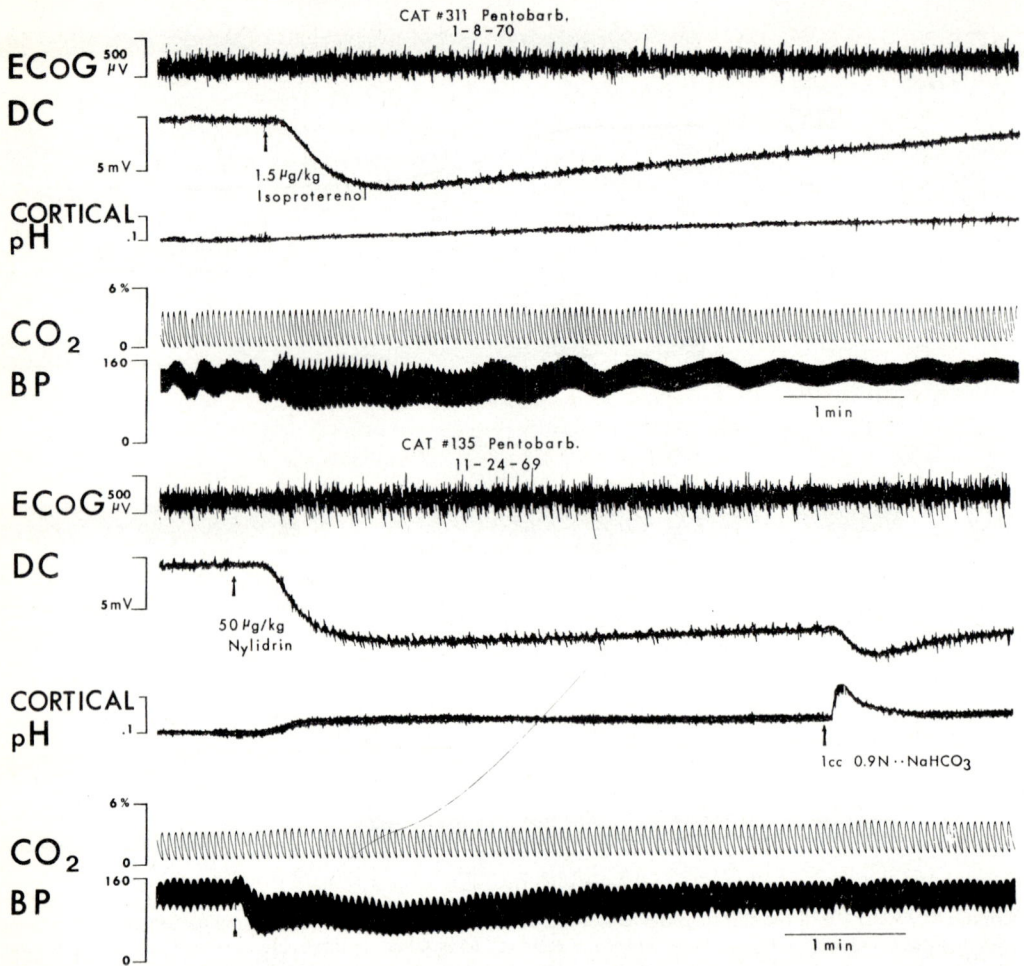

FIGURE 6    Compared with nylidrin the effect of isoproterenol is evanescent. At the peak of nylidrin shift (below), injection of NaHCO₃ into the brachiocephalic trunk produces cortical acidification identical to that in an earlier control record. [Decrease in cortical pH is denoted by upward deflection in the record.] The negative shift in DC level coincident with bicarbonate injection reflects an independent barrier electropotential varying with blood acid-base balance (Corrie *et al.*, 1969).

integrity (Rapoport, 1964). Figure 6 (cat No. 135) is an example of this phenomenon produced at the peak of a negative shift with nylidrin. It is identical to a control response thirty minutes earlier, and suggests that altered blood-brain barrier to bicarbonate is not an important factor in production of nylidrin SP shifts. Note the small additional DC deflection presumably related to blood alkalinization. Shifts linked with blood acid-base changes have been explored in other species (Tschirgi and Taylor, 1958; Held *et al.*, 1964). In cats by inducing *metabolic* acidosis, we have recently observed that SP moves positive by 10–25 millivolts for each 10 fold rise in [H$^+$] concentration (Corrie *et al.*, 1969). Finn and his coworkers (1968) made similar

observations with sodium gammahydroxybutyrate infusion. The *negative* potential slope following blood alkalinization is of similar magnitude and is independent of catecholamine related shifts. Mild alkalosis (pH 7.45–7.60) actually enhanced DC responses in the present experiments just as it potentiates activity at most beta adrenergic receptor sites (Baisset and Montastruc, 1967).

There were no brain impedance changes during the SP shifts with these amines. Depth of anesthesia had little bearing on the results. While animals breathed 100% oxygen, we delivered 4.4% halothane or gave supplemental pentobarbital (30–50 mg/kg) without effect on the aminergic SP response, even with profound electrocorticographic (ECoG) suppression. (BP < 60 mmHg). In awake cats, records identical to the tracing in Figure 4 have been obtained by telemetry, using an FM transmitter and Beckman biopotential electrodes (Methods-C3). In these preparations drugs were injected percutaneously into the saphenous vein.

## DISCUSSION

In our earlier experiments designed to control CBF in cats, DC responses (4–8 mv) developed when hypercarbic blood ($P_ACO_2$ 60–90 mmHg) was perfused at a constant rate through both carotid arteries. This evidence seemed to weaken the hypothesis causally relating SP shifts and CBF changes. However, we desired to make observations in a more nearly intact preparation. The lateral sinus outflow technique in monkeys (Methods–2) allows quantitation of CBF and instantaneous registration of blood flow change. This method measures only 50–70% of total CBF; therefore, validity of the results depends on the assumption that the region of the brain monitored remains constant (Rapela and Green, 1964).

The data in Figure 1, obtained using this technique, is in accord with that of other studies in which hypertension is shown to constrict medium sized, intracranial arteries, thus limiting small vessel pressure and CBF (James et al., 1969; Yoshida et al., 1966). Yet, DC shifts developed despite regulation of blood flow. In other preparations, intracarotid injections of epinephrine, at concentrations too meager to elevate blood pressure, transiently reduced CBF although the associated DC shifts persisted. This constrictor effect agrees with the results of several studies reviewed by Sokoloff (1959).

Besson et al. (1970) published certain illustrations in support of a blood flow—SP hypothesis, but without corroborating CBF records (Figures 8A, 9, 10A). We doubt that the DC shifts they depict have either a quantitative or temporal correlation with CBF change. An example is given of the large SP shift caused by $CO_2$, while blood pressure in a spinal cat hovers at 60 mmHg (Figure 8A). Harper and Glass (1965), however, demonstrated that $CO_2$ produces no increase in CBF at equivalent pressures. Besson and his coworkers (1970, Figure 9) have also shown a threefold increase in SP shift induced by epinephrine, as expiratory $CO_2$ drops from 4.8 to 1.0%. Based on our findings and the cardiovascular data of others (Baisset and Montastruc, 1967), the above observation could be attributed

solely to enhancement of beta adrenergic activity with hypocapneic alkalosis.

A fifty percent rise in blood pressure, evoked by epinephrine at the peak of a $CO_2$ run, (Besson et al., 1970, Figure 10A), should cause a proportionate CBF increase; for when CVR is sustained at a low level by 20% $CO_2$, CBF passively follows changes in BP. Since there was no DC shift with epinephrine, it was erroneously concluded that the drug did not further increase CBF. Moreover, we have often noted inactivation of the DC response to epinephrine when arterial pH falls below 6.9 with either $CO_2$ administration or acid infusion.

Anatomical factors in cats preclude direct CBF measurement. An arterial meshwork at the base of the skull, the rete mirabilis, supplying both intracranial and cervicopharyngeal structures (Davis and Story, 1943), could lead one to make injudicious assumptions about CBF based on carotid flow. Furthermore, the internal carotid artery is rudimentary. Cats generally have multi-channeled lateral venous sinuses not suitable for outflow measurements.

We know of just one other study which describes steady potential dynamics with direct recording across BBB using paired intracranial electrodes (Welch and Sadler, 1965). These authors found the ventricular surface of choroid plexus to be 14 mv positive to the lumen of an adjacent venule, and that both ouabain and potassium reduce this potential while low ventricular sodium increases it. Although we inserted micropipettes remote from identified active transport sites (Methods—3D), each provided a lead into volume conductor pathways on opposite sides of the vast blood-brain barrier surface. Placement of electrodes across the sagittal sinus in our experiments regularly permits recording for 1–6 hours, in contrast with the limited stability of choroid plexus recordings, typically 5–60 seconds (Welch and Sadler, 1965).

Circulating catecholamines probably act on either endothelium or perivascular glia to produce the effects observed in these experiments since their entry beyond this point is greatly restricted (Bertler et al., 1966).

Increased osmotic or hydrostatic pressure on one side of a membrane with fixed charge sets up an electrical potential proportional to rate of water flow (Wright and Diamond, 1969). These streaming potentials could be implicated in production of DC shifts with $CO_2$ and catecholamines, but direct proof is lacking. Others have implied a relationship between SP shifts and changes in tissue osmotic pressure (Kawamura et al., 1967). Sellers (1969) reported that epinephrine facilitates penetration of blood-borne viruses into the brain. Alpha blocking agents reversed this effect, while beta blockade did not. It was decided that hypertension alone promotes entry of these particles into brain by physically separating endothelial tight junctions.

There are several characteristics of the intra-cranial transvascular potential (ITVP) shift which distinguish it from local SP phenomena (e.g., spreading depression), and from the low amplitude transcortical shifts associated with dendritic de-polarization, such as those produced by epinephrine (Vanasupa et al., 1959). We suggest the ITVP shifts produced by $CO_2$ and beta amines may, indeed, reflect the varying metabolic control of active transport within the barrier.

Beta adrenergic drugs decrease membrane resist-ance, enhance bidirectional sodium flux, and stimulate active chloride transport in frog skin (Watlington, 1968). Presumably adenyl cyclase activity is increased by these compounds, accelerat-ing conversion of ATP to $3'-5'$ cyclic AMP. However, the effects of beta amines on membrane potential of smooth muscle vary with extracellular potassium concentration, hyperpolarizing by 12 mv at 1 mM $K^+$ and depolarizing by up to 5 mv at 10 mM $K^+$ (Somlyo and Somlyo, 1969). Provided similar receptor mechanisms are present in glia, and since interstitial potassium concentration is low in brain compared to that of plasma (Cserr, 1965), circulating beta amines could hyperpolarize glial membranes adjacent to the endothelium.

The negative shifts in ITVP produced by ad-ministering 10% $CO_2$ were not seen at the height of betaminergic responses (Figure 5). This finding suggests competitive interaction of these substances in closely related metabolic pathways. Although propranolol blocks the superficial receptor sites which bind beta adrenergic catecholamines, this drug should have no effect on the direct action of $CO_2$ within the cell.

The diversity of beta effects among species, both in direction and magnitude of action, is well known (Kaiser et al., 1964; Fleisch et al., 1970). In monkeys, some peculiarity of receptor template configuration may prevent the binding of both nylidrin and isoxsuprine. It is puzzling that negative SP shifts do not occur in dogs with either $CO_2$ or epinephrine, but potent beta agents have not yet been studied. Ouabain diminishes CSF production in cats but not in dogs (Oppelt et al., 1963), a discrepancy that may be symptomatic of more general differences between species regarding receptor inputs to high energy metabolic systems.

During the large negative shifts produced in monkeys by concussion (Meyer and Denny-Brown, 1955) the DC response to $CO_2$ is impaired (Marshall et al., 1968). These gross shifts provoked by injury have been equated with demarcation potentials between normal and killed tissue (O'Leary and Goldring, 1964). However, with restitution of function and recovery of steady potential, the $CO_2$ response returns. We have recently observed loss of betaminergic shifts in cats after brain trauma (unpublished observations). Could the transient increase in radioactive phosphate entry into brain observed following blow to the head (Cassen and Neff, 1960) be one of the reflections of derangement within a catecholamine dependent pathway?

Really compelling evidence for our contention that beta adrenergic amines modify transport across the blood-brain barrier can be secured only by radioisotope flux measurements and studies of CSF production during the electropotential shifts caused by these amines.

## SUMMARY

Beta adrenergic catecholamines and carbon dioxide produce large steady potential shifts in experiments conducted with cats and monkeys. The effects of these substances on brain SP are not simply additive; in fact, $CO_2$ induced shifts are blocked at the peak of an adrenergic response. Only those compounds having substituents which confer beta activity cause the negative SP shifts observed in this study. The beta-aminergic shift can be rapidly reversed by propranolol.

Such SP shifts are unrelated to changes in cerebral blood flow, brain impedance, or level of neuronal activity. Considerable evidence points to a direct effect of these compounds on active transport either across capillary endothelium or on astrocytic glial processes adjacent to the vascular lumen.

Paired microelectrodes which straddle the blood-brain barrier, even when placed within 1 mm of each other to record between sagittal sinus and subarachnoid space, yield results identical with those obtained by less direct methods. We suggest that these potential shifts are of BBB origin, and that they are sufficiently different from most other SP phenomena to be distinguished from them; the term 'intracranial transvascular potential (ITVP)' shift is proposed to describe them.

### ACKNOWLEDGMENT

We especially appreciate the indispensable technical con-tributions of Mr. Alvin Ziminsky to the conduct of these experiments.

## REFERENCES

Baisset, A., and Montastruc, P., 1967, [Hypocapnic gaseous alkalosis reinforces the beta circulatory effects of sympathomimetics], (in French), *C. R. Acad. Sci. (Paris)* **264**: 2529–2532.

Besson, J. M., Woody, C. D., Aleonard, P., Thompson, H. K., Albe-Fessard, D., and Marshall, W. H., 1970, Correlations of brain d–c shifts with changes in cerebral blood flow, *Amer. J. Physiol.* **218**: 284–291.

Bertler, A., Falck, B., Owman, C. H., and Rosengrenn, N. E., 1966, Localization of monoaminergic blood-brain barrier mechanisms, *Pharmacol. Rev.* **18**: 369–385.

Cassen, B., and Neff, R., 1960, Blood-brain behavior during temporary concussion, *Amer. J. Physiol.* **198**: 1296–1298.

Corrie, W. S., Hubbard, J. H., Thompson, H. K., and Marshall, W. H., 1969, Effects of metabolic acidosis on brain steady potential in cats, *Fed. Proc.* **28(2)**: 297.

Cserr, H., 1965, Potassium exchange between cerebrospinal fluid, plasma and brain, *Amer. J. Physiol.* **209**: 1219–1226.

Daniell, H. B., and Bagwell, E. E., 1969, Modification of the cardiovascular response to isoproterenol by pilocarpine, *Arch. int. Pharmacodyn.* **181**: 141–152.

Davis, D. D., and Story, H. W., 1943, The carotid circulation in the domestic cat, *Field Mus. Nat. Hist.* (Zool. ser.) **28**: 1–47.

Finn, H., Kao, F. F., Mei, S. S., and Harmel, M. H., 1968, CSF-blood potential in cats and its modification by sodium gamma-hydroxybutyrate, *Arch. int. Pharmacodyn.* **176**: 319–325.

Fleisch, J. H., Maling, H. M., and Brodie, B. B., 1970, Beta receptor activity in aorta. Variations with age and species, *Circulat. Res.* **26**: 151–162.

Harper, A. M., and Glass, H. I., 1965, Effect of alternations in the arterial carbon dioxide tension on the blood flow through the cerebral cortex at normal and low arterial blood pressures, *J. Neurol. Neurosurg. Psychiat.* **28**: 449–452.

Held, D., Fencl, V., and Pappenheimer, J. R., 1964, Electrical potential of the cerebrospinal fluid, *J. Neurophysiol.* **27**: 942–959.

James, I. M., Millar, R. A., and Purves, M. J., 1969, Observations on the extrinsic neural control of cerebral blood flow in the baboon, *Circulat. Res.* **25**: 77–93.

Kaiser, G. A., Ross, J., Jr., and Braunwald, E., 1964, Alpha and beta adrenergic receptor mechanisms in the systemic venous bed, *J. Pharmacol. exp. Ther.* **144**: 156–162.

Kawamura, H., Whitmoyer, D. W., and Sawyer, C. H., 1967, DC potential changes recorded between brain and skull in the rabbit after eating and drinking, *Electroenceph. clin. Neurophysiol.* **22**: 337–347.

Li, C. L., Bak, A. F., and Parker, L. O., 1968, Specific resistivity of the cerebral cortex and white matter, *Exp. Neurol.* **20**: 544–557.

Marshall, W. H., Ommaya, A. K., Richter, H., Thompson, H. K., and Woody, C. D., 1968, Relation of brain injury to slow potential changes accompanying hydrogen ion concentration changes in the blood, *Electroenceph. clin. Neurophysiol.* **24**: 190P.

Meyer, J. S., and Deny-Brown, D., 1955, Studies of cerebral circulation in brain injury. II. Cerebral concussion, *Electroenceph. clin. Neurophysiol.* **7**: 529–544.

Meyer, J. S., Ishikawa, S., and Lee, T. K., 1964, Electromagnetic measurement of internal jugular venous flow in monkeys. Effect of epilepsy and other procedures, *J. Neurosurg.* **21**: 524–539.

Mottschall, H. J., and Loeschske, H. H., 1963, [Measurement of transmeningeal potential of the cat with alterations of $p$ $CO_2$ and $H^+$ concentration in blood], (in German), *Pflugers Arch. ges. Physiol.* **277**: 662–670.

O'Leary, J. L., and Goldring, S., 1964, DC potentials of the brain, *Physiol. Rev.* **44**: 91–125.

Oppelt, W. W., Maren, T. H., Owens, E. S., and Rall, D. P., 1963, Effect of acid-base alterations on cerebrospinal fluid production, *Proc. Soc. exp. Biol.* (N.Y.) **114**: 86–89.

Rapela, C. E., and Green, H. D., 1964, Autoregulation of canine cerebral blood flow, *Circulat. Res.* **14-15**: Suppl. 1, 205–211.

Rapoport, S. I., 1964, Cortical pH and the blood-brain barrier, *J. Physiol.* (Lond.) **170**: 238–249.

Sellers, M. I., 1969, Studies on the entry of viruses into CNS of mice via circulation. Differential effects of vaso-active amines and $CO_2$ on virus infectivity, *J. Exp. Med.* **129**: 719–746.

Sokoloff, L., 1959, Action of drugs on the cerebral circulation, *Pharmacol. Rev.* **11**: 1–85.

Somlyo, A. V., and Somlyo, A. P., 1969, Pharmacology of excitation-contraction coupling in vascular smooth muscle and in avian slow muscle, *Fed. Proc.* **214**: 1634–1642.

Tschirgi, R. D., and Taylor, J. L., 1958, Slowly changing bioelectric potentials associated with the blood-brain barrier, *Amer. J. Physiol.* **195**: 7–22.

Vanasupa, P., Goldring, S., O'Leary, J. L., and Winter, D., 1959, Steady potential changes during cortical activation, *J. Neurophysiol.* **22**: 273–284.

Vaughan Williams, E. M., 1967, Central nervous system effect of beta adrenergic blocking drugs, *Ann. N.Y. Acad. Sci.* **139**: Article 3, 808–814.

Watlington, C. O., 1968, Effect of catecholamines and adrenergic blockade on sodium transport of isolated frog skin, *Amer. J. Physiol.* **214**: 1001–1007.

Welch, K., and Sadler, K., 1965, Electrical potential of choroid plexus of the rabbit, *J. Neurosurg.* **22**: 344–351.

Woody, C. D., Marshall, W. H., Besson, J. M., Thompson, H. K., Aleonard, P., and Albe-Fessard, D., 1970, Brain potential shift with respiratory acidosis in the cat and monkey, *Amer. J. Physiol.* **218**: 275–283.

Wright, E. M., and Diamond, J. M., 1969, An electrical method for measuring nonelectrolyte permeability, *Proc. Roy. Soc. B.* **171**: 203–225.

Yoshida, K., Meyer, J. S., Sakamoto, K., and Handa, J., 1966, Autoregulation of cerebral blood-flow; electromagnetic flow measurements during acute hypertension in the monkey, *Circulat. Res.* **19**: 726–738.

# 2-AMINE 4-PENTENOIC ACID (ALLYLGLYCINE): A PROPOSED TOOL FOR THE STUDY OF GABA MEDIATED SYSTEMS

GEORGINA RODRÍGUEZ DE LORES ARNAIZ
MARTHA ALBERICI DE CANAL and EDUARDO DE ROBERTIS

*Instituto de Anatomía General y Embriología, Facultad de Medicina,*
*Universidad de Buenos Aires, Buenos Aires, Argentina*

Structural and biochemical alterations produced by the convulsant drug 2-amine-4-pentenoic acid (allylglycine) in the cerebellar cortex are described. Most Purkinje cells show considerable shrinkage, condensation and darkening of the cytoplasm and nucleus and the basket synapses appear swollen and with loss of synaptic vesicles. This contrasts with the normal aspect of granular cells and glomerular synapses. In the cerebellum of the convulsant rat there is 32–35% inhibition of glutamic acid decarboxylase. Kinetic studies suggest that the inhibition is due to a competitive inhibition for the substrate. These results suggest that allylglycine may be used as a tool for the study in the central nervous system of inhibitory systems mediated by $\gamma$-aminobutyric acid.

In the past few years our laboratory has been engaged in studying the action of certain drugs which may induce seizure activity. In the convulsions induced by methionine sulfoximine, in addition to the known inhibition of glutamine synthetase (Peters and Tower, 1959; Sellinger and Weiler, 1963; Lamar and Sellinger, 1965) a strong inhibition of alanine aminotransferase and a lesser inhibition of glutamic acid decarboxylase (GAD) were found; on the contrary the acetylcholine system was unaffected. The ultrastructural changes found were mainly related to the non-aminergic (i.e., GAD rich) population of nerve endings (De Robertis, Sellinger, Rodríguez de Lores Arnaiz, Alberici and Zieher, 1967). Another convulsant studied which also affected the non-aminergic nerve endings was the 2-amine 4-pentenoic acid (allylglycine). Biochemically, the most significant result was the finding of an inhibition of GAD, during the period of convulsion, which was even greater by addition 'in vitro' of the drug. At the same time a decrease of 40 per cent in the concentration of $\gamma$-amino butyric acid (GABA) was found. These findings pointed toward a direct effect of allylglycine on the GABA system (Alberici, Rodríguez de Lores Arnaiz and De Robertis, 1969).

The experimental evidences that GABA may function as an inhibitory transmitter in the crustacean neuromuscular junction and as a central inhibitory transmitter in the vertebrate CNS have been extensively reviewed (Potter, 1968; Krnjević and Schwartz, 1968; Curtis, 1968). Since it is known that Purkinje neurons are inhibitory by way of GABA, the study of the action of the convulsant allylglycine has now been extended to the cerebellum. The morphological and biochemical alterations produced by allylglycine suggest that this drug may be an excellent tool for the study of GABA mediated systems.

Part of this investigation has been presented at the Symposium on 'Basic Mechanisms of the Epilepsies', Colorado Springs, November 1968 (De Robertis, Rodríguez de Lores Arnaiz and Alberici, 1969).

## METHODS

### Administration of allylglycine

DL-C-allylglycine (Sigma Chemical Co., St. Louis, Missouri) in 0.155 M NaCl solution was injected intraperitoneally into Wistar adult rats of about 100 g body weight at a dose of 150 mg/Kg. After a very short period of considerable excitation with running and jumping, convulsions followed by

This work has been supported by Grants of the Consejo Nacional de Investigaciones Científicas y Técnicas, Argentina and National Institutes of Health 5 RO1 NS 06953–05 NEUA. U.S.A.

rigidity were observed in most animals between 2 and 2.5 hours. At this time, the rats were decapitated and the cerebellum was removed in the cold room.

### Morphological studies

Slices of different portions of the cerebellar cortex were fixed in a mixture of glutaraldehyde and paraformaldehyde, then postfixed in 1% (w/v) osmium tetroxide in phosphate buffer, pH 7.4 and embedded in Epon 812. Thick sections of about $1\mu$ were stained according to Lane and Europa (1965) and observed with the light microscope. Thin sections were observed with a Siemens Elmiskop 1 Electron microscope. Controls from untreated rats were processed simultaneously.

### Assay of GAD

The whole cerebellum or the cerebellar cortex of rats undergoing convulsions and those of uninjected animals were homogenized individually at 10% (w/v) in bidistilled water for the assay of GAD. GAD was determined in triplicate according to the technique of Lowe, Robins and Eyerman (1958) by incubating 20 $\mu$l of buffer substrate mixture with 20 $\mu$l of a homogenate containing 2 mg fresh tissue. The final concentration of the components during the incubation were: potassium phosphate buffer, 0.1 M pH 6.4; pyridoxal phosphate, 0.5 mM and L-glutamate, 25 mM unless otherwise stated. The incubation was for 60 min at 37°C; the reaction was finished with 20 $\mu$l of 10% (w/v) trichloroacetic acid; after centrifugation, measurement of GABA was made on 10 $\mu$l aliquots as described by Lowe et al. (1958). To study the effect of allylglycine on GAD 'in vitro' the homogenate was from rat cerebral cortex and the buffer substrate mixture contained 2.5 or 5 mM allylglycine in final concentration. The blank tissue fluorescence for each homogenate was determined by adding the trichloroacetic acid before the homogenate. The enzyme activity was expressed in units per gram of fresh tissue. One unit is the amount of tissue which produces 1 $\mu$mole GABA per hour at 37°C.

## RESULTS

### Structural changes

The examination with the light microscope of the cerebellar cortex of rats undergoing allylglycine induced convulsions showed dramatic changes in the Purkinje cells. Among Purkinje cells having a normal aspect, others showed different degrees of retraction, condensation and darkening. Such irregularly shrunked cells were surrounded by clear spaces, the main dendrites were also very much condensed and the astrocytes found in the same layer showed some degree of swelling. In contrast, a rather normal appearance of the granular layer was observed (Figure 1).

At the electron microscope level, striking alterations of most of the constituents of the Purkinje cells were observed. The nucleus and cytoplasmic matrix showed a very dense condensation. The ribosomes appeared packed together and uniformly distributed, not in polysomes arrangements as in the control (Figure 2). The cisternae of the Golgi complex and endoplasmic reticulum showed some degree of dilatation (Figure 3). The main dendrites of the Purkinje cells showed similar changes, predominantly the extreme condensation (i.e., darkening) of the matrix. The base of the Purkinje cells appeared surrounded by distorted axons and terminals which in most cases had lost the synaptic vesicles or showed vacuolization and by clear spaces due to the swollen glial processes (Figure 3). In contrast, the glomerular regions of the granular layer were practically normal with the exception of a few processes situated at the periphery of the glomeruli (Figure 4).

### Effect of allylglycine on GAD activity

The assay of GAD in homogenates of the cerebellum from rats undergoing convulsions by allylglycine showed an inhibition of 32–35% with respect to the activity found in the controls (Table 1). A similar inhibition of GAD was found in homogenates from the cerebellar cortex. The extent of the inhibition was somewhat higher than that previously reported for the cerebral cortex; in all cases, the difference between control and treated rats was statistically significant.

Further studies were made on the inhibition of GAD 'in vitro' by using different concentrations of glutamate and 2.5 mM and 5.0 mM allylglycine. The results, plotted by the method of Lineweaver and Burk (1934) (Figure 5) supported our previous interpretation of a competitive inhibition by allylglycine, with respect to glutamate, for the active site on the enzyme. The Km for the control was $6.2 \times 10^{-3}$M while the Km values in the

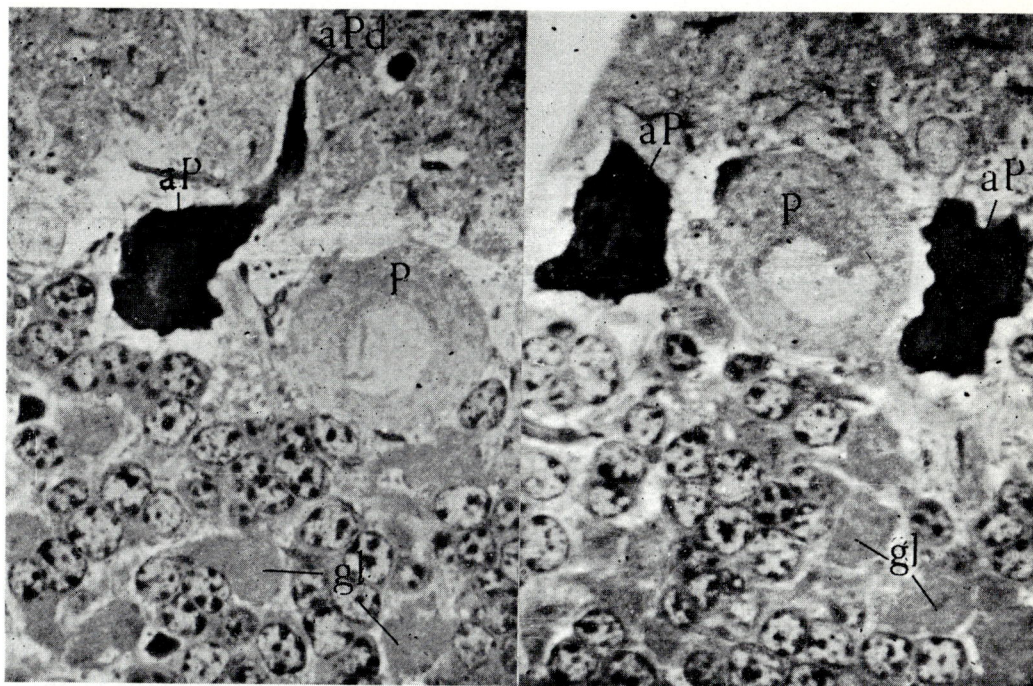

FIGURE 1   Light micrographs of the cerebellar cortex. Thick section embedded in Epon 812 (see Methods). Observe Purkinje (P) cells with a rather normal aspect among others showing great alterations (aP). In these there is a considerable shrinkage and darkening of the nucleus, cytoplasm and dendrites (aPd); the region of the basket synapses appears swollen. This contrasts with the normal aspect of the granular cells and glomerular synapses (gl) × 4,200.

TABLE 1

'In Vivo' action of Allylglycine on GAD Activity

|  | Control | Allylglycine | % Inhibition | P |
|---|---|---|---|---|
| Cerebellum | 33.7 ± 4.4 (3) | 21.9 ± 1.8 (4) | 35 | < 0.005 |
|  | 12.0 ± 1.1 (8) | 8.2 ± 1.2 (9) | 32† | < 0.001 |
| Cerebral cortex | 68.0 ± 4.2 (3) | 50.3 ± 2.3 (8) | 25‡ | < 0.001 |

GAD was determined in individual homogenates from cerebellum and cerebral cortex of untreated controls and allylglycine convulsant rats. Results are expressed as $\mu$moles GABA produced per hour per $g$ fresh tissue ± SD and as per cent inhibition with respect to the control. $P$ calculated by the student "t" for the difference between GAD activity in controls and injected is also included. In parenthesis is the number of homogenates assayed.
†Homogenates frozen overnight.
‡Data from Alberici et al. (1969).

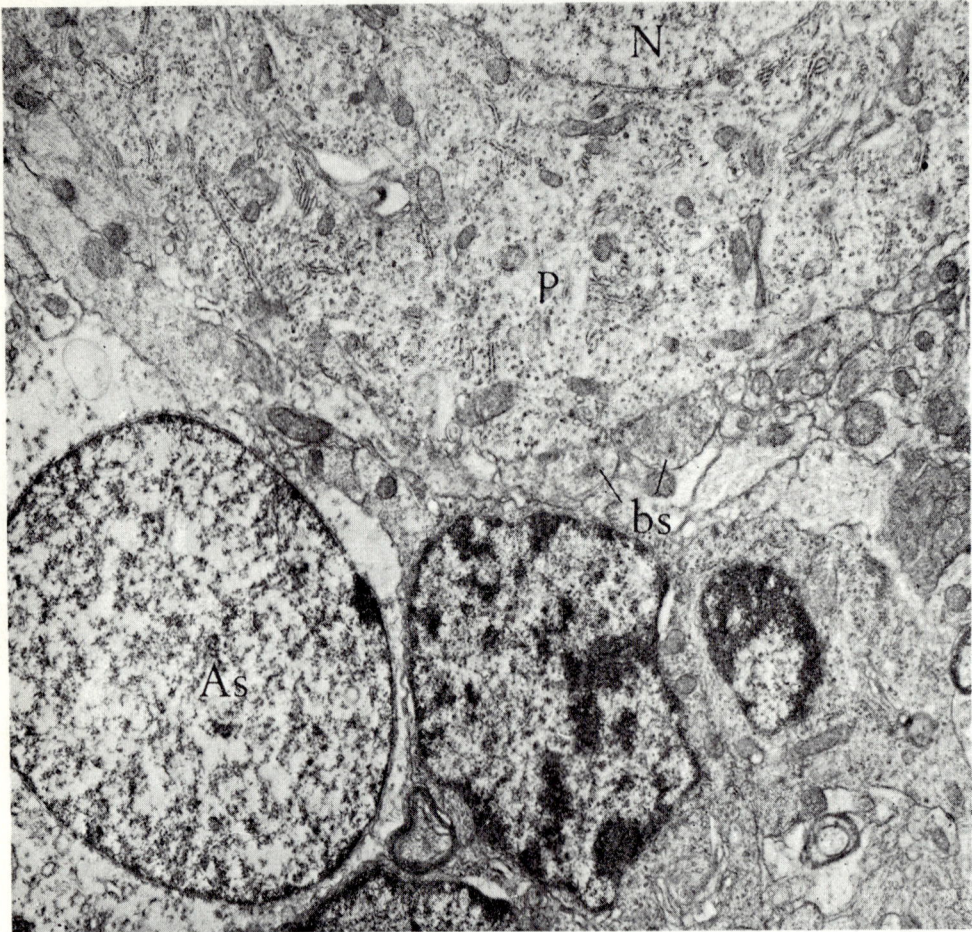

FIGURE 2    Electronmicrograph of the basal region of a Purkinje cell (*P*) in a control rat. The nucleus (*N*) and cytoplasm show the characteristic neuronal ultrastructure. The edge of the perikaryon is covered by the basket synapses (*bs*). Below there is one astrocyte (*As*) and granular cells. × 12,000.

presence of 2.5 mM and 5 mM inhibitor were respectively $1.6 \times 10^{-2}$M and $1.5 \times 10^{-2}$M.

## DISCUSSION

In a previous paper from this laboratory (Alberici et al., 1969) marked ultrastructural changes were observed in some nerve endings of the cerebral cortex (i.e. swelling, reduction in the number of synaptic vesicles, and vacuolization) of rats undergoing convulsions by allylglycine. Such alterations affected specially the GAD-rich fraction of nerve endings (Salganicoff and De Robertis, 1963) while the other fraction appeared less altered. This was

an indication of a selective action of the convulsant allylglycine on the non-aminergic synapses of the cerebral cortex (see De Robertis and Rodríguez de Lores Arnaiz, 1969).

More direct evidences of the selective action of allylglycine on certain neurons and synapses are presented. Using light and electron microscopy striking alterations of the perikaryon and large dendrites of Purkinje cells and on the axosomatic endings on the same neurons (i.e., basket synapses) were observed. In the glomerular regions of the granular layer, the endings of mossy fibers appeared completely normal and the only few altered axons were found in the periphery of the glomeruli, which may correspond to Golgi II axons. It is

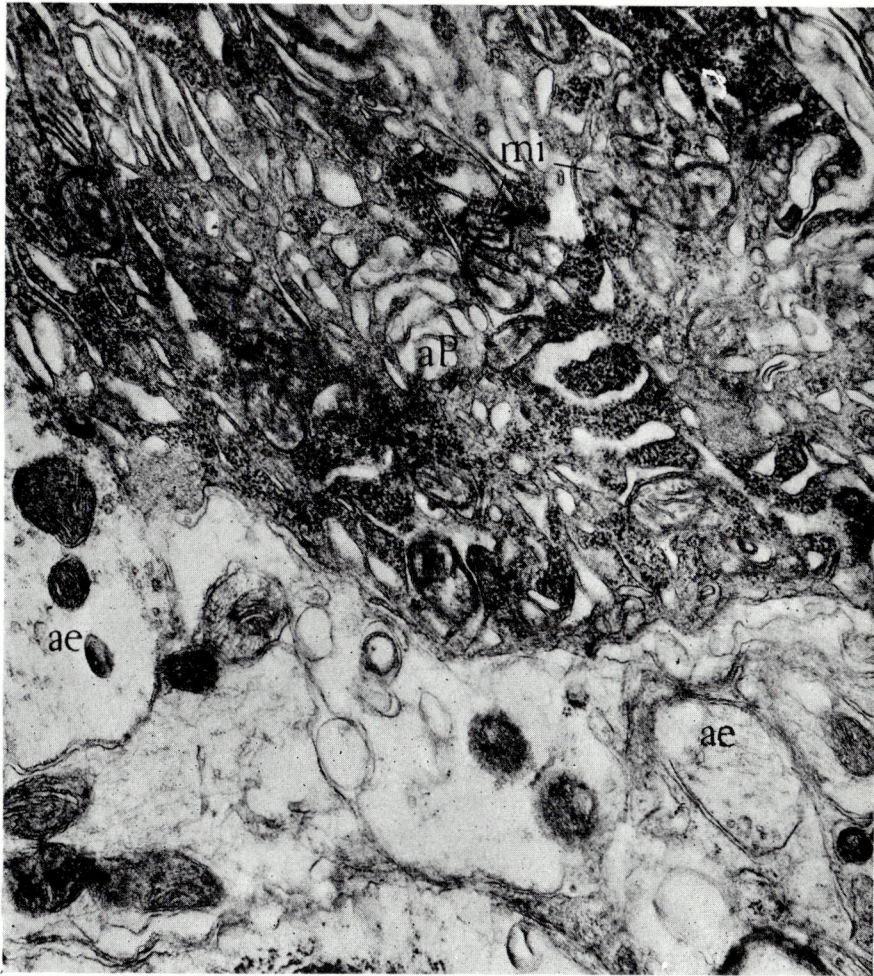

FIGURE 3    Electronmicrograph of the same region is in Figure 2 but from allylglycine convulsant rat. The altered Purkinje cell (*aP*) shows considerable condensation and darkening of the cytoplasmic matrix with dilatation of the cisternae of the Golgi complex and endoplasmic reticulum. Mitochondria (*mi*) look swollen. The basket synapses show considerable swelling and loss of synaptic vesicles; (*ae*) altered nerve ending. × 18,000.

interesting to remark that Eccles, Llinás and Sasaki, (1964) have shown that at the glomerulus the nerve terminals of mossy fibers make excitatory synapses while nerve terminals of Golgi II cells make inhibitory synapses on dendrites of granular cells.

Andersen, Eccles and Voorhoeve (1963) have given electrophysiological evidences that the dendrites of Purkinje cells receive mostly excitatory inputs from parallel and climbing fibers, while the perikaryon receives exclusively inhibitory synapses from basket cells and recurrent collaterals of Purkinje axons. Uchizono (1965) has shown that

the Purkinje perikaryon is surrounded by nerve terminals which contain elliptic (flattened) vesicles interpreted as inhibitory in nature. After the administration of allylglycine, in the basal region of Purkinje cells, the nerve terminals showed loss of synaptic vesicles and vacuolization.

We have already described an inhibition 'in vitro' of GAD by incubation with allylglycine which was prevented by the previous addition of the buffer-substrate mixture (Alberici *et al.*, 1969). It was also indicated that the inhibition of GAD by allylglycine is not by way of the cofactor, pyridoxal

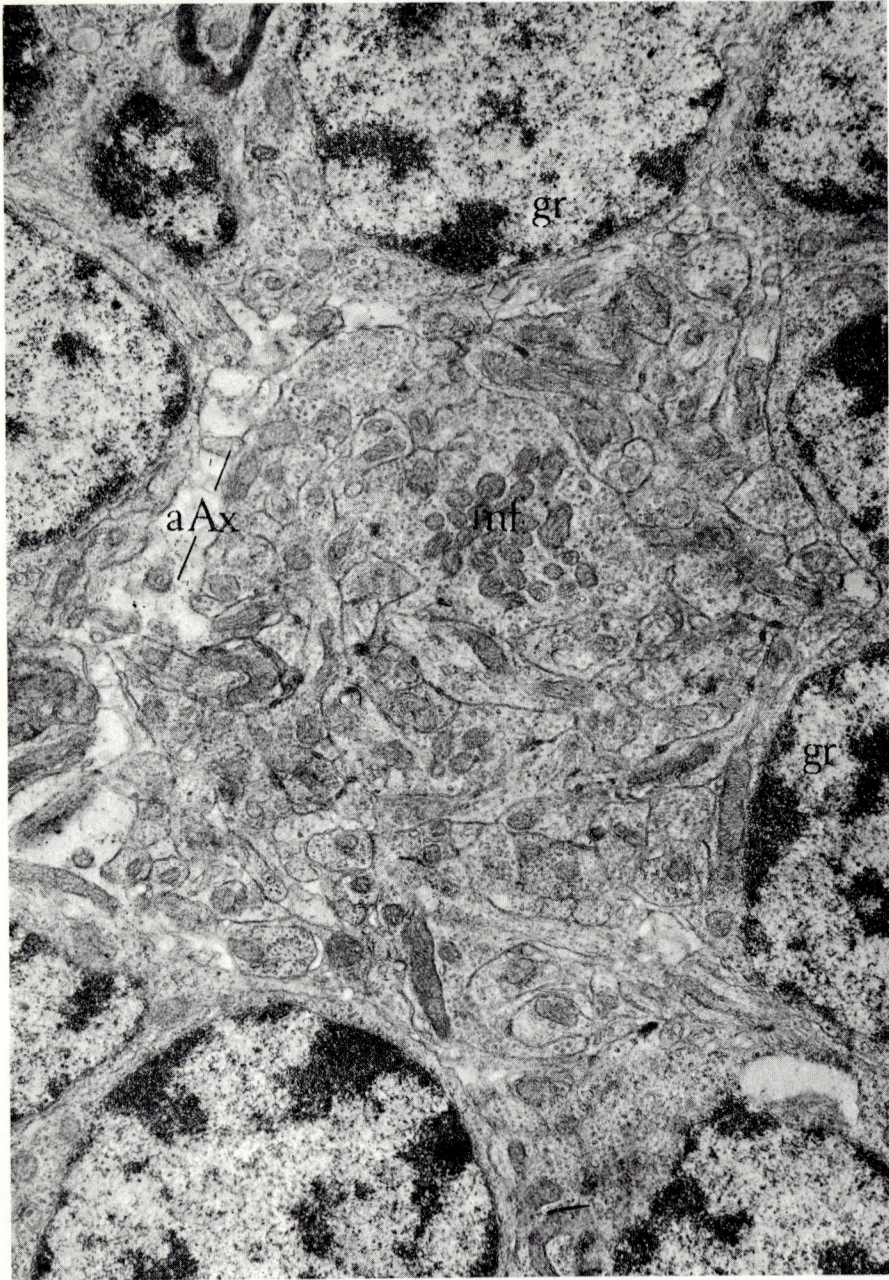

FIGURE 4  Electronmicrograph of a glomerular synapse of the cerebellum of an allylglycine convulsant rat. Observe the normal aspect of the granular cells (*gr*) and glomerulus; *mf* mossy fiber; *aAx* altered axon.

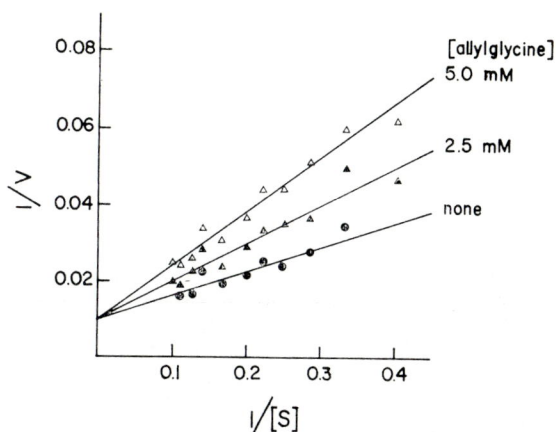

FIGURE 5 Lineweaver-Burk plot of the inhibition 'in vitro' of rat cerebral cortex GAD by allylglycine. $(S)$ is the concentration of glutamate. Velocity $(V)$ is expressed in $\mu$moles of GABA produced per hour per $g$ fresh tissue.

phosphate. This finding agrees with the fact that piridoxine given 'in vivo' before or simultaneously with allylglycine has no protective effect (McFarland and Wainer, 1965). On the other hand, the inhibition of GAD by pyridoxal phosphate antagonists is generally produced at low cencentrations (Killiam and Bain, 1957). The relatively high concentration of allylglycine needed to inhibit GAD suggested a competition for the substrate rather for pyridoxal phosphate. The results presented here on the activity of GAD as a function of the substrate concentration in the presence of the inhibitor (Figure 5) fit with the suggestion of a competitive inhibition by allylglycine with respect to glutamate for the active site of the enzyme.

In the above-mentioned paper, an inhibition of GAD 'in vivo' which was correlated with a decrease of GABA in the cerebral cortex was also described. In the present work a significant inhibition 'in vivo' of 32%–35% in GAD activity was found in the cerebellum.

With cell fractionation techniques it was already demonstrated that GAD is an enzyme contained in the presynaptic nerve endings (Salganicoff and De Robertis, 1963; Weinstein et al., 1963). Recently a high concentration of GAD within the Purkinje axon terminals in the dorsal portion of the Deiters' nucleus and in the nucleus interpositus, has been demonstrated (Fonnum, Storm-Mathisen and Walberg, 1970).

Our findings of a marked inhibition of GAD in the cerebellum of allylglycine convulsant rats together with the structural alterations observed

in the Purkinje cells support, the now widely recognized concept, that these neurons are inhibitory in nature (see Ito, 1968) and that they may function by way of GABA (Obata, Ito, Ochi and Sato, 1967; Roberts, 1968). The possibility of using allylglycine as a tool for the study of inhibitory systems mediated by GABA in the CNS seems of great interest for further physiological, biochemical and structural studies.

## ACKNOWLEDGMENTS

We are grateful to Dr. F. J. Barrantes for processing the sections for light microscopy and to Mrs. Lina Levi de Stein and Mrs. Alba Mitridate de Novara for their skilful technical assistance.

## REFERENCES

Alberici, M., Rodríguez de Lores Arnaiz, G., and De Robertis, E., 1969, Glutamic Acid decarboxylase inhibition and ultrastructural changes by the convulsant drug allylglycine, Biochem. Pharmacol. 18: 137–143.

Andersen, P., Eccles, J., and Voorhoeve, P. E., 1963, Inhibitory synapses on soma of Purkinje cells in the cerebellum. Nature (Lond) 199: 655–656.

Curtis, D. R., 1968, Pharmacology and neurochemistry of mammalian central inhibitory processes. In: Structure and Function of Inhibitory Neuronal Mechanisms, von Euler, C., Skoglund, S., and Söderberg, U., ed. pp. 429–456, Oxford: Pergamon Press.

De Robertis, E., and Rodríguez de Lores Arnaiz, G., 1969, Structural components of the synaptic region. In: Handbook of Neurochemistry, Lajtha, A., ed. pp. 365–392. New York: Plenum Press.

De Robertis, E., Rodríguez de Lores Arnaiz, G., and Alberici, M., 1969, Ultrastructural Neurochemistry. In: Basic Mechanisms of Epilepsies, Jasper, H. H., Ward, A. A., Pope, A., ed. pp. 137–158. Boston: Little, Brown & Co.

De Robertis, E., Sellinger, O. Z., Rodríguez de Lores Arnaiz, G., Alberici, M., and Zieher, L. M., 1967, Nerve endings in methionine sulphoximine convulsant rats, a neurochemical and ultrastructural study, J. Neurochem. 14: 81–89.

Eccles, J. C., Llinás, R., and Sasaki, K., 1964, Golgi cell inhibition in the cerebellar cortex, Nature (Lond.) 204: 1265–1266.

Fonnum, F., Storm-Mathisen, J., and Walberg, F., 1970 Glutamate decarboxylase in inhibitory neurons. A study of the enzyme in Purkinje cell axons and boutons in the cat. Brain Research 20: 259–275.

Ito, M., 1968, Two extensive inhibitory systems for brain stem nuclei. In: Structure and Function of Inhibitory Neuronal Mechanisms, von Euler, C., Skoglund, S. and Söderberg, U., ed. pp. 309–322. Oxford: Pergamon Press.

Killiam, K. F., and Bain, J. A., 1957, Convulsant hydrazides I: 'In vitro' and 'in vivo' inhibition of vitamin $B_6$ enzymes by convulsant hydrazides, J. Pharmacol. exp. Ther. 119: 255–262.

Krnjević, K., and Schwartz, S., 1968, The inhibitory transmitter in the cerebral cortex. In: *Structure and Function of Inhibitory Neuronal Mechanisms*, von Euler, C., Skoglund, S., and Söderberg, U., ed. pp. 419–427. Oxford: Pergamon Press.

Lamar, C., and Sellinger, O. Z., 1965, The inhibition *in vivo* of cerebral glutamine synthetase and glutamine transferase by convulsant methionine sulfoximine, *Biochem. Pharmacol.* **14**: 489–506.

Lane, B., and Europa, D. L., 1965, Differential staining of ultrathin sections of Epon-embedded tissues for light microscopy, *J. Histochem. Cytochem.* **13**: 579–582.

Lineweaver, J., and Burk, D., 1934, The determination of enzyme dissociation constants, *J. Amer. Chem. Soc.* **56**: 658–666.

Lowe, I. P., Robins, E., and Eyerman, G. S., 1958, The fluorometric measurement of glutamic decarboxylase and its distribution in brain, *J. Neurochem.* **3**: 8–18.

McFarland, D., and Wainer, A., 1965, Convulsant properties of allylglycine, *Life Sci.* **4**: 1587–1590.

Obata, K., Ito, M., Ochi, R., and Sato, N., 1967, Pharmacological properties of the postsynaptic inhibition by Purkinje cell axons and the action of γ-aminobutyric acid on Deiters neurons, *Exp. Brain Res.* **4**: 43–57

Peters, E. L., and Tower, D. B., 1959, Glutamic acid and glutamine metabolism in cerebral cortex after seizures induced by methionine sulphoximine, *J. Neurochem.* **5**: 80–90.

Potter, D. D., 1968, The Chemistry of inhibition in crustaceans with special reference to gamma-amino butyric acid. In: *Structure and Function of Inhibitory Neuronal Mechanisms*, von Euler, C., Skoglund, S., and Soderberg, U., ed., pp. 359–370. Oxford: Pergamon Press.

Roberts, E., 1968, Some biochemical physiological correlations in studies of γ-aminobutyric acid. In: *Structure and Function of Inhibitory Neuronal Mechanisms*, von Euler, C., Skoglund, S. and Söderberg, U., ed., pp. 401–418. Oxford: Pergamon Press.

Salganicoff, L., and De Robertis, E., 1963, Subcellular distribution of glutamic decarboxylase and gamma-aminobutyric alphaketoglutaric transaminase, *Life Sci.* **2**: 85–91.

Sellinger, O. Z., and Weiler, P., 1963, The nature of the inhibition in vitro of cerebral glutamine synthetase by the convulsant methionine sulphoximine, *Biochemical Pharmacology* **12**: 989–1000.

Uchizono, K., 1965, Characteristics of excitatory and inhibitory synapses in the central nervous system of the cat, *Nature* (Lond.) **207**: 642–643.

Weinstein, H., Roberts, E., and Kakefuda, T., 1963, Studies of subcellular distribution of γ-aminobutyric acid and glutamic decarboxylase in mouse brain, *Biochem. Pharmacol.* **12**: 503–509.

# UNIT ANALYSIS OF INPUTS TO CINGULATE CORTEX IN AWAKE, SITTING SQUIRREL MONKEYS I. EXTEROCEPTIVE SYSTEMS†

DAVID S. BACHMAN and PAUL D. MacLEAN

*Section of Comparative Neurophysiology and Behavior*
*Laboratory of Brain Evolution and Behavior*
*National Institute of Mental Health*
*Poolesville, Maryland*
*20837*

Exploration of the rostral cingulate cortex in awake, sitting squirrel monkeys failed to reveal any significant altera-tion of unit activity with photic, auditory, and somatic stimulation. As opposed to these negative findings, somatic stimulation of the hind limb activated 14 % of the tested units in the supracingulate cortex. The receptive fields were usually large and bilateral.

## INTRODUCTION

In preceding microelectrode studies on inputs to the limbic cortex of the squirrel monkey explora-tion has been carried out on all but the rostral cingulate and posterior orbital areas (Cuénod *et al.*, 1965; MacLean *et al.*, 1968; Yokota *et al.*, 1970; and Sudakov *et al.*, 1971). The present study deals specifically with the question of somatic, visual, and auditory inputs to the rostral cingulate cortex, including the supracallosal, pregenual, and subcallosal areas. The alleged relief of intractable pain by anterior cingulumotomy (Foltz and White, 1962; Ballantine *et al.*, 1967) and the impairment of immediate recall of visually presented words (Fedio and Ommaya, 1970) by anterior cingulate stimulation suggest the possibility of somatic and visual connections with the cingulate gyrus. There is the possibility that visual impulses might be transmitted to the cingulate cortex from the *nucleus opticus tegmenti* (Marg, 1964) *via* mammillo-thalamic pathways. In a preceding exploration of the posterior cingulate cortex only 5 of 249 units (2%) responded to photic stimulation (Cuénod *et*

al., 1965). Two of these were also excited by somatic stimulation. Preliminary findings of the present study have been published in an abstract (Bachman and MacLean, 1971). For comparative purposes the results of exploring the supracingulate cortex will also be described.

## METHODS

Twenty-one tracks were explored between frontal planes F20 and F5 in 3 adult squirrel monkeys prepared with a chronically fixed stereotaxic plat-form (MacLean, 1967). Extracellular unit record-ings were obtained with platinum-iridium micro-electrodes (Kinnard and MacLean, 1967). During experiments the monkey sat in a special restraining chair (MacLean *et al.*, 1968). The methods for applying and calibrating the strength of visual, auditory, and somatic stimuli, as well as for verify-ing histologically the loci of units were described in preceding publications (MacLean *et al.*, 1968; Sudakov *et al.*, 1971).

## RESULTS

### 1. Cingulate Cortex

Figure 1A shows the distribution of 106 units that

---

† Paper prepared in honor of Wade Hampton Marshall The results of somatic stimulation supply information about unit activity that confirms and extends an early evoked potential study by Dr. Marshall and co-workers (Woolsey, Marshall, and Bard, 1942).

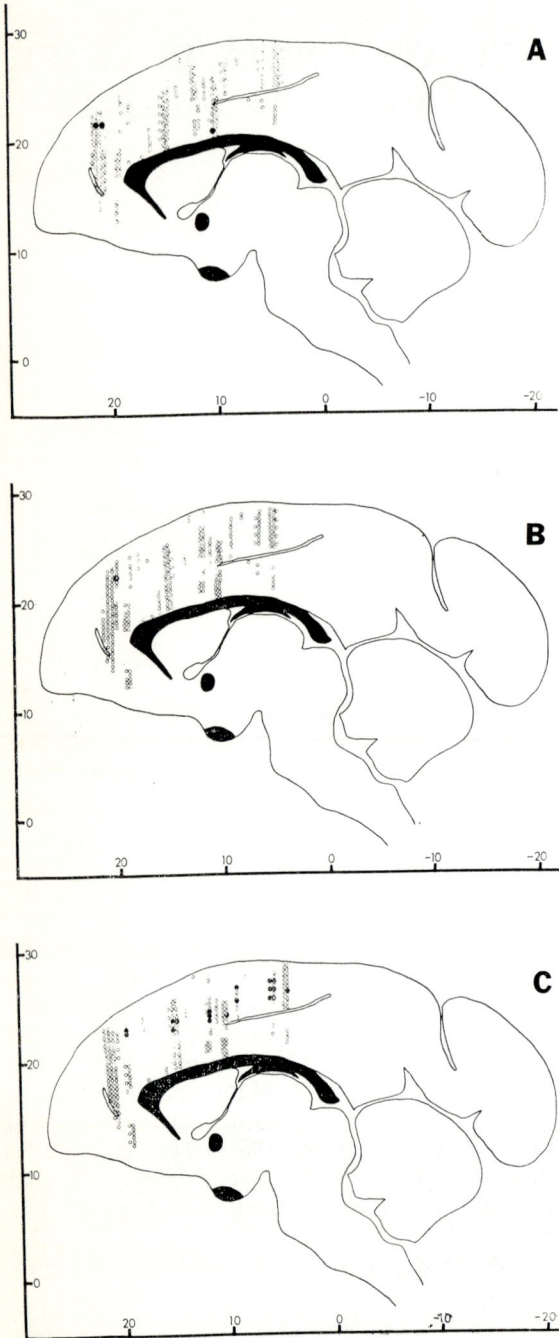

were tested during ocular illumination with the flash of a stroboscope and the light of a tungsten lamp. Only one unit was affected; it showed a decrease in its firing rate during stimulation with the tungsten light. None of the 101 units tested during the presentation of tones and clicks (Figure 1B) responded. The same was true for 84 units during somatic stimulation (Figure 1C) with light touch, pressure and pinprick.

## 2. *Supracingulate Cortex*

Of the 153 units examined in the supracingulate cortex (Figure 1A) one showed an increase, and another a decrease, in the rate of discharge following a stroboscopic flash. One unit that was activated by light was also inhibited by clicks (Figure 1B). One hundred and forty-two other units were unaffected by auditory stimulation.

As opposed to the negative findings on the cingulate cortex, 18 of 130 (14%) supracingulate units were responsive to somatic stimulation (Figure 1C and Figure 2). The receptive fields were large and usually bilateral, always involving the legs. Fourteen of the 18 units responded to light tactile stimulation, whereas pinprick and pressure were effective in all cases. The mean response latencies ranged from 40 to 185 msec.

## DISCUSSION

The negative findings of this study would suggest that the clinically observed alteration in the appreciation of pain and the deficit in the immediate recall of visually presented words (see 'Introduction') do not depend on somatic and visual projections to the cingulate cortex. The failure of cingulate units to respond to photic stimulation is

FIGURE 1  Loci of tested cingulate and supra-cingulate units plotted on a parasagittal diagram of the squirrel monkey's brain. Open circles, unresponsive units; filled circles, units responding to visual (A), auditory (B), or somatic (C) stimulation. Marginal scales in millimeters refer to atlas coordinates (see Plate E, Gergen and MacLean, 1962).

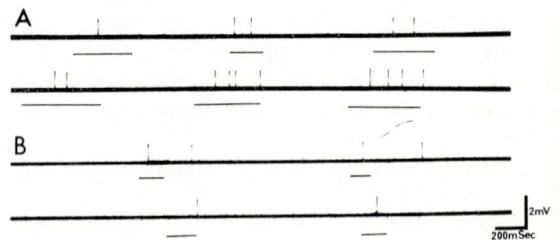

FIGURE 2  Responses of a unit in the supracingulate cortex to somatic stimulation by light touch (A) and pinprick (B). The bars underneath records correspond to signal of transducer.

evidence against a transmission of impulses from the *nucleus opticus tegmenti* by way of the mammillothalamic pathway.

Cingulumotomy has been claimed to attenuate withdrawal symptoms associated with morphine addiction (Foltz and White, 1957). It is well-known that stimulation of the anterior cingulate gyrus elicits a variety of autonomic changes (e.g., Smith, 1945; Ward, 1948; Dua and MacLean, 1964). The modification of withdrawal symptoms possibly reflects an alteration of the cingulate's participation in autonomic activity. Further evidence of the importance of this part of the limbic lobe in visceral functions is indicated by an extension of the present study in which we have observed cingulate unit and slow wave responses to vagal stimulation.

Except for the bilaterality of the receptive fields for most units, the findings on unit responses to somatic stimulation are in agreement with the evoked potential study by Woolsey, Marshall and Bard (1942). No evidence was found of somatic representation below the cingulate sulcus.

## SUMMARY

Exploration of the rostral cingulate cortex in awake, sitting squirrel monkeys failed to reveal any significant alteration of unit activity with photic, auditory, and somatic stimulation. As opposed to these negative findings, somatic stimulation of the hind limb activated 14% of the tested units in the supracingulate cortex. The receptive fields were usually large and bilateral.

## ACKNOWLEDGMENTS

We express our appreciation to Dr. Wade H. Marshall for his support of this work; and Lily Weinshilboum, Alvin C. Ziminsky, and John Lewis for their technical assistance.

## REFERENCES

Bachman, D. S., and MacLean, P. D., 1971, Unit study of exteroceptive inputs to cingulate cortex of squirrel monkeys, *Fed. Proc.* **30**: 434.

Ballantine, H. T., Cassidy, W. L., Flanagan, N. B., and Marino, R., 1967, Stereotaxic anterior cingulotomy for neuropsychiatric illness and intractable pain, *J. Neurosurg.* **26**: 488–495.

Cuénod, M., Casey, K. L., and MacLean, P. D., 1965, Unit analysis of visual input to posterior limbic cortex. I. Photic stimulation, *J. Neurophysiol.* **28**: 1101–1117.

Dua, S., and MacLean, P. D., 1964, Localization for penile erection in medial frontal lobe, *Amer. J. Physiol.* **207**: 1425–1434.

Fedio, P., and Ommaya, A. K., 1970, Bilateral cingulum lesions and stimulation in man with lateralized impairment in short-term verbal memory, *Exp. Neurol.* **29**: 84–91.

Foltz, E. L., and White, L. E., 1957, Experimental cingulumotomy and modification of morphine withdrawal, *J. Neurosurg.* **14**: 655–673.

Foltz, E. L., and White, L. E., 1962, Pain 'relief' by frontal cingulumotomy, *J. Neurosurg.* **19**: 89–100.

Gergen, J. A., and MacLean, P. D., 1962, *A Stereotaxic Atlas of the Squirrel Monkey's Brain (Saimiri sciureus)* U.S. Government Printing Office, Washington, D.C.

Kinnard, M. A., and MacLean, P. D., 1967, A platinum micro-electrode for intracerebral exploration with a chronically fixed stereotaxic device, *Electroenceph. Clin. Neurophysiol.* **22**: 183–186.

MacLean, P. D., 1967, A chronically fixed stereotaxic device for intracerebral exploration with macro- and microelectrodes, *Electroenceph. Clin. Neurophysiol.* **22**: 180–182.

MacLean, P. D., Yokota, T., and Kinnard, M. A., 1968, Photically sustained on-responses of units in posterior hippocampal gyrus of awake monkey, *J. Neurophysiol.* **31**: 870–883.

Marg, E., 1964, The accessory optic system, *Ann. NY Acad. Sci.* **117**: 35–52.

Smith, W. K., 1945, The functional significance of the rostral cingular cortex as revealed by its responses to electrical excitation, *J. Neurophysiol.* **8**: 241–255.

Sudakov, K., MacLean, P. D., Reeves, A., and Marino, R., 1971, Unit study of exteroceptive inputs to claustrocortex in awake, sitting, squirrel monkey, *Brain Res.* **28**: 19–34.

Ward, A. A., 1948, The cingular gyrus: area 24, *J. Neurophysiol.* **11**: 13–23.

Woolsey, C. N., Marshall, W. H., and Bard, P., 1942, Representation of cutaneous tactile sensibility in the cerebral cortex of the monkey as indicated by evoked potentials, *Bull. The Johns Hopkins Hosp.* **70**: 399–441.

Yokota, T., Reeves, A. G., and MacLean, P. D., 1970, Differential effects of septal and olfactory volleys on intracellular responses of hippocampal neurons in awake, sitting monkeys, *J. Neurophysiol.* **33**: 96–107.

# QUANTITATIVE SYNAPTIC CHANGES WITH DIFFERENTIAL EXPERIENCE IN RAT BRAIN

KJELD MØLLGAARD

*University of Copenhagen, Denmark*

MARIAN C. DIAMOND, EDWARD L. BENNETT,

MARK R. ROSENZWEIG and BERNICE LINDNER

*University of California, Berkeley*

Significant differences in size and number of synaptic junctions were found between littermate rats assigned at weaning (25 days of age) to enriched or impoverished environments and kept there for 30 days. The synapses measured were asymmetrical axodendritic junctions in the neuropil of layer III of the occipital cortex. Rats given experience in the enriched condition (EC) showed, in comparison to littermates in the impoverished condition (IC), synapses that averaged 52% greater in length but that were only 67% as numerous. The EC rats had more large synapses as well as fewer small synapses than did IC rats, so the EC size distribution could not have been derived simply by loss of small synapses from the IC distribution. The total area of synapses in the EC group, taking both size and number of contacts into account, was 40% greater than in IC. Thickness of cortex was 4.0% greater in EC than in IC, a value that compares closely with the 4.6% found in several previous 25-to-55 day EC–IC experiments. Problems of measurement and sampling in electron microscopy are considered. The results are discussed in relation to concepts of brain mechanisms of learning and memory storage, some aspects of which can now be studied directly.

Experience in enriched or impoverished environments causes changes in brain anatomy and chemistry as well as in behavior in the rat (Bennett, *et al.*, 1964 and 1970; Diamond *et al.*, 1966; Rosenzweig, 1971; Rosenzweig and Bennett, 1971); this report brings the anatomical research to the synaptic level with an electron microscopic study of changes in synaptic size and number.

Our group has taken progressive steps to understand the factors involved in bringing about structural changes in the rat cerebral cortex as a consequence of experience. In the initial anatomical experiments with 80 days of exposure to enriched or impoverished environmental conditions (EC and IC), measures of the cortical depth showed a 6.2% ($P < .001$) EC–IC difference in the occipital cortex (Diamond *et al.*, 1964). Further experiments demonstrated differences not only in cortical depth but also in glial number (14%, $P < .01$) (Diamond *et al.*, 1966), in perikaryon cross-sectional area

(12%, $P < .001$) and in cross-sectional area of neuronal nuclei (12%, $P < .01$) (Diamond, 1967). A next logical step in determining factors involved in the anatomical changes in the cortex was to examine synapses in the neuropil at the ultrastructural level.

It has been postulated that permanent changes may occur in synapses as a result of functional activity and that these changes may play some part in learning and memory (Kandel and Spence, 1968). According to Bloom and Iversen (1970) 'most morphologists accept as "synapses" junctions having only two essential features: 1. a specialized form of intercellular contact between presynaptic and postsynaptic processes, and 2. the presence of synaptic vesicles in the presynaptic terminal.' Our attention was focused on the first of these features. In the present experiment we quantified both the numbers and the lengths of the postsynaptic thickenings, the 'highly organized proteinaceous

## PROCEDURES

### Methods and Materials

Twelve pairs of male rats from the Berkeley $S_1$ strain were exposed from July 17 to August 15, 1969, to enriched or impoverished environments described briefly here. (For a fuller description see Rosenzweig *et al.*, 1971). From each litter, rats were assigned at random so that one animal went into a condition that was more complex or enriched than colony life and one was placed in a restricted or impoverished condition. In the enriched condition 12 rats lived together in a large cage with half a dozen stimulus objects placed in the cage every day out of a pool of 25 varied objects. In addition, these rats were removed from the cage each day for thirty minutes and placed in an open field maze with the barriers changed daily. The impoverished rats were placed singly in small cages in an isolation room; the IC rat could move about freely but it could see or touch no other rat. Food and water were available to both groups ad libitum.

The experimental conditions started after weaning at day 25 and lasted 30 days until day 55, when all the rats were sacrificed. The rats were handled as pairs through the complete anatomical procedures, from sacrifice to completion of measurement of synapses. Before sacrifice, the animals had been assigned code numbers, so that the anatomists did not know the experimental condition of any individual rat.

Following anesthesia with ether, the animal was perfused through the left cardiac ventricle with saline immediately followed by 4% glutaraldehyde in a 0.3 M sodium phosphate buffer (pH 7.4). (In more recent experiments than those reported here, we are using modified techniques of tissue preparation.) The brain was removed, and a small tissue block was excised from the right hemisphere utilizing a precisely ruled plastic 'T' square to insure the uniform removal of the anterior half of our usual sample of occipital cortex (Bennett *et al.*, 1964).

### Preparation of Material for Electron Microscopy

The excised tissue blocks were immersed for 10 hours in 4% glutaraldehyde in 0.3 M sodium phosphate buffer. After two hours in 1% osmium tetroxide in the same buffer, the blocks were dehydrated in graded ethanol solutions, transferred to propylene oxide and imbedded in araldite. Finally each tissue block was cut out, oriented and re-cemented to an araldite mount so that sections could be cut in the coronal plane. The orientation was verified by thick sections, and other tissue was cut away to leave a truncated pyramid whose center was one-third of the way down from the pial surface and therefore predominantly in layer III but extending slightly into layer IV. The boundary between layers III and IV is difficult to determine in the occipital cortex of the rat, although the boundary between layers IV and V is distinct.

The arrangement of the tissue block and pyramid is shown in Figure 1. Thin sections, 60–80 nm (600Å–800Å) were cut from the pyramid on a Porter-Blum MT-2 microtome with glass or diamond knives, depending on the hardness of the tissue. The sections were picked up on uncoated

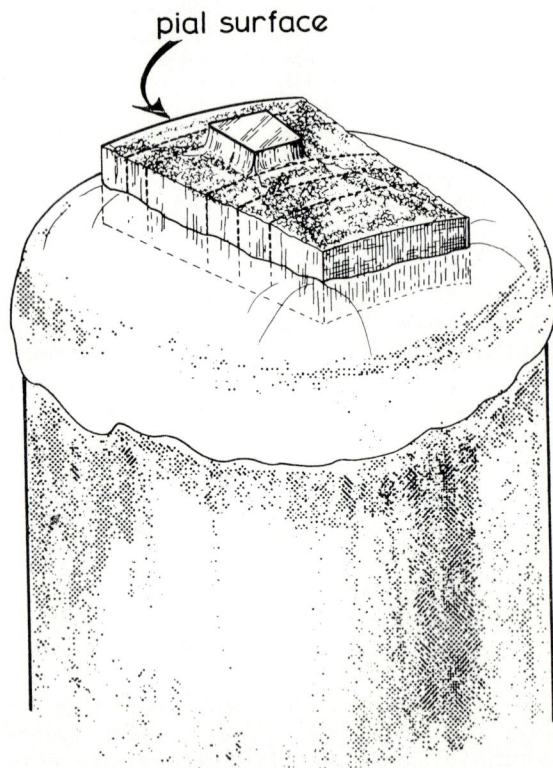

FIGURE 1 Tissue block mounted and prepared for sectioning. See description in text, above. The depth of the block, from the pial surface to the bottom of layer VI, is about 1 mm.

grids and double-stained with uranyl acetate and lead citrate (Reynolds, 1963). The material was examined with a Siemens Elmiskop I. The magnification was fixed at 14000× throughout the study. Since the magnification changes slightly with focusing, each series of pictures was started and finished with a photograph of a grating replica. As a result of changes in magnification, a small correction factor was used for the pictures of three of the twenty-four animals.

Only one section on each grid was chosen, so that successive sections through the same synapses would not be used. Within the chosen section, photographs were taken of as many areas of well-stained and well-fixed neuropil as possible. Since each section typically afforded only a few good pictures, many grids were examined for each animal. Satisfactory pictures were obtained from all animals in the experiment. All the micrographs were printed at a 3× magnification, giving a total magnification of 42000 in the 24.0×18.4 cm prints. Such a print then represents an area of 25 $\mu$m².

*Measurements*

It seemed to us important to make comparisons on synapses of the same morphological type. We decided to measure the axodendritic, asymmetrical synapses described by Colonnier (1968) as corresponding closely to Gray's type I synapses (Gray, 1959). The asymmetrical synapses are characterized by the presence of variable thick cytoplasmic postsynaptic opacities bordering the presynaptic membranes (Colonnier, 1968). (See Figure 2.) These characteristics served to identify most of the synapses measured. A vesicle population of spheroidal type is associated with asymmetrical synapses, and the great majority of dendritic spines are also associated with asymmetrical synapses. These secondary characteristics were used to identify some asymmetrical synapses where there was not a clear synaptic cleft, presumably because the plane of section was oblique to the plane of the junction. Only those junctions that

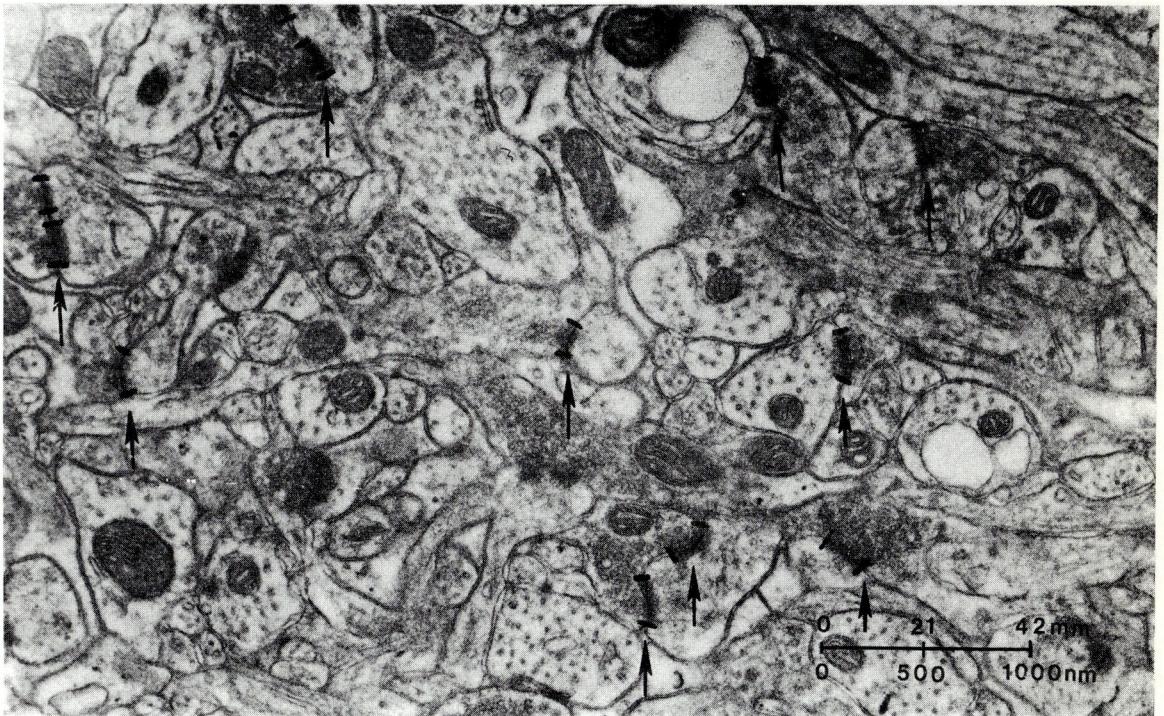

FIGURE 2 Electron micrograph of neuropil in layer III of rat occipital cortex, showing a variety of synapses. The arrows show asymmetric synapses that reached our criteria. The nanometer (nm) scale shows the dimensions of the neural structures. The millimeter (mm) scale refers to the dimensions of the photographs on which the measurements were made.

reached an acceptable criterion of clarity were measured.

The measurements of synaptic length were done on the postsynaptic thickenings. A magnifying glass was always used in measuring the length. Curved thickenings were divided in smaller parts and the sum of the parts was taken. In some cases a large synapse had one or more small discontinuities in the postsynaptic thickening; if the vesicles were uniformly distributed on the presynaptic side, the thickening was measured as one synapse in spite of discontinuities. Serial sections and reconstructions have shown that such perforations are normally present in larger synapses (Peters and Kaiserman-Abramof, 1969). At the magnification used, 4.2 mm equaled 100 nm (1000Å). The number of synapses was counted on each picture.

The tissue sections should afford random samples within the area chosen. Since all sections were processed as a single group, the only source of anatomical differences must be the difference in behavioral treatment, that is, the EC versus the IC environment. Since the purpose of this investigation was to compare the two groups of animals, no attempt was made to correct the results for shrinkage caused by fixation and imbedding.

Within the plates, we measured all synapses that met a criterion of clarity. Stereological sampling procedures (Weibel and Elias, 1967) would not have been appropriate since synapses are shown to vary in shape (Peters and Kaiserman-Abramof, 1969). The following precautions were taken to insure that no bias could enter during the processes of counting and measuring synapses: Prints from one pair of rats were read at a time, the EC and IC prints having been coded to conceal the identity of the animals, and the EC and IC prints of a pair were interspersed randomly. All of this work was done by a single person on coded prints, so that criteria of selection of synapses would not vary among groups.

## Preparation of Material for Light Microscopy

Cortical depth measures were taken on the left hemisphere in our usual way (Diamond, 1967); the tissue measured was directly opposite the tissue removed from the right hemisphere for electron microscopic study. Transverse frozen sections were cut at 20 $\mu$m and stained with a modified Nissl stain. The sections were projected with the use of a micro-slide projector and outlines of the brain were drawn (mag. 22.5 $\times$). On the drawings, lines were extended up from the corpus callosum to the dorsal surface of layer II. The lines were spaced 2 mm apart beginning immediately lateral to the elevation of the corpus callosum and extending to the widest dimension of the hemisphere. The individual lines were measured with a millimeter rule. Thus, the depth was measured through the whole dorsal surface of the cortex.

## RESULTS

### Synaptic Measurements

Usable electron microscope pictures were obtained from layer III of the occipital cortex of all 12 littermate pairs of rats raised in the enriched (EC) or impoverished (IC) environments. A total of 349 pictures was obtained; the number of pictures varied for each rat, as shown in Table 1. With two exceptions, 10 or more pictures were obtained for each animal. The number of identifiable asymmetric synapses per picture was determined. In all, 2,211 synapses were identified and measured; 1,405 from IC rats and 806 from EC rats. As shown in Table 1, the number of synapses per print was 35% fewer for the EC rats than for the IC rats. In 11 of the 12 pairs, the EC rat had fewer synapses per picture than did its IC littermate; in 9 of these pairs, the distribution differed at beyond .05 level of significance, as determined by the Kolmogorov-Smirnov test. In mean number of synapses per picture, IC exceeded EC significantly in 10 of the 12 pairs, as determined by t-tests. (See the P values in Table 1.) The one case of reversal, in which EC had more synapses per picture than IC, was not statistically significant.

The detailed distributions of synapses per plate are shown for three pairs of animals in Figure 3. The pairs chosen include the one case where EC had somewhat more synapses per plate than did IC (A), a case where the distributions for the pair were close to the means for the groups (B), and the case with the largest EC–IC difference (C). Figure 4 presents the distributions for the 12 animals of each group combined and shows graphically the group differences described in the paragraph above.

The lengths of the postsynaptic thickenings, as well as synaptic frequencies, distinguish EC from IC at the ultrastructural level. Table 2 shows that the mean length of synapses is greater for EC than IC in every pair, the smallest intrapair difference

TABLE 1

Distribution of Synapses per Picture

| Pair | Total Number of Pictures | Mean Synapses per Picture | Standard Error of Mean | Analysis of Variance | | | Kolmogorov-Smirnov | |
|---|---|---|---|---|---|---|---|---|
| | | | | Fs | df | p | $x^2$ | p |
| 1 EC | 23 | 6.3 | ±.6 | 6.9 | 1,40 | <.05 | 6.9 | <.05 |
| IC | 19 | 8.4 | ±.5 | | | | | |
| 2 EC | 10 | 5.1 | ±.6 | 1.5 | 1,19 | NS | 3.5 | NS |
| IC | 11 | 6.0 | ±.5 | | | | | |
| 3 EC | 12 | 4.5 | ±.7 | 17.8 | 1,28 | <.001 | 12.8 | <.01 |
| IC | 18 | 8.2 | ±.6 | | | | | |
| 4 EC | 12 | 5.5 | ±.6 | 5.1 | 1,17 | <.05 | 3.4 | NS |
| IC | 7 | 7.8 | ±.9 | | | | | |
| 5 EC | 16 | 3.9 | ±.2 | 16.8 | 1,30 | <0.01 | 10.1 | <.01 |
| IC | 16 | 6.9 | ±.7 | | | | | |
| 6 EC | 5 | 5.6 | ±.7 | 8.6 | 1,22 | <.01 | 6.5 | <.05 |
| IC | 19 | 9.5 | ±.6 | | | | | |
| 7 EC | 11 | 4.4 | ±.5 | 19.4 | 1,19 | <.001 | 10.8 | <.01 |
| IC | 10 | 8.1 | ±.7 | | | | | |
| 8 EC | 16 | 3.9 | ±.3 | 36.0 | 1,31 | <.001 | 15.8 | <.001 |
| IC | 17 | 7.8 | ±.6 | | | | | |
| 9 EC | 12 | 5.4 | ±.5 | 8.6 | 1,22 | <.01 | 10.7 | <.01 |
| IC | 12 | 8.2 | ±.8 | | | | | |
| 10 EC | 13 | 4.8 | ±.5 | 2.1 | 1,28 | NS | 1.2 | NS |
| IC | 17 | 3.9 | ±.4 | | | | | |
| 11 EC | 11 | 5.2 | ±.7 | 11.9 | 1,31 | <.01 | 11.9 | <.01 |
| IC | 22 | 8.2 | ±.5 | | | | | |
| 12 EC | 22 | 4.8 | ±.3 | 16.3 | 1,38 | <.001 | 8.4 | <.05 |
| IC | 18 | 7.1 | ±.5 | | | | | |
| All 12 EC | 163 | 4.94 | ±.16 | 101.9 | 1,347 | <.001 | 64.7 | <.0001 |
| All 12 IC | 186 | 7.56 | ±.20 | | | | | |

being 32% and the largest 67%, with an average difference of 52%. The standard error of the mean of lengths for each animal was about 4% of the mean length. In each pair, the mean synapse length was significantly greater for the EC than for the IC rat; P values, all less than .001, are shown to the right in Table 2. The mean measured length of all EC synapses on the photographs was 15.2 mm and that of the IC synapses was 10.0 mm; the equivalent calculated lengths were 362 nm and 238 nm respectively.

Length distributions for three pairs are given in Figure 5. The pairs are those with the smallest EC–IC difference (A), a difference in the middle of the range (B), and the largest EC–IC difference (C).

The cumulative frequency distribution of synapse lengths for all animals is shown in Figure 6. The significance of the difference in length distributions was evaluated by the non-parametric Kolmogorov-Smirnov test for each pair of rats, and it was highly significant in every case as is shown by the P values in Table 2. Consideration of such cumulative

## ADDENDUM TO TABLE 1

Complete Data for Individual Animals

| Pair | Number of Pictures with Each of the following Frequencies of Synapses: | | | | | | | | | | | | | | | Total Number of Pictures |
| --- | 1 | 2 | 3 | 4 | 5 | 6 | 7 | 8 | 9 | 10 | 11 | 12 | 13 | 14 | 15/16 | --- |
| 1 EC | 1 | | 3 | | 5 | 4 | 3 | 3 | 3 | | | | | 1 | | 23 |
| IC | | | | 1 | | 2 | 6 | 3 | | 4 | 1 | | 2 | | | 19 |
| 2 EC | | | 2 | 3 | 1 | 1 | 2 | 1 | | | | | | | | 10 |
| IC | | | 1 | | 4 | 2 | 2 | 1 | 1 | | | | | | | 11 |
| 3 EC | | 3 | 1 | 4 | 1 | | 1 | 1 | 1 | | | | | | | 12 |
| IC | | | | | | 2 | 2 | 3 | 5 | 2 | 2 | 1 | | | 1 | 18 |
| 4 EC | | | 2 | 3 | 2 | 1 | 2 | 1 | | 1 | | | | | | 12 |
| IC | | | | | 1 | 2 | | 1 | 2 | | | 1 | | | | 7 |
| 5 EC | | 1 | 5 | 6 | 3 | 1 | | | | | | | | | | 16 |
| IC | | | 2 | 1 | 3 | 3 | | 1 | 4 | | | 2 | | | | 16 |
| 6 EC | | | 1 | 2 | 1 | | 1 | | | | | | | | | 5 |
| IC | | | | | 1 | 2 | 3 | 3 | 1 | 1 | 3 | 1 | 2 | 2 | | 19 |
| 7 EC | | 2 | 1 | 3 | 3 | 1 | | 1 | | | | | | | | 11 |
| IC | | | | | 1 | 2 | 2 | | 2 | 2 | | 1 | | | | 10 |
| 8 EC | | 2 | 5 | 5 | 2 | 1 | 1 | | | | | | | | | 16 |
| IC | | | | 1 | 3 | 1 | 3 | 2 | 2 | 2 | 3 | | | | | 17 |
| 9 EC | 1 | | | 1 | 2 | 6 | 2 | | | | | | | | | 12 |
| IC | | | | | 1 | 1 | 5 | 2 | 1 | | 1 | | | | 1 | 12 |
| 10 EC | | 2 | 1 | 2 | 4 | 2 | | 2 | | | | | | | | 13 |
| IC | 1 | 2 | 4 | 3 | 5 | 1 | 1 | | | | | | | | | 17 |
| 11 EC | | | 2 | 3 | 3 | 1 | 1 | | | | 1 | | | | | 11 |
| IC | | | 1 | 1 | 5 | 3 | 3 | 2 | 2 | 2 | 2 | 2 | 1 | | | 22 |
| 12 EC | | | 6 | 5 | 4 | 2 | 5 | | | | | | | | | 22 |
| IC | | | | 2 | 2 | 3 | 5 | 2 | 2 | | 1 | 1 | | | | 18 |
| All 12 EC | 2 | 10 | 28 | 36 | 32 | 21 | 17 | 10 | 4 | 1 | 1 | | | 1 | | 163 |
| All 12 IC | 1 | 2 | 7 | 9 | 24 | 26 | 33 | 23 | 19 | 13 | 12 | 8 | 5 | 2 | 2 | 186 |

distributions for each pair of rats indicates, that with the size of the effect being studied, only about 20 synapses per rat need to be measured to obtain an intrapair distribution difference significant at beyond the .05 level; approximately 30 synapses per rat need be measured to reach the .01 level of significance.

Since EC rats show fewer asymmetrical synapses than IC, but the EC synapses are on the average larger, could the EC distribution be obtained simply by removal of smaller synapses from the IC distribution? To test this possibility, we prepared Figure 7 which presents distributions of the lengths of synapses in proportion to the relative numbers found in EC and IC when equal weight was given to the frequency distribution of each rat. From these calculations, when the area under the IC curve is normalized to 1000 synapses, then the area under the EC curve is normalized to 667 synapses. It is obvious that the EC rats have significantly fewer small synapses than the IC; that is, synapses measuring less than 10 mm (equivalent length 238

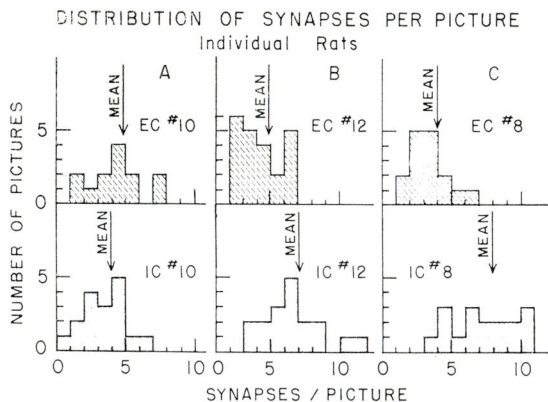

FIGURE 3 Distributions of numbers of asymmetric axodendritic synapses for rats of three EC–IC pairs. A shows the single pair in which the EC rat had more synapses per picture than IC; this difference was not significant. B shows the pair that most closely represents the mean effect; EC had 4.8 and IC 7.1 mean synapses per plate ($P < .001$). C presents the largest EC–IC difference.

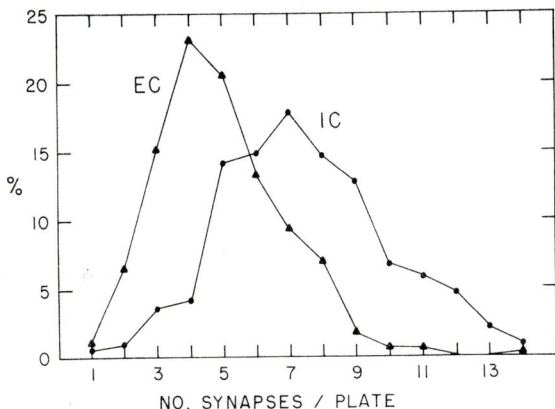

FIGURE 4 Distributions of numbers of asymmetric axodendritic synapses measured per photograph for all EC (Δ) and IC (O) rats.

nm); EC and IC have about an equal number of synapses of measured length 11 to 13 mm (equivalent length 262 to 310 nm), but the EC rats have significantly more synapses longer than 14 mm (333 nm). It might also be noted that only one EC synapse of 5 mm or less in length was measured, whereas 5% of the IC synapses were 5 mm (equivalent length 119 nm) or less. Every IC rat had one or more 5 mm synapses. On the other hand, only 13 per 1000 IC synapses were 22 mm or longer, whereas 84 per 667 EC synapses were 22 mm or over (525 nm) in length. That is, there are six times as many large synapses in EC than in IC, despite

the smaller total number in EC. Thus, the greater number of large synapses in EC demonstrates that the EC distribution cannot be derived from the IC distribution by elimination of small synapses.

Total area of EC and IC synapses was calculated, using the measured cross-section lengths and the numbers of junctions; that is, assuming the junctions to be roughly circular, we calculated the area for each size of cross-section and then multiplied

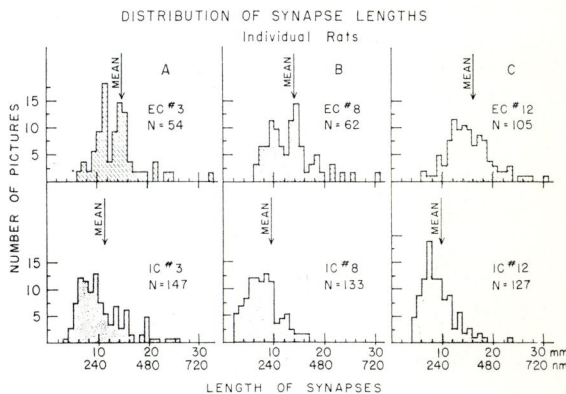

FIGURE 5 Distributions of lengths of postsynaptic thickenings for three EC–IC pairs. A shows the smallest EC–IC difference which was nevertheless significant at beyond the .001 level. B shows the pair that is most typical of the overall effect, and C presents the largest EC–IC difference.

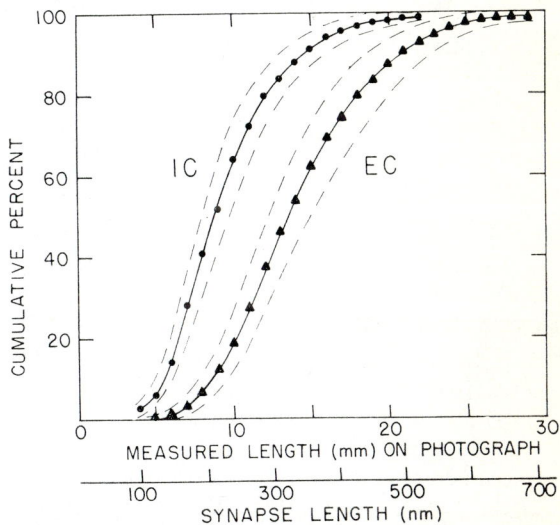

FIGURE 6 Cumulative frequency distribution of synaptic lengths for all EC (Δ) and IC (O) rats. The standard deviations are shown by dotted lines above and below the main curves. Lengths are given both in terms of actual measured length (mm) on electron microscope photographs and in equivalent synapse length (nm).

KJELD MØLLGAARD

## TABLE 2

The EC–IC Differences in Synaptic Length Distributions for Each Pair of Rats

| Pair | Total Number of Synapses | Mean Length of Synapses | ± Standard Error of Mean | Fs | Analysis of Variance df | p | Kolmorogov-Smirnov Test $x^2$ | p |
|------|------|------|------|------|------|------|------|------|
| 1 EC | 145 | 16.8 | ±.5 | 104 | 1,302 | <.001 | 63.7 | <.0001 |
| IC | 159 | 11.2 | ±.3 | | | | | |
| 2 EC | 51 | 15.7 | ±.6 | 43 | 1,115 | <.001 | 23.6 | <.0001 |
| IC | 66 | 10.8 | ±.4 | | | | | |
| 3 EC | 54 | 14.5 | ±.6 | 23 | 1,199 | <.001 | 30.8 | <.0001 |
| IC | 147 | 11.0 | ±.4 | | | | | |
| 4 EC | 66 | 13.7 | ±.5 | 30 | 1,119 | <.001 | 18.9 | <.0001 |
| IC | 55 | 9.9 | ±.4 | | | | | |
| 5 EC | 62 | 16.0 | ±.6 | 76 | 1,171 | <.001 | 36.7 | <.0001 |
| IC | 111 | 10.2 | ±.4 | | | | | |
| 6 EC | 28 | 13.2 | ±.8 | 30 | 1,206 | <.001 | 27.8 | <.0001 |
| IC | 180 | 9.3 | ±.2 | | | | | |
| 7 EC | 48 | 13.6 | ±.6 | 25 | 1,127 | <.001 | 25.8 | <.0001 |
| IC | 81 | 9.8 | ±.5 | | | | | |
| 8 EC | 62 | 13.8 | ±.6 | 62 | 1,193 | <.001 | 30.7 | <.0001 |
| IC | 133 | 9.3 | ±.3 | | | | | |
| 9 EC | 65 | 15.9 | ±.8 | 54 | 1,161 | <.001 | 30.5 | <.0001 |
| IC | 98 | 10.0 | ±.4 | | | | | |
| 10 EC | 63 | 15.3 | ±.7 | 46 | 1,128 | <.001 | 39.0 | <.0001 |
| IC | 67 | 9.7 | ±.4 | | | | | |
| 11 EC | 57 | 12.8 | ±.6 | 41 | 1,236 | <.001 | 28.5 | <.0001 |
| IC | 181 | 9.0 | ±.3 | | | | | |
| 12 EC | 105 | 16.0 | ±.4 | 156 | 1,230 | <.001 | 97.8 | <.0001 |
| IC | 127 | 9.7 | ±.3 | | | | | |
| All 12 EC | 806 | 15.2 | ±.2 | 803 | 1,2209 | <.001 | 463.6 | <.0001 |
| All 12 IC | 1405 | 10.0 | ±.1 | | | | | |

this by the number of cross-sections of that size. (We realize that the sampling tends to under-estimate the size of large synapses because fractions can be measured, whereas the small synapses are measured only if all or most of their diameters are included in a section; this will be considered in the Discussion.) From these calculations it appears that the total area of the synaptic thickenings in EC exceeds that in IC by 40% in the region studied. This value undoubtedly overestimates the difference because Peters and Kaiserman-Abramof (1969) found perforations in the postsynaptic thickenings, especially in the larger axodendritic contacts. In their serial reconstructions these 'holes' often amounted to one-third of the area of contact. In our observations, however, the perforations were usually small, often 3 or 4 mm on the photographs, and therefore not more than one-fifth the diameter of the larger synapses or only a few percent of the contact surface.

### Cortical Depths

There was a 4.0% difference ($P < .05$) in cortical

ADDENDUM TO TABLE 2

Complete Data for Individual Animals

| Pair | 3 | 4 | 5 | 6 | 7 | 8 | 9 | 10 | 11 | 12 | 13 | 14 | 15 | 16 | 17 | 18 | 19 | 20 | 21 | 22 | 23 | 24 | 25 | 26 | 27 | 28 | 29 | 30–40 | Total Number of Synapses |
|---|---|---|---|---|---|---|---|---|---|---|---|---|---|---|---|---|---|---|---|---|---|---|---|---|---|---|---|---|---|
| 1 EC | | 1 | 4 | 5 | 4 | 2 | 2 | 6 | 8 | 12 | 15 | 7 | 9 | 9 | 11 | 15 | 7 | 5 | 6 | 5 | 8 | 3 | 3 | | | 1 | | | 145 |
| IC | | | | | 12 | 22 | 15 | 19 | 19 | 9 | 13 | 12 | 11 | 1 | 3 | 4 | 3 | 2 | 1 | 2 | | 1 | 3 | | | | 2 | 5 | 159 |
| 2 EC | | 1 | 2 | 1 | 6 | 7 | 8 | 3 | 4 | 6 | 5 | 5 | 4 | 3 | 7 | 1 | 1 | 1 | 3 | 2 | | 4 | | | | | | | 51 |
| IC | | | | 3 | 6 | 7 | | 7 | 2 | 12 | 5 | 4 | 1 | 4 | 2 | 1 | 1 | 1 | 3 | 1 | | | 1 | | | | | | 66 |
| 3 EC | | 1 | 4 | 11 | 1 | 2 | 1 | 3 | 5 | 10 | 2 | 5 | 8 | 7 | 2 | 1 | 1 | 1 | 1 | 2 | | 1 | 1 | 1 | | | | 1 | 54 |
| IC | | | | | 18 | 17 | 14 | 19 | 11 | 8 | 5 | 10 | 4 | 9 | 2 | 1 | 2 | 7 | 1 | 1 | | 1 | 1 | | | | | | 147 |
| 4 EC | | | 3 | 2 | 1 | 4 | 6 | 4 | 7 | 5 | 3 | 7 | 5 | 6 | 3 | 3 | 3 | 2 | 2 | 1 | | 2 | | | | | | | 66 |
| IC | | | | 2 | 10 | 6 | 6 | 7 | 6 | 6 | 1 | 2 | 3 | 2 | | 1 | | | | | | | | | | | | | 55 |
| 5 EC | | 3 | 7 | 12 | 10 | 2 | 3 | 3 | 3 | 8 | 5 | 2 | 3 | 4 | 4 | 3 | 4 | 9 | 3 | 1 | 3 | | | 1 | 1 | 1 | | | 62 |
| IC | 1 | 3 | 4 | | | 7 | 12 | 15 | 7 | 10 | 7 | 6 | 3 | 5 | 2 | 1 | 2 | 1 | | 1 | 1 | | | 1 | | 1 | | | 111 |
| 6 EC | | 3 | 1 | | | | 5 | | 4 | 5 | 2 | 5 | 3 | 3 | 2 | 1 | 1 | 1 | | | 1 | | 1 | 1 | | | | | 28 |
| IC | 1 | 3 | 4 | 19 | 36 | 34 | 19 | 19 | 10 | 6 | 7 | 5 | 5 | 3 | 3 | 3 | 1 | | | | | 1 | | | | | | | 180 |
| 7 EC | | 2 | 5 | 4 | 1 | 2 | 4 | 4 | 8 | 4 | 3 | 5 | 6 | 1 | 1 | 1 | 2 | 2 | 1 | 2 | 1 | 1 | | | | | 1 | | 48 |
| IC | | | | | 17 | 13 | 7 | 8 | 5 | 5 | 1 | 3 | 4 | | 1 | 3 | | 2 | | | | | | | | | | | 81 |
| 8 EC | 7 | | 9 | 11 | 2 | 4 | 4 | 7 | 6 | 4 | 3 | 7 | 9 | 4 | 1 | 2 | 3 | 1 | | 2 | | 1 | | 1 | | | | 1 | 62 |
| IC | | | | | 16 | 16 | 15 | 17 | 15 | 5 | 6 | 7 | 3 | 2 | 2 | 2 | | | | | | | | | | | | | 133 |
| 9 EC | | | 5 | 1 | 2 | 3 | 2 | 5 | 4 | 6 | 7 | 1 | 6 | 2 | 2 | 3 | 2 | 1 | 5 | 5 | 2 | 1 | 2 | | 1 | | | 2 | 65 |
| IC | | | | 8 | 19 | 6 | 11 | 12 | 7 | 10 | 8 | | 2 | 5 | 2 | | | | 1 | 1 | | 1 | | | | | | | 98 |
| 10 EC | | | 2 | 7 | 1 | 2 | 5 | 6 | 5 | 1 | 8 | 6 | 2 | 4 | 5 | 2 | 4 | 3 | 2 | 1 | 1 | 1 | 1 | 1 | 1 | 1 | 1 | 2 | 63 |
| IC | | | | | 11 | 9 | 8 | 11 | 6 | 4 | 3 | 2 | | | 1 | 1 | | 1 | | | | | | | | | | | 67 |
| 11 EC | | | 13 | | 2 | 7 | 5 | 6 | 6 | 5 | 2 | 8 | 3 | 2 | 1 | 3 | 2 | | 2 | | | 2 | | | 1 | 1 | 1 | 1 | 57 |
| IC | | | | 30 | 30 | 25 | 23 | 18 | 14 | 10 | 4 | 3 | 3 | 2 | 1 | | 1 | | | | | | | | | | | | 181 |
| 12 EC | | | 6 | 11 | 2 | 1 | 1 | 5 | 3 | 8 | 12 | 10 | 11 | 10 | 7 | 9 | 8 | 4 | 3 | 2 | 2 | 3 | | 1 | 1 | 1 | | 1 | 105 |
| IC | | | | | 15 | 24 | 15 | 15 | 10 | 11 | 3 | 7 | 2 | 3 | 1 | 2 | 1 | | | | | 1 | | | | | | | 127 |
| All 12 EC | 1 | 18 | 1 | 5 | 16 | 29 | 38 | 52 | 63 | 74 | 67 | 63 | 69 | 55 | 46 | 44 | 37 | 29 | 25 | 23 | 17 | 19 | 8 | 4 | 4 | 3 | 3 | 12 | 806 |
| All 12 IC | | | 64 | 123 | 200 | 186 | 153 | 167 | 112 | 96 | 63 | 61 | 41 | 36 | 20 | 17 | 10 | 14 | 5 | 5 | 3 | 4 | 2 | 1 | | 1 | 1 | 1 | 1405 |

depth between the EC and IC rats in this experiment, which is close to the 4.6% mean found in four previous 25-to-55-day EC–IC experiments. The data for the 12 pairs are shown in Table 3. A high positive correlation (0.79, $P < .001$) was found between cortical depth differences and synaptic

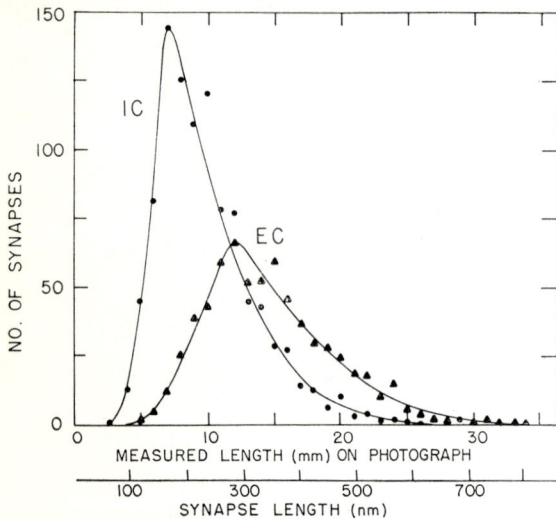

FIGURE 7 Frequency distribution of length of 1000 synapses from IC rats (O) and 667 synapses from EC rats (Δ). The totals differ because, as discussed in the text, EC rats have two-thirds as many synapses in the area measured as did IC rats.

TABLE 3

Percentage Differences in Cortical Depth

| Pair No. | Percent Difference, EC–IC |
|---|---|
| 1 | 3.1 |
| 2 | −5.9 |
| 3 | −2.8 |
| 4 | 0.4 |
| 5 | 10.8 |
| 6 | −0.4 |
| 7 | 2.5 |
| 8 | 1.8 |
| 9 | 15.0 |
| 10 | 9.9 |
| 11 | 3.6 |
| 12 | 9.2 |
| Mean = | 4.0 |
| S.D. = | 0.16 |
| $P$ | < .05 |

length differences. The correlation was not significant between the differences in cortical depth and the number of synapses.

## DISCUSSION

Consideration of certain aspects of the experimental conditions and the measures will help to relate these results to other cerebral effects and to suggest how far they may be extrapolated. Factors to be considered in successive sections of the Discussion are these: (a) the behavioral conditions and animal subjects, (b) the cortical region studied, (c) conceptualizations concerning cerebral changes in learning and memory storage, (d) problems of two-dimensional sampling in ultrastructural measurements, (e) meaning of synapse size, (f) meaning of synapse number, (g) combined effects of alterations in synapse size and number, (h) possible relations of synaptic area to cortical AChE.

### Behavioral Conditions and Subjects

The enriched and impoverished environmental conditions were employed because they have previously been demonstrated to bring about changes in a number of anatomical measures (cortical thickness, glial/neuronal ratio, size of perikaryon and of neuronal nucleus), in neurochemical measures (activities of the enzymes AChE and ChE, and RNA/DNA ratio), and performance in several behavioral tests (Lashley III maze, visual reversal discrimination, Hebb-Williams test). Control experiments demonstrate that these rather gentle and natural treatments do not produce stress (Riege and Morimoto, 1970; Rosenzweig and Bennett, 1969). The age period (30 days of exposure beginning at weaning at about 25 days of age) has produced clear effects with most of the measures listed above. The exact age is not critical, however, since rather similar results are obtained whether exposure is from 60 to 90 days of age or from 25 to 55 days as in the present experiment. Rats were used since most of the previous work has been done with them, but we have done similar experiments with mice and gerbils; all three species show rather similar cerebral effects of differential experience on the measures studied previously (Rosenzweig and Bennett, 1969), so presumably the synaptic changes will occur in other species than the rat.

## Cortical Region Studied

The synaptic measures were taken in the occipital cortex, the region that has yielded larger effects on most measures than have other cortical regions (Rosenzweig *et al.*, 1969). While this area includes the visual projection region, vision is not necessary to produce the previous effects; they have been found in experiments conducted in the dark or with blinded rats (Rosenzweig *et al.*, 1969). This region is known to receive inputs from sensory modalities other than vision, but it is not yet known why it seems particularly sensitive to effects of differential experience.

Measurements have been concentrated on layer III of the cortex in this experiment, although they will later be extended to other layers. Layer I was avoided because it does not change in depth as a consequence of EC–IC experience (Diamond *et al.*, 1964; Walsh *et al.*, 1969). It has been suggested that plastic changes might be found more readily in the supragranular layers II and III than in lower layers because the supragranular layers have evolved more recently and are the last to develop embryologically (Altman, 1966). The reasons for measuring asymmetric synapses within the neuropil were given earlier in methods. Within the neuropil, the large majority of the synapses measured had their postsynaptic membranes on dendritic spines and the rest, on small dendritic branches.

Further work will be necessary to test whether the observed changes in synaptic size and number also occur in other types of synapses, in other layers of the cortex, and in other cortical regions. In advance of obtaining such information, we may nevertheless consider briefly the possible general functional significance of changes in synaptic size and number.

## Concepts about Cerebral Changes in Learning and Memory Storage

There has long been speculation that long-term memory storage might involve changes of synaptic size and number. Almost as soon as the neuron doctrine has been enunciated, Tanzi (1893) speculated that learning might involve the growth of new neuronal terminals, and Ramón y Cajal (1904) supported this hypothesis. Many subsequent investigators (e.g., Hebb, 1949) have accepted the likelihood that learning occurs through formation of new synaptic connections. The opposite possibility, that learning involves selective elimination of neurons and/or synaptic connections, has received less attention, although Ramón y Cajal felt that in embryological development there is a selection among somewhat random connections, '... due to atrophy of certain collaterals and the progressive disappearance of disconnected or useless neurons' (Ramón y Cajal, 1929). This has been supported by later work (Jacobson, 1970). Hebb (1949) and others have supposed that little used connections may atrophy even in the adult. Recently Dawkins (1971) has suggested that the relatively high continual rate of deaths of neurons does not occur at random but may be a mechanism of memory storage. Other workers (such as Eccles, 1965) have supposed that learning and memory involve 'growth just of bigger and better synapses that are already there, not growth of new connections.' It should also be mentioned that some workers (such as Sperry, 1965) doubt whether learning and memory involve any morphological changes; perhaps there are 'only changes in physiological resistance and conductance ... or various endogenous properties of neurons and glia.'

## Problems of Two-dimensional Sampling

Certain problems and limitations of our two-dimensional sampling of the very complex three-dimensional neuropil demand at least brief discussion.

The sampling of small unit areas of neuropil does not provide any indication of the position of a synapse with regard to the postsynaptic cell, yet this position is of major importance for the relative effectiveness of a synapse in determining the activity of the postsynaptic cell. That is, the closer to the cell body and to the axon hillock the synapse is, the larger the role it will play in determining the excitation or inhibition of impulses. An estimation of the position of the synapses would require serial sections and reconstructions, an effort far beyond the scope of the present study. We can only hope that random sampling has ensured that material from the EC and IC groups is comparable in this respect.

It should be noted that our sampling is not strictly two-dimensional since the sections are cut about 70 nm thick, and this is almost as great as the lengths of the smallest synapses that we find (about 100 nm). As a simplifying assumption, we can think of the synapse as a circular plate (see Figure 8). It might well be asked why we do not find synaptic cross-sections that appear still smaller

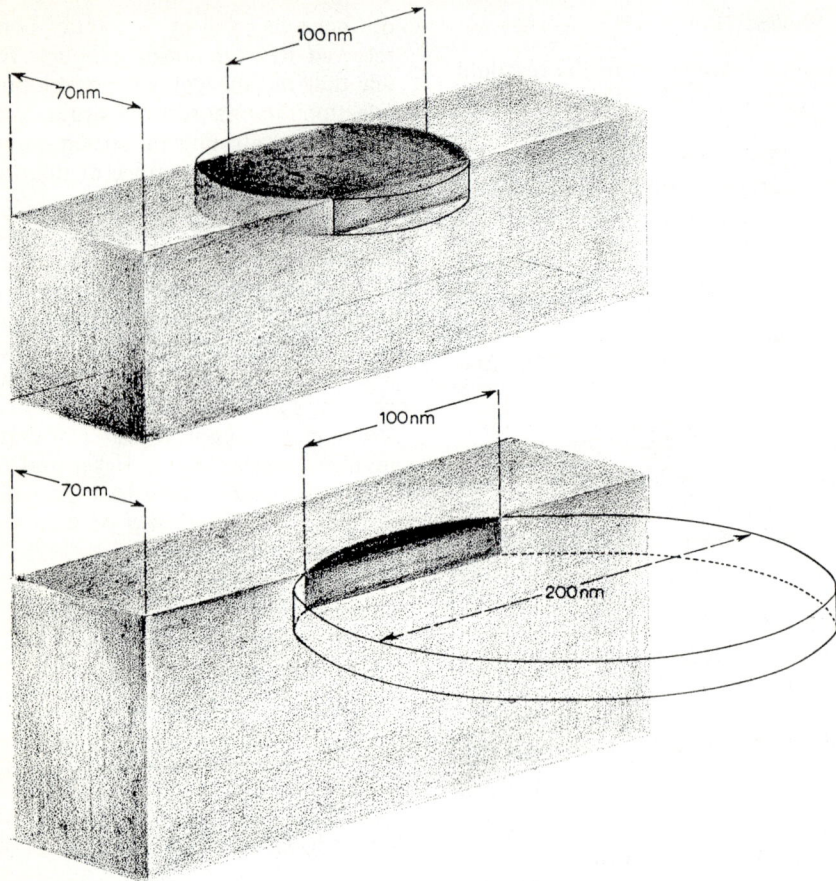

FIGURE 8    Diagram illustrating sections (the vertical gray blocks) through small and large synapses (represented by the horizontal discs). See discussion in text.

than 100 nm, since some tissue sections probably cut only the periphery of a synaptic plate (as in the lower portion of Figure 8). The thickness of the section provides an answer, since if the fraction of the synapse does not occupy a good part of the thickness of the section, then it will probably be superimposed on other structural material which will obscure it. For this reason, a small part of the rim of a large synapse will not be measurable and will not be confused with a small synapse. The sampling underestimates the size of large but not of small synapses, since any fraction of a large synapse that occupies the width of a section is measurable, whereas a small synapse can be measured only if all or most of its diameter occupies the section.

## Meaning of Synaptic Size

Our data demonstrate the presence of bigger synaptic junctions in EC than in IC rats. Are bigger junctions also better in terms of effectiveness? The intensity of postsynaptic current increases with the local concentration of transmitter and of available receptor molecules (Katz, 1966). Let us assume that the amount of transmitter chemical-receptor during transmission is proportional to the area of membrane in which the receptors are localized. In that case an increase in the area of the synaptic membrane would increase the amount of transmitter-receptor complex formed, thus increasing peak postsynaptic current and thereby increasing the effectiveness of the synapse in determining the activity of the postsynaptic cell.

## Minimal Size of Synapses?

Our observations indicate that synapses in the region measured do not occur smaller than about 100 nm, although somewhat smaller ones could have been measured in the sections obtained. This led us to ask whether there may in fact be a minimal possible synaptic size. Let us assume that synaptic vesicles must contact the presynaptic membrane for release of their contents by exocytosis. (This is a prevailing although not necessary assumption; see Robertson, 1970.) It would seem unlikely that any synapse would be so small as to allow space for only a single vesicle to contact the membrane at a given time. (The vesicles in our preparation are about 30 to 40 nm in diameter, which is about the size of ACh vesicles measured by others.) A synaptic plate with room for a central vesicle and a surrounding ring of six vesicles might well be the next larger size of circular synapse that would be functionally efficient. This would be about 100 nm in diameter, and this is in fact the diameter of the smallest junctions that we have found.

*Meaning of Synaptic Number*

Concerning synaptic number, this study coupled with another from our laboratory suggests that both increase and decrease of synapses may occur in response to differential experience. The data of the present experiment show the EC rats to have significantly fewer asymmetrical synapses than IC rats in the neuropil of layer III of the occipital region. On the other hand, we have data from a study conducted in collaboration with Dr. Albert Globus (in preparation) that demonstrate a positive EC–IC difference in number of dendritic spines (and thus presumably of synaptic contacts) on pyramidal cells in the occipital cortex of rats. Globus made counts of spines on EC–IC littermate pairs of rats from four experiments with similar behavioral conditions to those of the present experiment. On basal dendrites, there was an 8.6% EC–IC difference ($P < .01$) in the number of spines per unit of length along the dendrite. On the other dendrites (apical, oblique and terminal branches), there was no EC–IC difference per unit of length. The counts of spines by light microscopy on cells stained by the Golgi-Cox method cannot be assigned to cortical layers for direct comparison with the present electron microscopic results. The synapses measured were largely on dendritic spines, as noted above. It is possible that EM counts in other cortical layers will reveal an increase in the number of junctions, or the relation between EM synaptic counts and Golgi-Cox spine counts may be more complex; only further research can solve this question.

The unexpected finding that enriched experience leads to a diminution in numbers of axodendritic contacts can be related to changes in ontogeny. Mugnaini (1970) and others have noted that 'many of the synaptic connections seen at early stages of histogenesis are lost during ontogeny.' Our results indicate that this can occur also in the post-weaning period. Of course, the difference between EC and IC animals does not show clearly whether EC rats lose contacts or IC rats gain them. We will want, as in others of our experiments, to study also rats kept in the standard colony (SC) conditions as a baseline group. A developmental study of rats in EC, SC and IC will be required to show the changes in size, form and number of various sorts of synapses as consequences of differential experience.

*Combined Alterations in Synaptic Size and Number*

A few other workers have also reported changes in both size and number of synaptic junctions as a function of differential experience. Since quantitative electron microscopy of synapses has only begun, since some reports are only preliminary, since small numbers of subjects are typically used, and since experimental conditions have differed, it is not surprising that results reported to date do not show much consistency.

Cragg (1967, 1968, 1969) has made several reports on measures of axon terminals in rats kept permanently in the dark or removed for various periods of light exposure. In 1967 Cragg reported that when dark-reared rats were exposed to daylight, there were opposite changes in the upper and lower halves of the occipital cortex. In the upper half the terminals became larger and less numerous, whereas in the lower half they became smaller and more numerous. His latest report deals with axon terminals in the geniculate nucleus. These are 15% greater in diameter in dark-raised rats, but 34% less in density per unit of tissue volume. When length of synaptic contacts were measured, these 'did not show any regular differences between the light- and dark-exposed rats of successive pairs' (p. 61). Fifková (1970) measured axodendritic synapses in layers II–IV of occipital cortex of rats after 6 weeks of unilateral lid suture. She reported that synapses in the hemisphere contralateral to the sutured eye were 20% fewer in number but 7.5% larger in cross-section than in the control hemisphere. Foote, Aghajanian and Bloom (1970) counted synapses in layer I in rats raised for the first 6 weeks in the dark or under a normal light cycle. The most reproducible difference was that the dark-raised showed 30–40% more junctions than did the controls (Akert *et al.*, 1970). While these studies of visual stimulation show a variety of results, they do indicate an inverse relation between changes in size and number of synapses (or terminals), which is what the present data also show.

The difference of effects between upper and lower halves of cortex reported by Cragg emphasizes the necessity of relating results to depth within the cortex. In this regard, the altered thickness of cortex as a result of EC–IC treatment might be a source of error in positioning the region sampled. That such error did not occur is demonstrated by the fact that even the three pairs that showed greater depth of cortex in IC than in EC (see Table 3) nevertheless yielded typical results in both synaptic length and number (see Tables 1 and 2).

Our data show clearly greater synaptic size and lesser synaptic number in EC as compared to IC rats. These measures come only from layer III and we realize that investigation of other layers may show different results, as Cragg has observed between the upper and lower halves of cortex. On the basis of present results we would like to suggest the hypothesis that the IC brain represents a more immature condition with many small synapses that are not committed to functional circuits. Learning and experience in EC could modify this 'pluripotency,' increasing the effectiveness of functional pathways and suppressing connections that yield undesirable results.

*Can Differences in Synaptic Area Be Related to Differences in Cortical AChE?*

It is tempting to try to relate the EC–IC differences in synaptic area in this experiment to EC–IC differences in cortical acetylcholinesterase (AChE) activity that we have found in other studies (Bennett *et al.*, 1964; Rosenzweig *et al.*, 1971), but a good many problems and unknowns lie in the way of doing so. Favoring the attempt is the fact that AChE is found especially richly at synaptic junctions. But the 40% greater synaptic area observed in EC versus IC in the present experiment would not be expected to lead to a comparable difference in total AChE activity for the following reasons:

a) AChE is found all along the neuronal membrane, although not as richly as at synapse (De Robertis, 1971). In order to estimate the increase in

AChE with increase in synaptic area, one would need to know both the relative concentrations of AChE at junctions and along non-specified membrane and also the relative areas of all cholinergic synapses and all other membrane. These values are not known, but calculations with various plausible values indicate that EC might be expected to have at most only a few percent more total AChE activity than IC, within layer III of the occipital cortex.

b) The chemical analyses have been done on samples that included all of the cortex, so that effects in all layers are averaged together. Until synaptic changes are studied in all of the layers of the cortex, it will be premature to try to relate differences in synaptic area to differences in AChE, although we hope eventually to be able to do so.

## SUMMARY

Twelve littermate pairs of male $S_1$ rats were placed in an enriched environmental condition (EC) or an impoverished condition (IC) at 25 days of age and were kept there for 30 days. Measurements of asymmetrical axodendritic synapses in layer III of occipital cortex showed 52% greater mean length in EC than in IC. Number of synapses per unit area of neuropil was 35% less in EC than in IC. Both differences were highly significant statistically. The EC distribution of synapse lengths cannot be derived from IC simply by elimination of small contacts, since EC has significantly more large junctions than IC. The total synaptic area in the region measured was calculated to be about 40% greater in EC than in IC. Thickness of cortex was 4.0% greater in EC than IC, but typical synaptic effects were found even in the three pairs in which IC cortex was thicker than EC. Problems of measurement and sampling in electron microscopy were considered. The results were discussed in relation to concepts of brain mechanisms of learning and memory storage.

ACKNOWLEDGMENTS

This research was supported by National Science Foundation Grant GB-8011 and by the U.S. Atomic Energy Commission. Dr. Møllgaard received a research stipend from the University of Copenhagen and during this research was visiting lecturer in the Department of Physiology-Anatomy, University of California, Berkeley.

Consultation with Professor Peter Satir during the course of this research is deeply appreciated.

# REFERENCES

Akert, K., Gray, E. G., and Bloom, F. E., 1970, Structure of specialized junctions, *Neuroscience Research Program Bulletin* **8**: 336–360.

Altman, J., 1966, *Organic foundations of animal behavior*. Holt, Rinehart & Winston, New York.

Bennett, E. L., Diamond, M. C., Krech, D., and Rosenzweig, M. R., 1964, Chemical and anatomical plasticity of brain, *Science* **146**: 610–619.

Bennett, E. L., Rosenzweig, M. R., and Diamond, M. C., 1970, Time courses of effects of differential experience on brain measures and behavior of rats. In: *Molecular approaches to learning and memory*, Byrne, W. L., ed. Academic Press, New York, pp. 55–89.

Bloom, F. E., and Iversen, L. L., 1970, Macromolecules in synaptic function, *Neurosciences Research Program Bulletin* **8**: 325.

Colonnier, M., 1968, Synaptic patterns on different cell types in the different laminae of the cat visual cortex, An electron microscope study, *Brain Research* **9**: 268–287.

Cragg, B. G., 1967, Changes in visual cortex on first exposure of rats to light: Effect on synaptic dimensions, *Nature* **215**: 251–253.

Cragg, B. G., 1968, Are there structural alterations in synapses related to functioning?, *Proceedings of the Royal Society, Series B* **171**: 319–323.

Cragg, B. G., 1969, The effects of vision and dark-rearing on the size and density of synapses in the lateral geniculate nucleus measured by electron microscopy, *Brain Research* **13**: 53–67.

Dawkins, R., 1971, Selective neurone death as a possible memory mechanism, *Nature* **229**: 118–119.

De Robertis, E., 1971, Molecular biology of synaptic receptors, *Science* **171**: 963–971.

Diamond, M. C., 1967, Extensive cortical depth measurements and neuron size increases in the cortex of environmentally enriched rats, *Journal of Comparative Neurology* **131**: 357–364.

Diamond, M. C., Krech, D., and Rosenzweig, M. R., 1964, The effects of an enriched environment on the histology of the rat cerebral cortex, *Journal of Comparative Neurology* **123**: 111–119.

Diamond, M. C., Law, F., Rhodes, H., Lindner, B., Rosenzweig, M. R., Krech, D., and Bennett, E. L., 1966, Increases in cortical depth and glia numbers in rats subjected to enriched environment, *Journal of Comparative Neurology* **128**: 117–125.

Eccles, J. C., 1965, Possible ways in which synaptic mechanisms participate in learning, remembering and forgetting. In: *The anatomy of memory*, Kimble, D. P., ed. Science and Behavior Books, Inc., Palo Alto, California, p. 97.

Fifková, E., 1970, The effect of monocular deprivation of the synaptic contacts of the visual cortex, *Journal of Neurobiology* **1**: 285–295.

Foote, W. E., Aghajanian, R. J., and Bloom, F. E., unpublished. Cited by Bloom, F. E., 1970, Correlating structure and function of synaptic ultrastructure. In: *The Neurosciences, Second study program*, Schmitt, F. O., ed. pp. 729–747, Rockefeller University Press, New York.

Gray, E. G., 1959, Axo-somatic and axo-dendritic synapses of the cerebral cortex: An electron microscope study, *Journal of Anatomy* **93**: 420–432.

Hebb, D. O., 1949, *The organization of behavior*. Wiley, New York.

Jacobson, M., 1970, Development, specification, and diversification of neuronal connections. In: *The neurosciences, Second study program*, Schmitt, F. O., ed. pp. 116–129. Rockefeller University Press, New York.

Kandel, E. R., and Spence, W. A., 1968, Cellular neurophysiological approaches in the study of learning, *Physiological Review* **48**: 65–134.

Katz, B., 1966, *Nerve, muscle, and synapse*. McGraw-Hill, New York.

Mugnaini, E., 1970, The relation between cytogenesis and the formation of different types of synaptic contact, *Brain Research* **17**: 169–179.

Peters, A., and Kaiserman-Abramof, I. R., 1969, The small pyramidal neuron of the rat cerebral cortex, The synapses upon dendritic spines, *Zeitschrift für Zellforsch.* **100**: 487–506.

Ramon y Cajal, S., 1904, *Textura del sistema nervioso del hombre y de los vertebrados*. Nicolás Moya, Madrid, Spain.

Ramón y Cajal, S., 1929, *Studies on vertebrate neurogenesis*. C. C Thomas, Springfield, Illinois.

Reynolds, E. S., 1963, The use of lead citrate at high *pH* as an electron-opaque stain in electron microscopy, *Journal of Cell Biology* **17**: 208–212.

Riege, W. H., and Morimoto, H., 1970, Effects of chronic stress and differential environments upon brain weights and biogenic amine levels in rats, *Journal of Comparative and Physiological Psychology* **71**: 396–404.

Robertson, J. D., 1970, The ultrastructure of synapses. In: *The neurosciences, Second study program*, Schmitt, F. O., ed., pp. 715–728. Rockefeller University Press, New York.

Rosenzweig, M. R., 1971, Effects of environment on development of brain and of behavior. In: *Biopsychology of development*, Tobach, E., ed., pp. 303–342. Academic Press, New York.

Rosenzweig, M. R., and Bennett, E. L., 1969, Effects of differential environments on brain weights and enzyme activities in gerbils, rats, and mice, *Developmental Psychobiology* **2**: 87–95.

Rosenzweig, M. R., Bennett, E. L., and Diamond, M. C., 1971, Chemical and anatomical plasticity of brain: Replications and extensions, 1970. In: *Macromolecules and behavior*, 2nd edition, Gaito, J., ed. Appleton-Century-Crofts, New York.

Rosenzweig, M. R., Bennett, E. L., Diamond, M. C., Wu, S.-Y., Slagle, R. W., and Saffran, E., 1969, Influences of environmental complexity and visual stimulation on development of occipital cortex in rat, *Brain Research* **14**: 427–445.

Sperry, R. W., 1965, Embryogenesis of behavioral nerve nets. In: *Organogenesis*, DeHaan, R. L., and Ursprung, H., eds., pp. 161–186. Holt, Rinehart & Winston, New York.

Tanzi, E., 1893, I fatti e le induzioni nell'odierna istologia del sistema nervoso, *Riv. Sperim. de Fren. e de Med. Leg. XIX:* 419–472.

Walsh, R. N., Budtz-Olsen, O. E., Penny, J. E., and Cummins, R. A., 1969, The effects of environmental complexity on the histology of the rat hippocampus, *Journal of Comparative Neurology* **137**: 261–266.

# AN INTERDISCIPLINARY ANALYSIS OF AGGRESSION

## ROBERT M. DORMAN

*Philadelphia, Pa., U.S.A.*

Plato was one of the most influential ancients to recognize the creative power of aggression. *The Dialogues*, through the dynamic character of Socrates, provide direct testimony for the creativeness of aggression in the individual. In the final lines of the *Euthyphro*, following Socrates' total destruction of Euthyphro's "knowledge" of divinity, Socrates' invective is:

> What are you doing my friend? Will you leave, and dash me down from the mightly expectation I had of learning from you what is holy and what is not, and so escaping from Meletus' indictment? I counted upon showing him that now I had gained wisdom about things divine from Euthyphro, and no longer out of ignorance made rash assertions and forged innovations with regard to them, but would lead a better life in the future. (#1, P.185)†

From the instructiveness of sarcasm to teaching through bitter anger, Socrates leads us through his defense in the Apology:

> It is these people, gentlemen, the disseminators of these rumors, who are my dangerous accusers . . . And what is more, they approached you at the most impressionable age, when some of you were children or adolescents, and they literally won their case by default, because there was no one there to defend me. And the most fantastic thing of all is that it is impossible for me even to know and tell you their names . . . All these people, who have tried to set you against me out of envy and love of slander—and some too merely passing on what they have been told by others—all these are very difficult to deal with. (#2, P.5).

Twenty-five hundred years ago Western Philosophy taught us that the development of knowledge and the dissolution of enslaving beliefs is by no means a non-passionate process, but is rather a personal, courageous and aggressive one. We have forgotten some of this teaching because human aggression contains such a long history of barbarism that we have generally given it a destructive meaning. It is to exactly this paradox of aggression—used constructively in searching philosphical discourse, and destructively as in genocide—that this paper is addressed.

In the hundred years following Darwin, many

† Numbered references immediately following text.

schools of thought concerned with the genesis of human aggression have emerged. Presently, the most heated controversy seems to rage between "Environmentalists" and "Instinct Theorists." The former tend to explain almost all human adaptive behavior in terms of cultural conditioning, while the latter emphasize man's genetic endowment. An introduction into these divergent points of view might juxtapose Nietzsche (who was heavily influenced by Darwin, #3) with Cultural Relativist Ruth Benedict in order to clarify the evolution of current arguments.

Nietzsche saw man as an highly aggressive animal who could only flourish by engaging in the violent behavior inherited from his animal predecessors. Believing that humans were evolved from predators, Nietzsche emphasized aggression as vital to a healthy psyche.

> We thoroughly misunderstand the beast of prey and the man of prey (Cesare Borgia, for example); we thoroughly misunderstand "nature" as long as we seek a "diseased condition" at the bottom of these healthiest of all tropical monsters and growths. Or, even worse, as long as we seek an inborn "hell" in them, as almost all moralists have heretofore done. (#4, P.105).

As a scholar of Greek philosophy, Nietzsche adopted Socrates' aggressive method of delivery, believing that instinctive behavior was an integral part of philosophy. About the thought process itself he says:

> After keeping an eye on and reading the lines of the philosophers for a long time, I find that I must tell myself the following: The largest part of conscious thinking must be considered instinctual activity, even in the case of philosophical thinking. (#4, P.3).

Discussing the creative uses of aggression he urges:

> The ability to endure and sustain, enduring gratitude and enduring vengefulness—both only toward one's equals; subtlety in requital and retaliation; a subtly refined concept of friendship; a certain need to have enemies (as outlets for the passions: envy, quarrelsomeness and wantonness—basically, in order to be capable of being a good friend): . . . (#4, P.205).

Nietzsche held that the natural outlet for man's

aggression is other men, and that philosophies negating this principle must lead to internalized aggression and the subsequent destruction of the individual and society.

> To refrain from wounding, violating, and exploiting one another, to acknowledge another's will as equal to one's own: this can become proper behavior, in a certain coarse sense, between individuals when the conditions for making it possible obtain (namely the factual similarity of the individuals as to power and standard of value, and their co-existence in one great body). But as soon as one wants to extend this principle, to make it the *basic principle of society*, it shows itself for what it is; the will to negate life, the principle of dissolution and decay. (#4, P.201).

We see then that Nietzsche was arguing for the intelligent and subtle control of a vital biological force, needing reason to channel it toward the constructive ends of societies and individuals. According to Nietzsche, if a society does not allow or teach the constructive use of aggression, it will be internalized by individuals toward self-destructive ends, or externalized destructively as in Fascism (which Nietzsche loathed (#3)). It would be of interest now to follow this line of thinking into Ruth Benedict's work, as she claimed to be influenced by Nietzsche.

In *Patterns of Culture* (#5) Benedict's basic assumption that human behavior can be explained almost entirely in terms of specific cultural attitudes and *not* instincts, is most interestingly derived from her affinity to Nietzsche. Her categorization of American Indian cultures in the chapter on the Pueblos of New Mexico came from Nietzsche's study of Greek tragedy.

> The basic contrast between the Pueblos and the other cultures of North America is the contrast that is named and described by Nietzsche in his studies of Greek tragedy. He discusses two diametrically opposed ways of arriving at the values of existence. The Dionysian pursues them through the annihilation of the ordinary bounds and limits of existence. The desire of the Dionysian, in personal experience or in ritual, is to press through it toward a certain psychological state, to achieve excess ... The Apollonian distrusts all this, and has often little idea of the nature of such experiences. He finds ways to outlaw them from his conscious life ... In Nietzsche's fine phrase, even in the exaltation of the dance he 'remains what he is, and retains his civic name!' (#5, P.78-79).

It must be noted, however, that Benedict's bond with Nietzsche breaks completely, in that Nietzsche included development of both outlooks on life within the same individual in order to experience life fully, while Benedict saw them as representing opposing cultures, with a bias in favor of the Apollonian. While Nietzsche also favored the Apollonian concern for moderation, he believed Dionysian passion was essential for growth and development—denial of its expression leading to internalized aggression and destruction of the individual. Conversely, Benedict neither believed that both forms of behavior were necessary within the same individual, nor that Dionysian (aggressive and violent) experience was beneficial to the individual or society. In effect, Benedict maintained that the need for and performance of aggression (seen as destructive) is a function of whether or not it is taught by the specific culture in which one lives.

In order to contrast the destructive and violent Dionysian way of life with the non-violent moderation of the Apollonian, Benedict writes in great detail about the Pueblo people, the Zuni in particular, whose culture "has not disintegrated like that of all the Indian communities outside of Arizona and New Mexico." (#5, P.57).

> Just as according to the Zuni ideal a man sinks his activities in those of the group and claims no personal authority, *so also he is never violent.* (emphasis mine) Their Apollonian commitment to the mean in the Greek sense is never clearer than in their cultural handling of emotions. Whether it is anger or love or jealousy or grief, moderation is the first virtue. The fundamental tabu upon their holy men during their brief periods of office is against any suspicion of anger. Controversies, whether they are ceremonial or economic or domestic, are carried out with an unparalleled lack of vehemence. (#5, P.106).

So then, the Zuni ideal teaches moderation and complete non-violence, unlike that of the western plains Indians, whose Dionysian ideal often sent them out in search of violent experience:

> The most conspicuous of these is probably their practice of obtaining supernatural power in a dream or vision ... On the western plains they sought these visions with hideous tortures. (#5, P.81).

But is the picture of non-violence that Benedict paints of the Pueblos consistent with the evidence that she cites?

In introducing us to the ancestral origins of the Zuni Indians, the author states that:

> An early people, the Basket makers, had lived there so long before that we cannot calculate the period of their occupancy, and they were supplanted, and perhaps largely exterminated, by the early Pueblo people. (#5, P.57-58).

It would not seem that possible ancestral occupancy through extermination would particularly suggest the development of a non-violent culture, however, evidence concerning contemporary Pueblo ideology and behavior is given and is more telling.

Explaining the great importance and time bound up in ritual, the author states:

The Zuni are a ceremonious people, a people who value sobriety and inoffensiveness above all other virtues. Their interest is centered around their rich and complex ceremonial life ... No field of activity competes with ritual for foremost place in their attention. Probably most grown men among the western Pueblos give to it the greater part of their waking life. (#5, P.59-60).

With this all-important emphasis on religious ceremony in mind, the following passage is most interesting:

One of the obligations that rest on every priest or official during the time when he is participating in religious observances is that of feeling no anger. But anger is not tabu in order to facilitate communication with a righteous god who can only be approached by those with a clean heart. (#5, P.61).

The author tells us that priests or officials (all of the males at frequent intervals) must not feel anger, yet anger is quite acceptable to facilitate spiritual communication during ceremony. Despite what the Zuni may think, it would seem that they have not rejected anger, but ritualized it—ritual of course being their major field of endeavor.

In so far as the Zuni are portrayed by the author as rejecting torture ("Torture was even more consistently rejected"), (#5, P.90) it is difficult to understand what she considers the following evidence to indicate.

The Pueblos do not understand self torture. Every man's hand has its five fingers, *and unless they have been tortured to secure a sorcery confession* they are unscarred (emphasis mine) (#5, P.91).

Witchcraft among the Pueblos, like so many of their situations, is an anxiety complex. They vaguely suspect one another; and if a man is sufficiently disliked, witchcraft is sure to be attributed to him (#5, P.122).

Likewise, self-torture seems indicated by the following:

The priests, again, in their retreat before their altars sit motionless and withdrawn for eight days, summoning the rain (#5, P.64).

In the Cactus Society, a warrier cult, they dashed about striking themselves and each other with cactus-blade whips ... (#5, P.91).

And with respect to childhood initiation:

The Pueblo practice of beating with stripes is likewise without intent to torture. The lash does not even draw blood. Far from glorying in any such excess, as the Plains Indians do, a Zuni child whipped at adolescence or earlier, at the tribal initiation, may cry out and even call for his mother when he is struck by the initiating masked gods.

The adults repudiate with distress the idea that the whips might raise welts ... The fact that it is the same act that is used elsewhere for self-torture has no bearing upon the use that is made of it in this culture. (#5, P.91).

If these passages do not clearly indicate ritualized torture and violence, then the author has dramatically failed to offer an acceptable alternative explanation. If the Zuni reject the idea of torture, then why do they engage in such practices as often they do, or even at all? Neither the probability that they have ritualized almost all of it, nor their *belief* that they do not *intend* such extremes, lessens the fact that they do in fact torture with apparent frequency. If it is argued that the Zuni do not torture because they don't *believe or say they intend torture*, rather only the observance of religious rites, then the author is employing an often used but frequently irrelevant standard, for what a given people *believe or say* they are doing quite often bears no resemblance to what they *are* doing.

Benedict seems to be sharing "logic" with the Inquisitors, who also did not believe in torture, but rather the teachings of love, brotherhood and nonviolence. As a mild example of what has been used to convert people to these doctrines, Daniel Mannix recounts:

If the garotte failed, the Inquisitors then had recourse to the strappado. The victim's hands were tied behind him and he was then hoisted up by his wrists to a pulley on the ceiling. For maximum effect, the prisoner was raised slowly and allowed to stand tiptoe for the length of time it required to say a psalm Miserere three times slowly ... If this failed, he was raised almost to the ceiling and suddenly allowed to drop but brought up with a sharp jerk that dislocated his arms at the shoulders. (#6, P.46).

This wasn't torture either, simply an example of man's attempt to bring "misguided" people back to the paths of "righteousness". Ruth Benedict and the Cultural Relativists are forced to accept this reasoning if their major explanations of human behavior are *based* upon examining the *ideologies* within various cultures.

Benedict's unfortunate emphasis upon ideological interpretation of Zuni life did not allow her to place enough emphasis on the fact that the Zuni pay a terrific price for any "moderation" that may exist between individuals in daily encounters. In explaining the Zuni value of "sobriety and inoffensiveness", Ester Goldfrank points out the extreme use of fear in the society's control of individuals. Emphasizing the "omnipresence of ideas of witchcraft" affecting the growing child, Goldfrank writes:

As he grows older he learns that even those nearest him, even his parents may be witches, and that he himself may be possessed without becoming aware of it. He learns that, once possessed, he cannot lay the evil power aside, nor can he expect effective help from relatives or friends. Only the medicine-men may be able to exorcise the "bad spirit". At times—and most frequently these were times of drought or disaster—not even they would come to his aid. The accused was then tried and tortured before a body of his peers and his priests and, in most cases, executed by the War chief and his assistants, since the need for trial was practically cause for conviction. (#7, P.523).

It is perhaps ironic that Benedict's example of a society in which non-violence and cooperation are "valued", is in fact one that has apparently no understanding that it teaches destructive aggression, leading, as Nietzsche suggested, to destructive aggression internalized through fear and externalized in irrational persecution.

There are three essential points upon which this discussion has converged. First, the author has failed to provide evidence that her example of a non-violent society is anything but violent, and that her dichotomization of model Indian cultures, portrayed as violence or non-violence oriented, is fundamental. Second, despite the fact that a culture may teach the belief that one should be or is moderate and non-violent, it can easily include the teaching of behavior that is excessive and violent, under the guise of ritualization. And last, if the Relativist position is adopted, no truly functional meaning of aggressive behavior can be developed, in that no standard is provided against which even slightly divergent forms of behavior can be measured. For this reason in particular, it is imperative that we go on to look for such a standard.

Less than twenty years ago, two behavioral scientists, Niko Tinbergen and Konrad Lorenz, founded the science of ethology, in order to seek objective standards for the evaluation of normal animal behavior. Their assumption is that behavior, as an aspect of physiology, is subject to laws of physiological evolution, and that by studying the normal behavior of species on all levels of evolutionary complexity, trends in physiology (structural and functional interaction) bearing upon human development might emerge. (#8). Since adaptive mutation has usually increased physiological complexity in the development of new species, logic would suggest that information concerning the *value of behavior patterns in terms of natural selection* would be most significant. There are several principles which are fundamental to understanding this approach.

To begin, behavior must be understood as physiological function. It must be realized that an organism's gross responses (behavior in external environment) are an aspect of a continuum of stimulations and responses occurring within the organism. As John Dewey argued:

> The older dualism between sensation and idea is repeated in the current dualism of peripheral and central structures and functions; the older dualism of body and soul finds a distant echo in the current dualism of stimulus and response. Instead of interpreting the character of sensation, idea and action from their place and function in the sensorimotor circuit, we still incline to interpret the latter from our preconceived and preformulated ideas of rigid distinctions between sensations, thoughts and acts ... More specifically what is wanted is that sensory stimulus, central connections and motor responses shall be viewed, not as separate and complete entities in themselves, but as divisions of labor, functioning factors, within the single concrete whole ... (#9, P.253).

If we fail to realize that organismic behavior always includes the integration of physiological activity, we engage in misleading dualisms. Bodily structures become separated from their functions and physiological functions are walled off from overt behavior. For example, if thinking is not considered behavior, cognitive function must be separate from the integration of cerebral function, which is ultimately bound up with cerebral structure. To accept that thinking is not behavior and a physiological event, forces the unfortunate conclusion that thinking must be a non-physiologic process occurring in an organism, which *is* a physiological unit.

Implicit in this dualism is the belief that learning from environment means only learning from *external* environment. Since learning of any kind must be accompanied by changes in the nervous system (aspect of internal environment) which cannot occur without information from other somatic sources within the organism (muscular, cardiac and endocrine changes, etc.) learning must also be in part derived from *internal* environment.

> There are many kinds of learning, ranging from the simplest modifications of innate behavior to the most complex, symbolic transactions seen in the reasoning of man. All, however, are characterized by an enduring change in the behavior of the organism, perhaps a permanent change. From a biological point of view, the change in behavior must be a change in the functioning of the nervous system, and, if it is a permanent change, perhaps it is also a change in the structure of the nervous system. (#10, P.80).

Disallowing learning from internal environment separates physiological modification of the nervous system from the rest of somatic physiology (interacting with external environment) which in effect makes learning from any source impossible. If learning does not equally include nervous system modification through information from internal environment, as well as information from external environment, learning cannot be behavior.

Relying too heavily on abstract dualisms allows some theorists to argue that animal behavior is not applicable to human behavior in that humans "learn" solely from "environment", while "lower" animals with only limited learning capacities behave mainly on the level of "programmed" behavior.

> The notable thing about *human* behavior is that it is learned. Everything a human being does as such he has had to learn from other human beings. From any dominance of biologically or inherited predetermined reactions that may prevail in the behavior of other animals, man has moved into a zone of adaptation in which his behavior is dominated by learned responses. It is within the dimension of culture, the learned, the man-made part of the environment that he grows, develops, and has his being as a behaving organism. Whatever other recondite elements may be involved in his behavior, and whatever the limits that his genetic constitution may set upon his learning capacities, this is the conclusion of behavioral sciences— the sciences concerned with the study of the origins of man's behavior. (#11, P.12).

It is precisely this fundamental misunderstanding of behavior, learning, and environment, as well as the misconception of the evolutionary process to which it leads, that the ethologists are trying to avoid. Unless behavior, learning and environment are seen as aspects of the same continuum, the evolution of anatomical structure becomes separated from physiology, which is both logically and scientifically incomprehensible.

With this explanation of physiological activity in mind, we can proceed into the ethological position that while structure and function are biologically inseparable, the development of different and complex patterns of adaptive behavior, produced by environmental advantage, *preceded gross structural changes*. This is not a contradictory point of view if we accept the fact that in any given species there are individuals which will survive more efficiently than others in different environments. Combining the theories of natural selection and Mendellian genetics, we realize that random mutations in physiology are selected in or out by their survival value within a given environment. For this reason major phylogenetic changes in physiology would in fact follow selective advantages of individuals' increased efficiency to survive in an environment. If we can allow that a minor structural change due to random mutation can permit a major functional change in survival advantage of an individual, it can be concluded that major structural change within a species is preceded by major functional ones in individuals. Reversing this argument does not appear to be correct in that major changes in individual structure due to random

mutation are usually lethal. Interdisciplinary support for ethological theory comes from the following:

> Harvard University's Ernst Mayr is biology's authority in the field of systematics—that is the evolutionary classification of species. At a symposium in 1958 he said, "On the whole it seems correct to state, as Lorenz has emphasized, that behavior movements often precede phylogenetically the special structures that make these movements particularly conspicuous"..."A shift into a new niche or adaptive zone is, almost without exception, initiated by a change in behavior. The other adaptations to the new niche, particularly the structural ones, are acquired secondarily." (#12, P.13).

Unless the laws of evolution, which dictate that changes in physiology must have survival advantage do not apply to human animals, the locus of a standard for evaluating human behavior might be rooted in an evaluation of the survival value ascribed to specific developmental trends pervading vertebrate evolution. To believe that human evolution is fundamentally different from the rest of animal evolution is an unfortunate position. In doing so one would have to explain away the entire field of comparative physiology, which can trace the origins of our own basic physiology through the development of species (#13, #14).

The suggestion that it is only our brain that has undergone an essentially different process is just as unlikely. While it is true that each species has developed its own selective physiological advantages, it appears equally true that all of these advantages are based upon survival value and it must be assumed that major changes in the human brain have also been a product of survival value. To accept this argument and dismiss the importance of interspecies (including human) similarities of cerebral structure and function is to disregard entirely too much neurobiological research, which very frequently develops practical theory and technique on animals, before development along similar lines is attempted in humans. If it is argued that the development of the neo-cortex and its functional relationships with phylogenetically older central and peripheral structures makes human behavior unequatable with that of other animals, then we must proceed into whatever scientific evidence is available to support or disprove this possibility.

Compelling neurophysiological theory suggests that development of phylogenetically new cerebral structures has not *necessarily* shifted the control of function so much as integrated new structures with old in order to allow increasingly complex adaptive responses. In a popular work, Paul D. MacLean suggests:

> There are those who argue that one has no right to apply

behavioral observations on animals to human affairs, but it is to be emphasized that man has inherited the basic structure and organization of three brains, the two oldest of which are quite similar to those of animals. For purposes of discussion I refer to the three prototypes as reptillian, old mammalian and new mammalian. Despite their great differences in structure and chemistry all three brains must interconnect and function together. One might imagine that man's brain has evolved somewhat like a house to which wings and superstructure are added. (#15, P.25).

It is implicit in this statement that while each major addition to the evolving brain has been integrated with the older structures, phylogenetically old structures have to a great degree maintained their physiological identities. Evidence for this hypothesis is provided by several sources.

Studying functional anatomical relationships of species on varying levels of evolutionary complexity provides a partial clue as to whether new structures have taken over function from old ones. In discussing the interrelation of the neo-cortex with the limbic cortex and hypothalamus (found even in the fish) vis-à-vis survival functions (gross behavior patterns necessary to individual and/or species survival, eg. eating, sex, etc.) MacLean says that the:

> . . . old limbic cortex is basically primitive compared with the new cortex. From this it might be inferred that it continues to function at an animalistic level in animals as in man. Also, in marked contrast to the new cortex it has strong connections with the hypothalamus, which plays a basic role in integrating the performance of mechanisms involved in self-preservation and the preservation of the species . . .
>
> Clinical and experimental findings suggest that the lower part of the ring (ring of the limbic system) fed by the amygdala (a nucleus within the limbic system) is primarily concerned with emotions and behavior that insure self-preservation. Its circuits are kept busy with the selfish demands of feeding, fighting and self-protection. (#5, P.29).

Further evidence supporting the integrity of function in older cerebral structures comes from surgical removal of newer ones. Bard and Macht present evidence from a decorticate cat preparation:

> The next day when the door of her cage was rattled she began to tremble violently, meowed plaintively, circled the cage frantically and exhibited widely dilated pupils and maximal piloerection over back and tail. This appeared to be a definite and rather full exhibition of fright. (#16, P.65).

Eliciting full affective reponses from decorticate animals would correlate with field data confirming spontaneous aggression in chickens (#17), doves and many other bird species, (#18) since the avian brain has very little cortex at all.

Another experimental technique shedding light on the localization of function is intracerebral electrical stimulation.

> We know from observations of people with brain damage, from electrical stimulation of regions of the brain in patients undergoing brain surgery while conscious, and particularly from the celebrated experiments on cats by Walter R. Hess of Zurich, that in mammals the chief center for regulation of drives lies in the part of the brain stem called the diencephalon. (#19, P.57).

A relatively substantial account of aggressive reactions elicited by diencephalic (phylogenetically old) and limbic system stimulation can be found in work from numerous laboratories (#20, #21, #22). And still other laboratories provide evidence indicating subcortical control of states of consciousness through the technique of pharmacological stimulation (#23, #24, #25, #26). In effect, current experimental evidence seems to indicate that primary modulation of survival functions lies within old instead of new cerebral structures.

It must be realized, however, that while some cerebral structures are considered phylogenetically older than others, this is not to say that they have undergone no evolutionary changes. For example, with the evolution of the cortico-spinal system, main control over motor function shifted from the extra-pyramidal system to the cerebral cortex (#27, chapt. 23). But as considerable evidence does not suggest that the newest areas of the cerebral cortex have taken predominant control over basic survival functions, it is also not clear that the increased complexity of cerebral integration of function has drastically changed numerous survival functions, from the lowest to the highest vertebrates. Evidence in support of this hypothesis will be offered both from laboratory studies and from ethological field studies, but first a few terms must be discussed.

Lorenz, in *Evolution and Modification of Behavior,* published after his *On Aggression,* attempts to clarify the concept of "innate", which is contained in the term "instinctive". While rejecting "that there are, in the machinery of behavior, quite considerable self-contained units into which learning does not enter, and which, in the hierarchical organization of appetitive behavior, are intercalcated with links that are adaptively modifiable by learning", (#28, P.102) he offers the following:

> The chances of random modification being adaptive are not greater than those of mutation being so; the probability is assessed as $10^{/8}$ by geneticists. Any modifiability which regularly proves adaptive, as learning indubitably

does, presupposes a programming based on phylogenetic-ally acquired information (#28, P.104).

T. C. Schneirla in criticizing Lorenz' books tells us that:

> Lorenz' critics dispute neither his emphasis on ". . . the great fact of adaptiveness in behavior . . . ," as he implies, nor on correlations between genotypes (empirically described species genetics) and phenotypes (individual patterns developed). Rather, they reject much of his evidence as partial and unreliable, and question on these and other grounds the assumptions he applies dogmatically to these problems. (#29, P.62).

In effect, Lorenz' critics do not seem to take issue with the assumption that adaptive behavior of individuals within a given species is ultimately tied up in species genetics. They do argue that his extrapolations from this assumption are unjustified on the grounds of inadequate and unreliable evidence at best. It is imperative, then, that we examine some of the most relevant evidence pertaining to the hypothesis that specific survival functions have remained relatively unchanged, despite evolutionary developments in the integration of function, as this hypothesis is central to Lorenz' thinking.

Although it is not disputed that the function of heterosexual mating behavior at *least* performs the survival function of perpetuating species through the genetic advantages of sexual dimorphism, it is important to stress the obvious. The function of heterosexual mating behavior, despite increasing structural and functional complexities, has never left its procreative survival function throughout evolutionary development. With the additional argument that sexual intercourse as an isolated behavioral unit is performed by all species throughout mammalian evolution with far more similarity than dissimilarity, we seem forced to conclude that no matter what the capacities for modification a species' physiology may have, modifiability of certain behavior patterns (empirically defined units of physiological interaction involving internal and external environment) has been specifically and fundamentally limited by their survival function.

If we further define "learning" as an organism's physiological capacity to modify behavior patterns *potentially* advantageous to survival functions (species gross behavioral adaptation due to natural selection), then *increased* capacities for learning have not necessarily changed numerous survival functions, nor led to the *use* of significantly different behavior patterns necessary to survival functions performed by species on many levels of evolution. The implication is that wherever a particular species' behavior patterns involve survival functions discernable in many other

species, the probability of learning basically governing behavior patterns is small. Some other process common to the development of such species must be guiding the appropriate coordination of behavior patterns with specific survival functions. If this were not the case, individual survival, species survival, or behavior identifiable as similar through the development of many species would be highly improbable. More directly, if specific human behavior (e.g. sexual intercourse) can be shown as similar to survival function behavior of other species, it is unlikely that increased human learning capacities are primarily responsible for governing such human behavior; rather, learning probably plays a subservient role. Moreover, it seems probable that such evolutionarily developed mechanisms (other than classical conditioning or trial and error learning) common to many species including man probably lead individuals into behavior patterns appropriate to survival functions. To further defend this hypothesis, information on the evolutionary development of aggression is most appropriate.

Predatory behavior and defense against predation, involving violent physical attack between species, is generally defined as interspecific aggression by ethologists, who use the term intraspecific aggression to describe a violent (physical contact) or threatened (gestured) attack by an animal, usually directed toward a member of its own species. The reasoning which separates interspecific from intraspecific aggression in the ethologist's mind is based upon the different survival functions observed to be related to possible categorically different forms of behavior. Interspecific aggression serves the major survival functions of gathering food to nourish the organism, and individual or group defense against predation. Intraspecific aggression serves the survival functions of territorial defense, establishment of dominance hierarchies, sexual fighting and other survival functions which have been well documented by direct field observation. (#12, #30, #31, #32, #33, #34, #35). In this light, if aggression where it occurs on various levels of evolution is understood in terms of survival functions, then the arguments mentioned above against individual species primarily learning (as defined) similar behavior patterns should apply to aggression. It remains then to deal with specific criticsms against these observations and hypotheses as well as their application to human aggression.

In criticizing Lorenz' *On Aggression*, (#32) Ashley Montagu states:

> For a serious student of animal behavior Dr. Lorenz appears to be singularly ill-informed on the temperaments of prehuman primates. It is not "irascibility" which is the term most frequently used to describe the temperaments

of the prehuman primates by those who know them best, but "amiability". The field studies of Schaller on the gorilla, of Goodall on the chimpanzee, of Harrison on the orangutan, as well as those of others, show these creatures to be anything but irascible. All the field observers agree that these creatures are amiable and quite unaggressive, and there is not the least reason to suppose that man's prehuman primate ancestors were in any way different. Captured monkeys and apes in zoos and circuses are not the best examples from which to deduce the behavior of such creatures under natural conditions. (#36, P.12).

Montagu's references citing the "amiability" of primates do not seem to bear him out, for "the field studies of Schaller on the gorilla . . ." say that:

The small number of overt social interactions was a most striking aspect of intragroup behavior. The most frequently noted interactions were dominance (at .23 times per hour of observation) . . . The relative infrequency of interaction can probably be attributed to the following circumstances. Competition for food and mates provided little basis for strife since forage was abundant and sexual behavior not prominent. The members of the group were alert to the possibilities of aggressive encounters, and subordinate animals tended to circumvent issues before they materialized. (#37, P.341).

There was considerable variation in the intensity of aggressive behavior which gorillas exhibited toward each other and toward such intruders as man. In the order of increasing intensity, the responses included: (1) an unwavering but usually brief stare, sometimes with furrowed brow and slightly pursed lips; (2) a jerk of the head or a snap in the direction of the offending animal; (3) an incipient charge, indicated by a light forward lunge of the body, occasionally without moving the feet, but usually accompanied by one or two abrupt steps; (4) a quadrupedal bluff charge over a distance of from 10 to 80 feet; and (5) physical contact in the form of biting or wrestling. (#37, P.354).

"Goodall on the chimpanzee" says:

Instances of attack were seldom observed, and mature males were seen fighting on only one occasion, when William took a bite of fresh meat on which Huxley, an old male, was feeding.

However, this one observation is justifiably described by Goodall as it infers extensive fighting knowledge:

Huxley, with a scream, seized William and bit him on the shoulder, and J.B., a large and powerfully built mature male chased him from the tree. After a few moments both males rejoined Huxley, who reached out and hit William four or five times on the scrotum, while J.B. bit his shoulder. J.B. then hit and pulled at William's scrotum; the latter screamed loudly but did not try to escape or retaliate. No signs of injury were visible after the encounter.

Other instances of attack behavior are described elsewhere . . .

When an excited male bounds towards a subordinate the latter sometimes runs off rapidly. It may hurl itself out of a tree, or leap into a tree if on the ground . . .

At other times the subordinate animal, instead of running away, remains where it is or moves toward the male and makes an appeasement gesture . . . (#38, P.466-467).

It is also significant that Goodall observed and defined Lorenz' "displacement activities" and "redirected activities".

When confronted by an unusual situation and apparently uncertain as to how to act, for instance when upset by my presence and/or when inclined both to flee and to attack, chimpanzees may show certain responses that will be summarized under the headings referred to as "displacement activities" and "redirected activities". These terms are classificatory devices only and apply to activities that appeared to me to be irrelevant to the situation and out of context (displacement activities) or else to be redirected on to an object or objects other than the initial eliciting stimulus. (#38, P.467-468).

Although Montagu cites *Primate Behavior* as a major reference indicating primate amiability, he neglects to cite Hall and Devore's suggestion in the same volume that:

The nature of the dominance functions of one α male, or of a group of central hierarchy males, is the clue that leads to an adequate understanding of the major aspects of baboon social organization. In addition, however, the complex relationships stemming from these dominance functions must also be fully studied as they affect the whole group. For example, threat behavior we now know can have at least four forms according to the kinds of social interaction with which it is associated. One of the commonest of these is called, in ethological terms, "redirection of aggression" (Bastock, Morris and Moynihan, 1953), corresponding to "transferred threat" (Altmann, 1962). (#39, P.64).

Also omitted are Simmonds' observations of aggression and dominance in the bonnet macaque. Concerning the behavior of one male Simmonds writes:

Throughout the study he threatened more and more dominant males until he was displacing the number 2 male fairly regularly. He was not, however, part of the central core of dominant males at the end of the study. They tended to threaten him jointly, and their combined dominance exceeded this. (#40, P.185).

In short, while all articles in *Primate Behavior* (indeed a basic source) indicate the importance and occurrence of aggression in gorillas, chimpanzees, rhesus and bonnet macques, lemurs, langurs, and gibbons, only the orangutan has not been included as aggressive due to inadequate field observation.

While it is true that a great deal of primate social

behavior is unaggressive, it is equally untrue to suggest that aggression does not play a fundamental role in primate interaction. It is unfortunate that critics such as Montagu, (#36) Carrighar, (#44) and Zuckerman (#42) accuse Lorenz and his following of basing their conclusions on the behavior of caged animals, or of flagrantly misrepresenting the ideas of others. Particularly amusing is a Zuckerman criticism of *On Aggression*. Sir Solly Zuckerman, whose baboon behavioral studies are widely reported ("..., (1932) field observations of chacma behavior were confined to a few days in the eastern Cape Province, South Africa, by far the bulk of his data coming from detailed observations on *p. hamadryas* in the Regent's Park Zoo, London.") (#33, P.53) attacks Lorenz as follows:

> But it is a different matter when he cites the presumed data of others ... Who, for example, has seen "small apes ... leaping after a tiger or a leopard"? And to how many readers does Lorenz address the sentence "anybody who has ever seen ... the male chimpanzee defending his band or family with self-sacrificing courage"? Has this ever been seen? (#42, P.93).

Although I have not found the documentation for chimpanzee defense by males of their group, the following has been seen:

> Flight to the trees provides bonnet macaques the main escape from predation. During such flights the males drop to the rear and are the first to come into contact with the predators. (#40, P.196).

And in *African Genesis*, Ardrey recounts an observation of an incident involving baboons made over fifty years ago by naturalist Eugene Marais in South Africa.

> It was still dusk. The troop had just returned from the feeding grounds and had barely time to reach its scattered sleeping places in the high piled rocks behind the fig tree. Now it shrilled in terror. And Marais could see the leopard. It appeared from the bush and took its insolent time. So vulnerable were the baboons that the leopard seemed to recognize no need for hurry. He crouched just below a little jutting cliff observing his prey and the problems of the terrain. And Marais saw two male baboons edging along the cliff above him.
>
> The two males moved cautiously. The leopard, if he saw them, ignored them. His attention was fixed on the swarming, screeching, defenseless horde scrambling among the rocks. Then the two males dropped. They dropped on him from a height of twelve feet. One bit at the leopard's spine. The other struck at his throat while clinging to his neck from below. In an instant the leopard disemboweled with his hind claws the baboon hanging to his neck and caught in his jaws the baboon on his back. But it was too late. The dying disemboweled baboon had hung on just long enough and had reached the leopard's jugular vein with his canines. (#30, P.83).

In a slightly different vein Montagu argues:

> But teeth are no more an armament than is the hand, and it is entirely to beg the question to call them so. Virtually all the members of the order of primates, other than man, have large canine teeth, and these animals, with the exception of baboons, are predominantly vegetarians, and it is because they are vegetarians that they require large canine teeth; that they may, on occasion, serve a protective purpose is entirely secondary to their main function, which is to rip and shred the outer coverings of plant foods. (#36, P.4).

Even if it can be believed that huge canine teeth particularly assist a vegetarian, rather than aid in current species predation (baboons and chimpanzees) (#38, #43) or possible predation by direct ancestors and current species if ecological environment changes; it is not believable that such dentition was not equally evolved for defense against predation. It is also highly unlikely that the enlarged canine teeth evolved for reasons other than predation or defense against predation. The only known pure vegetarians (outside of some primates) possessing enlarged canines are some species of deer, who have very small or no antlers, which when well developed in other deer species serve the function of defense (#44). Concerning our pre-human ancestors, Anthropologist Sherwood Washburn states that:

> In all the apes and monkeys the males have large canine teeth ... This is an efficient fighting mechanism, backed by very large jaw muscles. I have seen male baboons drive off cheetahs and dogs, and according to reliable reports male baboons have even put leopards to flight ... All the evidence from living monkeys and apes suggests that the male's large canines are of the greatest importance to the survival of the group, and that they are particularly important in ground-living forms that may not be able to climb to safety in the trees. (#45, P.9).

It is interesting that Montagu's suggestion that "teeth are no more an armament than is the hand" seems quite correct, because recent evidence suggests that it is exactly the evolution of the human hand, with its relatively large topographical representation in the cerebral cortex, which may have been incorporated into certain survival functions that used to employ large canine teeth. In giving evolutionary significance to findings of Neurophysiologist Wilder Penfield concerning cerebral localization of function, Washburn suggests that:

> ... Certain parts of the brain have increased in size much more than others. As functional maps of the cortex of the brain show, the human sensory motor cortex is not just an enlargement of that of an ape. The areas for the hand, especially the thumb, in man are tremendously enlarged, and this is an integral part of the structural base that makes the skillful use of the hand possible. The selection

pressures that favored a large thumb also favored a large cortical area to receive sensations from the thumb and to control its motor activity. (#45, P.12-13).

And it seems that selective pressures such as defense and predation, were integrally accounted for in human evolution through the development of tool-using behavior.

> The small, early man-apes lived in open plains country, and yet none of them had large canine teeth. It would appear that the protection of the group must have shifted teeth to tools early in the evolution of the man-apes and long before the appearance of the forms that have been found in association with stone tools. The tools of Sterkfontein and Olduvai represent not the beginnings of tool use, but a choice of material and knowledge in manufacture which, as is shown by the small canines of the man-apes that deposited them there, derived from a long history of tool use. (#45, P.9).

Further support for tool using in predation and defense pre-dating man comes from Montagu.

> It is known that some primates will defensively, and sometimes offensively, throw stones and other materials at intruders. It is probable that the precursors of early man used stones in this same manner. This is an extra-corporeal, an instrumental, use of an object. A stone used for such a purpose is the crudest of all tools—but it is a tool. (#46, P.10).

If anthropogical data indicating the use of weapons by pre-human primates is considered in the light of the principles of animal evolution vis á vis survival functions, it is difficult to see how Montagu can say:

> . . . but most tools of prehistoric man, from his earliest days, were most certainly not designed to serve as weapons. (#36, P.5).

On what grounds can it be argued that man's use of tools did not have major significance in predation and defense, when such use of tools is indicated in our pre-human forerunners? As Derek Freeman empha-sizes, the discoveries of Dart, Robinson, Broom, Leakey and Bartholomew indicate the use of tools by our million-year-old ancestors, the Australopi-thecines. Why these tools were used as weapons is best summed up by Freeman.

> The limb and pelvic bones of the Australopithecines indicate that these homonids walked erect. This is an important fact, for as Darwin and others have stressed, the attainment of bipedal locomotion is a crucial stage in the transition from ape to man, for by freeing the hands it makes possible the evolution of the employment and manufacture of weapons and tools. A concomitant pro-cess in this evolutionary development is a reduction in the size of the teeth and of the facial skeleton. This condition we also find in the Australopithecinae, whose teeth are

small and of human conformation. Furthermore, the Australopithicinae were typically found (Bartholomew and Birdsill, 1953) in association with the mammalian fauna of the open grasslands of southern Africa, on which fauna they carnivorously preyed. (#47, P.114).

It is also likely that early man preyed upon members of his own species. Despite Montagu's asser-tion that:

> There is, however, no record of any people, prehistoric, nonliterate, or anywhere in the annals of human history, who made a habit of killing their fellow men in order to dine off them. (#36, P.12).

Freeman offers the following:

> Cannibalism, which is but one expression of man's carniv-orous nature, has been reported for almost all parts of the world, and, on the evidence of palaeoanthropology, was probably once a universal practice. Blanc (1961) has shown that a range of fossil skulls, from mid-Pleistocene onwards, are characterized by a careful and symmetrical incising of the periphery of the *foramen magnum* to produce an opening which, on comparative evidence, was made with the purpose of extracting the brain for eating. (#47, P.113).

And Oakley states:

> There is no doubt that *Pithecanthropus* of the Pekin Caves half a million years ago ate the brains of his fellow men; but, as one of the speakers said, these may have been men of the same species yet of neighboring groups. How can we tell? Even at this early time level must we rule out the possibility of ritualized behavior? The earliest use of what were probably missle-stones was not long after the beginning of hunting, perhaps a million years ago or more.

But if it is argued that pre-human and early human use of weapons for predation and defense was integral in the evolution of the increased complexities and capabilities of the human organism, then must it be argued that we are genetically endowed killers pro-grammed to perform acts of destructive violence? No, as T. C. Schneirla suggested, "it is a long way from genes to behavior." It is argued here that where pre-human survival function behavior indicates simil-arity with our own evolution, it is not likely that our increased capacities for modification of behavior have fundamentally changed our participation in survival functions from those of other animals. As aggressive behavior in the forms of intraspecific dominance hierarchies, predation, defense, sexual fighting and territoriality exist on most levels of vertebrate evolu-tion, including our own, it is unlikely that each species which participates in these survival functions in strikingly similar ways, developed behavioral simil-arities as a primary function of *learning*. It would seem then that there must be some evolutionary mechanisms working throughout the development of species which form the matrix into which learning is integrated. To argue that human aggression is solely a function of culture, etc. seems to omit entirely the common aspects of aggression throughout most of vertebrate evolution.

If we can accept human aggression as solely a cultural adaptation, then we must argue that direct pre-human forerunners also possessed culture, and so must current primates, chickens, and fish. Interspecific aggression (predation or defense against predation) is accomplished with claws and canines in panthers, canines in baboons, weapons in prehominids and hominids—where are the fundamental differences? Why should it be assumed that intraspecific aggression between nations is essentially different from the functions of territoriality in some primates, or that dominance hierarchies in military and business establishments are different in kind from dominance behavior in baboon societies? While it is true that our species has an enormously increased capacity for modifying behavior, our general modes of aggressive behavior still seem to revolve around archaic survival functions. The development of human society has in part obscured the uses to which we put aggressive survival functions; however, behavior patterns equatable with other species' behavior have not been deleted.

Should we infer that the evolution of human aggression is essentially destructive at this point in our development? To do this one must negate the value of philosophical discourse, exploring the unknown, social ordering, protection against attack, developing competence in the face of adversity, and this list grows long. While it is true that on almost no other level of evolution do members of the same species kill each other through organized conflict, humans in fact do. Is this our "programmed" destiny? Such a belief does not allow for species modification of behavior through learning, or explain behavior in terms of adaptive advantage. It is not likely that with our increased abilities to be successfully aggressive we must as a species have gained great advantage from aggression? If it is true that evolutionarily developed mechanisms usually inhibiting intraspecific killing occur in other species it would seem probable that we too should have them. But we do kill each other on an individual and a mass level. Perhaps as Lorenz suggests, our loss of huge canine teeth and sharp claws has led us to a loss of mechanisms inhibiting killing (#32). The above arguments indicating the "replacement" of teeth and claws as weapons by the "brain and hand", would seem to go against this suggestion.

If we as a species had lost all inhibitions against killing, the human history of destructive violence would probably be much worse than it already is. If almost all other potential intraspecific killing on the part of other vertebrates has been inhibited under normal conditions (with significant exception; #49, #50), we should probably be able to see such a mechanism within humans. There seem to be strong inhibition mechanisms that generally keep us from killing our own children, parents, brothers and sisters, and members of our own immediate groups. This does not mean that such possible inhibition mechanisms developed through evolution of species have not been in part overridden by some forms of environmental conditioning, or learning. Given conditions of starvation, plague, overcrowding and the break-up of individual small groups, aggression as a survival function *might* even include killing members of one's own species. When dominant members of a group "recognize the threat of destruction" from another group, often leaders can direct the development of aggression in group members through threatening banishment from the group, violence to an individual's family, or even death to the individual. Most valuable in this cotext are Stanley Milgram's studies on human obedience, indicating an overwhelming tendency for men to obey recognized authority (#51). It is in effect very obvious that humans and other animals *can* be conditioned to perform a great variety of absurd and bizarre acts. The fact that we *can* force other individuals and ourselves to go without sleep, starve to death, remain celebate, and engage in torture does not mean that there are no evolutionarily developed mechanisms guiding learning to include eating, sleeping, sexuality and constructive aggression.

Unless we make a serious enquiry into whatever laws we can find governing the evolution of species survival functions and be willing to equate them with man when applicable, it does not seem likely that we will find a working definition for any normal human behavior. If we refuse to develop the possibility that the learned aspects of human aggressive behavior are essentially rooted in evolutionary mechanisms directing survival function, and simply believe that culture can condition aggression in or out, we will probably maintain our historically hit and miss control over whether our aggression destroys us or is employed toward constructive ends.

### REFERENCES

1. Plato. *Euthyphro*. In *The Collected Dialogues of Plato*. Eds.: E. Hamilton and H. Cairns, New York, Random House, 1964. Pp. 169-185.

2.  Plato. *Socrates' Defense (Apology)*. In The Collected Dialogues of Plato. Eds.: E. Hamilton and H. Cairns, New York, Random House, 1964. Pp. 3-26.

3.  Kaufmann, W. *Nietzsche*. Cleveland, World Pub. Co., 1964.

4.  Nietzsche, F. *Beyond Good and Evil*. Chicago, H. Regnery Co., 1955.

5.  Benedict, R. *Patterns of Culture*. Boston, Houghton Mifflin, 1959.

6.  Mannix, D.P. *The History of Torture*. New York, Dell Pub., 1967.

7.  Goldfrank, E.S. Socialization, personality, and the structure of Pueblo society (with particular reference to the Hopi and Zuni). American Anthropologist, 47: 516-539, 1945.

8.  Tinbergen, N. On war and peace in animals and man. Science, 160: 1411-1418, 1968.

9.  Dewey, J. The reflex arc concept in psychology. In *Philosophy, Psychology and Social Practice*. Ed.: J. Ratner, New York, Capricorn Books, 1965. Pp. 252-266.

10. Dethier, V.G. and Stellar, E. *Animal Behavior: Its Evolutionary and Neurological Basis*. Englewood Cliffs, N.J., Prentice-Hall, 1965.

11. Montagu, M.F.A. Introduction in: *Man and Aggression*. Ed.: M.F.A. Montagu, New York, Oxford Univ. Press. 1968, Pp. vii-xiv.

12. Ardrey, R. *The Territorial Imperitive*. New York, Athenium, 1966.

13. Krieg, W.J.S. *Functional Neuroanatomy*. 3rd Edition, Bloomington, Ill., Brain Books, 1966.

14. Romer, A.S. *The Vertebrate Body*. Philadelphia, W.B. Saunders Co., 1964.

15. MacLean, P.D. Alternative neural pathways to violence. In *Alternatives to Violence*. Ed.: L.K.Y. Ng, New York, Time-Life Books, 1968, Pp. 24-33.

16. Bard, P. and Macht, M.B. The behavior of chronically decerebrate cats. In *Neurological Basis of Behavior*. Eds.: G.E.W. Wolstenholme and C.M. O'Connor, Boston, Little-Brown, 1958, Pp. 55-75.

17. Guhl, A.M. The social order of chickens. Sci. Amer., Feb., 1956. In *Psychobiology: The Biological Basis of Behavior*. Readings from Scientific American, San Francisco, W.H. Freeman & Co., 1967. Pp. 113-116.

18. Lorenz, K. *King Solomon's Ring*. New York, T.Y. Crowell Co., 1961.

19. Von Holst, E. and Saint Paul, U. Electrically controlled behavior. Sci. Amer., March, 1962. In *Psychobiology: The Biological Basis of Behavior*. Readings from Scientific American, San Francisco, W.M. Freeman & Co. 1967, Pp. 56-65.

20. Delgado, J.M.R. Free behavior and brain stimulation. Int. Rev. Neurobiol. 6c: 349-449, 1964.

21. Delgado, J.M.R. *Physical Control of the Mind: Toward a Psychoactive Society*. New York, Harper & Row, 1969.

22. Siegel, A. and Flynn, J.P. Differential effects of electrical stimulation and lesions of the hippocampus and adjacent regions upon attack behavior in cats. Brain Res., 7: 252-267, 1968.

23. Jouvet, M. The states of sleep; a neuropharmacological approach. Paper read at Assn. for Res. in Nerv. Ment. Dis., New York, Dec., Pp. 86-126, 1965.

24. Jouvet, M. The states of sleep. Sci. Amer., Feb.; 62-72, 1967.

25. Brody, J.F., DeFeudis, P.A. and DeFeudis, F.V. Effects of micro-injections of L-glutamate into the hypothalamus on attack and flight behavior in cats. Nature, 224: 1330, 1969.

26. Bandles, R.J. Facilitation of aggressive behavior in rats by direct cholinergic stimulation of the hypothalamus. Nature, 224: 1035-1036, 1969.

27. Gatz, A.J. *Manter's Essentials of Clinical Neuroanatomy and Neurophysiology*. Philadelphia, F.A. Davis Co., 1967.

28. Lorenz, K. *Evolution and Modification of Behavior*. Chicago, Univ. Chicago Press, 1967.

29. Schneirla, T.C. Instinct and aggression. In *Man and Aggression*, Ed.: M.F.A. Montagu, New York, Oxford Univ. Press, 1968, Pp. 59-64.

30. Ardrey, R. *African Genesis*. New York, Dell Pub., 1969.

31. Carthy, J.D. and Ebling, F.J. Eds.: *The Natural History of Aggression*. New York, Academic Press, 1966.

32. Lorenz, K. *On Aggression*. New York, Harcourt, Brace & World, 1966.

33. De Vore, I. Ed.: *Primate Behavior*. New York, Holt, Rinehart & Winston, 1965.

34. Fletcher, R. *Instinct in Man*. New York, Schocken Books, 1966.

35. Jay. P.J. Ed.: *Primates Studies in Adaptation and Variability*. New York, Holt, Rinehart & Winston, 1968.

36. Montagu, M.F.A. The new litany of "Innate Depravity" or original sin revisited. In *Man and Aggression*. Ed.: M.F.A. Montagu, New York, Oxford Univ. Press, 1968, Pp. 3-17.

37. Schaller, G.B. The behavior of the Mountain Gorilla. In Primate Behavior. Ed.: I. De Vore, New York, Holt, Rinehart & Winston, 1965, Pp. 324-367.

38. Goodall, J. Chimpanzees of the Gombe Stream Reserve. In *Primate Behavior*. Ed.: I. De Vore, New York, Holt, Rinehart & Winston, 1965, Pp. 425-473.

39. Hall, K.R.L. and De Vore, I. Baboon social behavior. In *Primate Behavior*. Ed.: I. De Vore, New York, Holt, Rinehart & Winston, 1965, Pp. 53-110.

40. Simonds, P.E. The bonnet macaque in South India. In *Primate Behavior*. Ed.: I. De Vore, New York, Holt, Rinehart & Winston, 1965, Pp. 175-196.

41. Carrighar, S. War is not in our genes. In *Man and Aggression*. Ed.: M.F.A. Montagu, New York, Oxford Univ. Press, 1968, Pp. 37-50.

42. Zuckerman, S. The human beast. In *Man and Aggression*. Ed.: M.F.A. Montagu, New York, Oxford Univ. Press, 1968, Pp. 91-95.

43. Hamburg, D. Recent evidence on the evolution of aggressive behavior. Engineering and Science, XXXIII, No. 6: 15-24, 1970.

44. Ulmer, F. Curator of Philadelphia Zoological Gardens, Personal communication.

45. Washburn, S.L. Tools and human evolution, Rep. from Sci. Amer., Sept. 1960. San Francisco, W.H. Freeman & Co., 1960, Pp. 1-16.

46. Montagu, M.F.A. Introduction in *Culture and the Evolution of Man*. Ed.: M.F.A. Montagu, New York, Oxford Univ. Press, 1967, Pp. vii-xii.

47. Freeman, D. Human aggression in anthropological perspective. In *Natural History of Aggression*. Eds.: J.D. Carthy and F.J. Ebling, New York, Academic Press, 1966, Pp. 109-120.

48. Oakley, K.P. Discussion of Human aggression in anthropological perspective, by Freeman, D. In *Natural History of Aggression*. Eds.: J.D. Carthy and F.J. Ebling, New York, Academic Press 1966.

49. Ardrey, R. *Social Contract*. New York, Athenium, 1970.

50. Ewer, R.F. *Ethology of Mammals*. Chapt. 6, New York, Plenum Press, 1968.

51. Milgram, S. Some conditions of obedience and disobedience to authority. Int. J. Psych., 6: 259-276, 1966.

## SELECTED BIBLIOGRAPHY NOT CITED IN TEXT

Adorno, T.W., et. al., *The Authoritarian Personality*. New York, Harper and Row, 1950.

Andersson, B., et. al. An appraisal of the effects of diencephalic stimulation of conscious animals in terms of normal behavior. In *Neurological Basis of Behavior*. Eds.: G.E.W. Wolstenholme & C.M. O'Connor, Boston, Little Brown, 1958, Pp. 76-84.

Asimov, I. *The Genetic Code*. New York, New Amer. Lib. of World Lit., 1963.

Bard, P. and Mountcastle, V.B. Some forebrain mechanisms involved in expression of rage with special reference to suppression of angry behavior. In *Basic Readings in Neuropsychology*. Ed.: R.L. Isaacson, New York, Harper & Row, 1964, Pp. 110-180.

Barnett, S.A. On the hazards of analogies. In *Man and Aggression*. Ed.: M.F.A. Montagu, Oxford Univ. Press, 1968.

Barnett, S.A. Grouping and dispersive behavior among wild rats. In *Aggressive Behavior*. Eds.: S. Garattini and E.B. Sigg, New York, Wiley & Sons, 1969.

Beatty, J. Taking issue with Lorenz on the Ute. In *Man and Aggression*. Ed.: M.F.A. Montagu, Oxford Univ. Press, 1968. Pp. 111-115.

Berkowitz, L. Simple views on aggression: an essay review. *American Scientist*, 57: 372-382, 1969.

Berrill, N.J. *Sex and The Nature of Things*. New York, Dodd, Mead & Co., 1953.

Bliss, E.L. and Ailion, J. Response of neurogenic amines to aggregation and strangers. *J. Pharm. Exper. Ther.*, 168 (2): 258-263, 1969.

Bonner, D.M. and Mills, S.E. *Heredity*. Englewood Cliffs, N.J., Prentice-Hall, 1964.

Boulding, K.E. Am I a man or a mouse—or both? In *Man and Aggression*. Ed.: M.F.A. Montagu, Oxford Univ. Press, 1968, Pp. 83-90.

Broderick, A.H. *Man and His Ancestry*. Greenwich, Conn., Fawcett Pub., 1964.

Broom R. The ape men. *Sci. Amer.*, Nov., 1949.

Brown, R. The authoritarian personality and the organization of attitude. In *Social Psychology*. New York, Free Press—Macmillan & Co., 1965, Pp. 477-546.

Calhoun, J.B. Population density and social pathology. *Sci. Amer.*, 128: No. 7: 571-603, 1963.

Carder, B. and Berkowitz, K. Rat's preference for earned in comparison with free food. *Sci.*, 167: 1273-1274, 1970.

Carpenter, C.R. The Howlers of Barro Colorado Island. In *Primate Behavior*. Ed.: I. De Vore, New York, Holt, Rinehart & Winston, 1965, Pp. 250-291.

Christian, J.J. Social subordination, population density and mammalian evolution. *Sci.*, 168: 84-90, 1970.

Crow, J.F. Ionizing radiation and evolution. Reprinted from: *Sci. Amer.*, Sept., 1959.

Dobzhansky, T. *Heredity and the Nature of Man*. New York, New Amer. Lib., 1960.

Eimerl, S. and De Vore, I. *The Primates*. New York, Life-Time Books, 1965.

Ellis, H. *Psychology of Sex*. New York, New American Lib., 1960.

Engles, F. *Socialism: Utopian and Scientific*. New York, International Pub. Co., 1935.

Fisher, A.E. Chemical stimulation of the brain. *Sci. Amer.*, June, 1964.

Fisher, J. Interspecific aggression. In *The Natural History of Aggression*. Eds.: J.D. Carthy and F.J. Ebling, New York, Academic Press, 1966, Pp. 66-74.

French, J.D. The reticular formation. *Sci. Amer.*, May, 1957.

Freud, S. *Beyond the Pleasure Principle*. New York, Liveright Pub., 1963.

Funkenstein, D.H. The physiology of Fear and anger. *Sci. Amer.*, May, 1965.

Gajdusek, C. Physiological and psychological characteristics of stone age man. *Engineering and Science: Biological Bases of Human Behavior*, XXXIII, No. 6: 25-71, 1970.

Gilula, M.F. and Danials, D.N. Violence and man's struggle to adapt. *Sci.*, 164: 396-405, 1969.

Gorer, G. Ardrey on human nature: animals, nations and imperitives. In *Man and Aggression*. Ed.: M.F.A. Montagu, Oxford Univ. Press, 1968, Pp. 74-82.

Gorer, G. Man has no "killer" instinct. In *Man and Aggression*. Ed.: M.F.A. Montagu, Oxford Univ. Press, 1968, Pp. 27-36.

Gregory, R.L. *Eye and Brain*. New York, McGraw-Hill, 1966.

Hall, K.R.L. Aggression in monkey and ape societies. In *The Natural History of Aggression*. Eds.: J.D. Carthy and F.J. Ebling, New York, Academic Press, 1966, Pp. 51-64.

Herrick, C.J. *The Evolution of Human Nature*. New York, Harper & Bros., 1961.

Hess, E.M. Imprinting in animals, *Sci. Amer.*, March, 1958.

Howell, C.F. *Early Man*. New York, Time-Life Books, 1965.

Hunsperger, R.W. and Bucher, V.M. Affective behavior produced by electrical stimulation in the forebrain and brainstem of the cat. In *Structure and Function of the Limbic System*, Progress in Brain Research, Vol. 27, Eds.: R.W. Adey and T. Tolkizane, Elsevier, 1967, Pp. 103-127.

Huxley, J. *Evolution in Action*. New York, New Amer. Lib., 1964.

Jay, P. The Common Langur of North India. In *Primate Behavior*. Ed.: I. De Vore, New York, Holt, Rinehart & Winston, 1965, Pp. 197-249.

Kermani, E.J. "Aggression"–biophysical aspects. *Dis. Nerv. Syst.*, 30: 407-414, 1969.

Kleitman, N. Patterns of dreaming, *Sci. Amer.*, Nov., 1960.

Klopper, A. Discussion of the physiological background of aggression. In *The Natural History of Aggression*. Eds.: J.D. Carthy and F.J. Ebling, New York, Academic Press, 1966, Pp. 73-76.

Leakey, M.D. Stone artifacts from Swartkrans. *Nature*, 225: 1222-1225, 1970.

Levine, S. Sex differences in the brain. *Sci. Amer.*, April, 1966.

Lorenz, K. The evolution of behavior. *Sci Amer.*, Dec., 1958.

Lorenz, K. Ritualized fighting. In *The Natural History of Aggression*.

MacLean, P.D., et. al. Experiments on localization of genital function in the brain. *Trans. Amer. Neurol. Assn.*, 1959.

MacLean, P.D. Contrasting functions of limbic and neocortical systems of the brain and their relevance to psychosomatic medicine. In *Biological Foundations of Emotion*. Ed.: E. Gellhorn, Glenview, Ill., Scott, Foresman & Co., 1968, Pp. 73-106.

Magoun, H.W. *The Waking Brain*. Springfield, Ill., Thomas Pub., 1963.

Maier, N.R.F. and Schneirla, T.C. *Principles of Animal Psychology*. New York, Dover Pub., 1964.

Mark, V.H., et. al. Brain disease and violent behavior. *Neuro-Ophthalmology*, IV: 282-287, 1968.

Mason, J.W. The scope of psychoendocrine research. *Psychosom. Med.*, XXX, No. 5: 565-575, 1968.

Masters, W.H. and Johnson, V.E. *Human Sexual Response*. Boston, Little & Brown, 1966.

Matthews, L.H. Overt fighting in mammals. In *The Natural History of Aggression*. Eds.: J.D. Carty and F.J. Ebling, New York, Academic Press, 1966, Pp. 23-38.

Medawar, P.B. *The Future of Man*. New York, New Amer. Lib., 1961.

Monnier, M. and Tissot, R. Correlated effects in behavior and electrical brain activity evoked by stimulation of the reticular system, thalamus and rhinencephalon in the conscious animal. In *Neurological Basis of Behavior*. Eds.: G.E.W. Wolstenholme and C.M. O'Connor, Boston, Little-Brown, 1958, Pp. 105-119.

Morris, D. *The Naked Ape*. New York, McGraw-Hill, 1967.

Morris, D. *The Human Zoo*. New York, McGraw-Hill, 1969.

Moruzzi, G. and Magoun, H.W. Brain stem reticular formation and activation of the EEG. In *Basic Readings in Neuropsychology*. Ed.: R.L. Isaacson, New York, Harper & Row, 1964, Pp. 153-280.

Moruzzi, G. Sleep and instinctive behavior. *Arch. Ital. Biol.*, 107: 175-217, 1969.

National Commission on the Causes and Prevention of Violence. *Violent Crime: The Challenge to Our Cities*. New York, Braziller Inc., 1969.

Nietzsche, F. *The Genealogy of Morals*. Garden City, N.Y., Doubleday, 1956.

Nietzsche F. *Philosophy in the Tragic Age of the Greeks*. Chicago, Regnery Co., 1962.

Penfield, W. The role of the temporal cortex in recall of past experience and interpretation of the present. In *Neurological Basis of Behavior*. Eds.: G.E.W. Wolstenholme and C.M. O'Connor, Boston, Little-Brown, 1958, Pp. 149-174.

Portman, A. and Grene, M. *Beyond Darwinism. Commentary*, Vol. **45**: 31-41, 1965.

Preston, J.B., et. al. The motor cortex—pyramidal system: patterns of facilitation and inhibition on motoneurons innervating limb musculature of cat and baboon and their possible adaptive significance. In *Neurophysiological Basis of Normal and Abnormal Motor Activities*. Eds.: M.D. Yahr and D.P. Purpura, New York, 1967, Pp. 61-72.

Pribram, K.H. and Kruger, L. Functions of the "Olfactory Brain." In *Basic Readings in Neuropsychology*. Ed.: R.L. Isaacson, New York, Harper & Row, 1964, Pp. 212-252.

Reynolds, V. and Reynolds, F. Chimpanzees of the Bondongo Forest. In *Primate Behavior*. Ed.: I. De Vore, New York, Holt, Rinehart & Winston, 1965, Pp. 368-424.

Schaller, G.B. Behavioral comparisons of the apes. In *Primate Behavior*. Ed.: I. De Vore, New York, Holt, Rinehart & Winston, 1965, Pp. 474-482.

Scott, J.P. and Fredricson, E. The causes of fighting in mice and rats. Physiol. Zool., XXIV: 273-309, 1951.

Scott, J.P. That old-time aggression. In *Man and Aggression*. Ed.: M.F.A. Montagu, New York, Oxford Univ. Press, 1968, Pp. 51-58.

Sheehan, E.R.F. Conversation with Konrad Lorenz. *Harper's Magazine*, 236, #1416: 69-77, 1968.

Simons, E.L. The early relatives of man. *Sci. Amer.*, July, 1969.

Southwick, C.H., et. al. Rhesus monkeys in North India. In *Primate Behavior*. Ed.: I. De Vore, New York, Holt, Rinehart & Winston, 1965, Pp. 111-159.

Sperry, R.W. Neurology and the mind-brain problem. In *Basic Readings in Neuropsychology*. Ed.: R.L. Isaacson, New York, Harper & Row, 1964, Pp. 403-429.

Storr, A. *Human Aggression*. New York, Athenium, 1968.

Tinbergen, N. The curious behavior of the stickleback. *Sci. Amer.*, Dec., 1952.

Tinbergen, *Animal Behavior*. New York, Time-Life Books, 1965.

Tolman, E.C. Cognitive maps in rats and men. *Psychol. Rev.*, 55: 189-208, 1948.

von Koenigswald, G.H.R. *The Evolution of Man*. Ann Arbor, Mich., Univ. Michigan Press, 1962.

Walker, D. *Rights in Conflict: The Violent Confrontation of Demonstrators and Police in the Parks and Streets of Chicago During the Week of the Democratic National Convention of 1968*. Philadelphia, Braceland Bros., 1968.

Washburn, S.L. and Hamburg, D.A. The Study of primate behavior. In *Primate Behavior*. Ed.: I. De Vore, New York, Holt, Rinehart & Winston, 1965, Pp. 1-15.

Washburn, S.L. and De Vore, I. The social life of Baboons. *Sci. Amer.*, Dec.. 1961.

Welch, B.L. and Welch, A.S. Aggression and the biogenic amine neurohumors. In *Aggressive Behavior*. Eds. S. Garattini and E.B. Sigg, New York, Wiley & Sons, 1969.

Williams, D. Neural factors related to habitual aggression. Brain, 92: 503-520, 1969.